Green Energy and Technology

More information about this series at http://www.springer.com/series/8059

Gianfranco Pistoia · Boryann Liaw
Editors

Behaviour of Lithium-Ion Batteries in Electric Vehicles

Battery Health, Performance, Safety, and Cost

Springer

Editors
Gianfranco Pistoia
National Research Council
Rome
Italy

Boryann Liaw
Department of Energy Storage and Advanced Vehicles
Idaho National Laboratory
Idaho Falls, ID
USA

ISSN 1865-3529 ISSN 1865-3537 (electronic)
Green Energy and Technology
ISBN 978-3-319-69949-3 ISBN 978-3-319-69950-9 (eBook)
https://doi.org/10.1007/978-3-319-69950-9

Library of Congress Control Number: 2017964589

Printed on acid-free paper

This Springer imprint is published by the registered company Springer International Publishing AG part of Springer Nature
The registered company address is: Gewerbestrasse 11, 6330 Cham, Switzerland

Contents

Lithium-Ion Battery Design for Transportation

Alvaro Masias

Abstract This chapter will discuss the technical requirements and status of applying lithium-ion batteries to electrified vehicles. It will begin by introducing the principles of vehicle propulsion, electrified features, powertrain design, and the resulting battery chemistry applicability. An understanding of vehicle needs will enable a discussion on lithium-ion battery pack design. Once the basic layout of pack design is understood, it is necessary to appropriately size a pack to meet its intended vehicle function relative to various drive cycles and other requirements. A review of current lithium-ion technology and applicability for automotive applications will then follow. This chapter will describe existing cell energy and power performance in the context of international performance targets. The various features of cell design for automotive will also be discussed along with a review of current safety testing standards and regulations. Finally, an examination of existing commercialized products will show how the vehicle, pack and cell design principles described are implemented in actual production vehicles.

Keywords Electrified vehicles · Cell design · Battery pack design
Pack performance targets · Safety

1 Introduction

Lithium-ion batteries are enabling a new generation of electrified vehicles to be commercialized by global automakers. A variety of governments including the USA, European Union, China, and Japan have announced increasingly strict fuel economy regulations for their respective markets. The modern fossil fuel powered automobile has been the subject of continuous engineering improvement for over one hundred years [1]. Comparatively, modern electrified automobiles are a

A. Masias (✉)
Ford Motor Company, 2101 Village Road, Dearborn, MI 48121, USA
e-mail: amasias@ford.com

G. Pistoia and B. Liaw (eds.), *Behaviour of Lithium-Ion Batteries in Electric Vehicles*, Green Energy and Technology,
https://doi.org/10.1007/978-3-319-69950-9_1

relatively new technology, yet their potential for petroleum displacement makes them a key component of virtually all automakers' current and future product portfolios.

The status of applying lithium-ion batteries to vehicles and the technical requirements involved will be discussed. As background, the concepts of vehicle propulsion, electrified features, powertrain design, and the resulting battery chemistry applicability will be introduced. A discussion on battery pack design will be enabled by this understanding of the vehicle needs. After the pack design, layout basics are understood, it will be possible to suitably size a pack to meet the designed vehicle features relative to different cycles.

A discussion of the suitability of lithium-ion technology for automotive applications will then follow. While considering various global performance targets, the current performance of cell energy and power will be reviewed. Next will be a survey of current safety testing standards and regulation considering the various features of cell design for automotive applications. How the identified vehicle, pack and cell design principles are implemented in vehicle production will be examined by reviewing existing commercialized electrified vehicles.

2 Vehicles

2.1 Vehicle Propulsion

In a conventional automobile, the propulsion power is provided solely by the engine, whereas in an electric vehicle the battery/motor delivers all traction power. In a hybrid electric vehicle, the traction force is provided by a mixture of the engine and motor/battery which must be efficiently managed by the vehicle [2, 3]. The traction power required to propel a vehicle must exceed that which is simply required by kinematics to accelerate the vehicle mass due to the additional forces for rolling resistance, aerodynamic drag, and elevation change:

$$F_{\text{Road Load}} = F_{\text{Kinematic}} + F_{\text{Rolling Resistance}} + F_{\text{Aerodynamic Drag}} + F_{\text{Elevation}}$$

$$F_{\text{Road Load}} = ma + mgC_{\text{RR}} + \frac{1}{2}\rho C_{\text{D}} A v^2 + mg \sin\theta$$

where

m mass of vehicle
a acceleration of vehicle
g acceleration of free fall due to gravity
C_{RR} coefficient of rolling resistance between tires and road surface
ρ density of ambient air
C_{D} coefficient of draft of the vehicle

A cross-sectional area of vehicle
v speed in the direction of travel
θ angle of road grade

The power required to propel the vehicle depends on this force and the vehicle velocity according to:

$$P_{\text{Road Load}} = v(F_{\text{Road Load}}) = mav + mgC_{\text{RR}}v + \frac{1}{2}\rho C_{\text{D}}Av^3 + mgv\,\sin\theta$$

Note this power has terms with linear and cubic relations to velocity. In a uniform acceleration, the power demand increases nearly linearly with time peaking at the end of the acceleration period. It is also notable that power varies as the cube of velocity, helping to account for the significantly increased power levels observed at highway speeds when compared to city driving patterns.

Empirically, it can be determined that most vehicles have strongly correlated acceleration 0–100 kph times (seconds), peak power (kW), and weights (kg). Vehicle properties such as aerodynamic drag, mechanical grip, and engine/motor performance can result in a complicated correlation. However, as a first-order approximation, it can be empirically estimated that:

$$T = A * \frac{W}{P}$$

where:

T Time (s)
A $0.68\,\text{kW} * \text{s}/\text{Kg}$
W Curb Weight (Kg)
P Power (kW)

Using the fourth-generation Toyota Prius (US Model Year 2009–2015) as an example yields W = 1380 kg, P = 100 kW, resulting in an estimated 0–100 kph time of 9.5 s, which was confirmed by Motor Trend Magazine in 2012 [4].

During braking, retarding force is provided through a combination of electrical (driving the traction motor backwards to establish a generator) and mechanical (friction disks or drums creating waste heat) loads. In this case, the braking force required is less than the force needed to decelerate the vehicle mass since the rolling resistance and aerodynamic drag forces also act to slow the vehicle:

$$F_{\text{Braking}} = F_{\text{Kinematic}} - F_{\text{RollingResistance}} - F_{\text{AerodynamicDrag}} - F_{\text{Elevation}}$$

$$F_{\text{Braking}} = ma - mgC_{\text{RR}} - \frac{1}{2}\rho C_{\text{D}}Av^2 - mg\,\sin\theta$$

Likewise, the power required to brake the vehicle is the product of this force and the vehicle velocity:

$$P_{\text{Braking}} = v(F_{\text{Braking}}) = mav - mgC_{\text{RR}}v - \frac{1}{2}\rho C_{\text{D}}Av^3 - mgv\sin\theta$$

Taking into account again the approximate proportionality of power to velocity, the braking power required declines approximately linearly with time in a uniform deceleration. Thus, for braking, the peak power regenerative braking power requirement occurs at the beginning of the braking event. Braking is partially assisted by the aerodynamic drag and rolling resistance forces.

However, as a practical matter, during braking the power expended typically far exceeds that experienced while accelerating and delivered by the combination of an engine and/or electric motor. Modern automobiles are expected to brake from speed in a fraction of the time and distance that they took to achieve the same speed. As a result, a simple kinematic study of these two vehicle events shows that the powers involved must vary significantly, with the excess power available to brake serving as a safety feature in the case of any engine or throttle malfunction.

2.2 *Electrified Vehicle Features*

There is a variety of electrified vehicle types with no universally accepted definition. We will use various performance features which the continuum of increasing electrification confers onto a vehicle as a means to classify the various vehicle types (see Fig. 1) [5].

The electrified vehicles which support external plugs to transfer electrical energy on board are the plug-in hybrid electric vehicle (PHEV) and the electric vehicle (EV). Although PHEVs and EVs have been around as concepts for some time, it is only through the recent performance revolution of lithium-ion batteries that they are becoming viable as vehicle technologies. The families of vehicles which do not

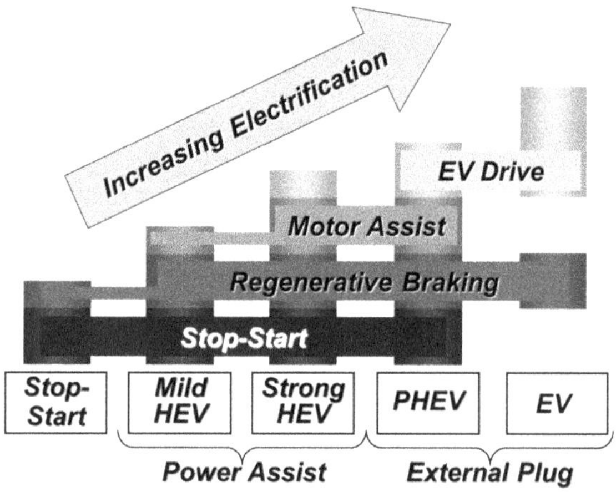

Fig. 1 Various types of electrified vehicles and defining features [5]

support external plugs are known as stop-start hybrids (S/S) and hybrid electric vehicle (HEV). The contribution of lithium-ion batteries to these vehicles (S/S and HEVs) can be described as evolutionary when compared to existing battery types.

A variety of automotive vehicle functions lend themselves well to electrification. As described in Fig. 1, increasing levels of electrification (quantity of electrical energy and power on board) enable different features.

Stop–Start: During engine idle times, such as when stopped at a traffic light or coasting, a traditional internal combustion engine vehicle can consume significant fuel to maintain power to auxiliary (non-traction) vehicle loads such as heating/cooling, infotainment, and illumination. By powering these features from the electrical system on board, a vehicle is able to turn off the engine completely, thereby leading to fuel savings. This feature typically denotes the first level of electrification given the limited power (≤ 10 kW) and energy (≤ 100 Wh) needs to be involved.

Regenerative Braking: Traditional automobiles will convert a vehicles kinetic energy into waste friction and heat under braking to decelerate. Incorporating an electrical system on board that can harness some or all of this electrical power allows for a significant efficiency boost by recuperating the energy used to accelerate the vehicle. Typical braking events are of a short duration (≤ 5 s), leading to relatively high power demands for the energies involved. As a result, it is common to size a battery system to capture only a portion (approximately 80–90%) of the braking power electrically and have a mechanical friction system work in concert to absorb the very high initial power pulse.

Motor Assist: Internal combustion engine fuel consumption will vary as a function of the engine speed and torque. A traditional automobile engine will have its operating point of engine speed and torque primarily determined by the driver throttle demand. However, a vehicle with electrified motor assist can divide the driver torque demands of the vehicle between the engine and battery-driven electric motor, thereby allowing the engine controller to shift the engine operating point to a more efficient region of the speed/torque engine map. The majority of strong hybrids have battery/motor electrical power sized around 30–40 kW and 300–400 Wh (useable) to achieve this motor assist function for significant portions of the various driving cycles used around the world.

EV Drive: During EV drive, the entire vehicle traction and auxiliary power demands are supported by the electrified systems on board. As a result, this function can require significantly increased electrical power and energy to achieve. EV drive energy consumption can range widely with typical sedan values from 95 to 240 Wh/km depending on the particular vehicle (weight, aerodynamics, road load, etc.) and driving cycle characteristics (rates of acceleration and deceleration, speed, idle time, etc.) [6].

The electrified power demands of EV drive can also vary widely based on the same vehicle characteristics as energy, ranging from ~ 15 kW (low-speed urban) to >100 kW (aggressive highway). Due to these increased electrical loads, most HEVs can only achieve limited EV drive range under moderate (urban or highway coasting) driving scenarios. PHEVs are most often differentiated from HEVs by the increased electrical range (energy) and power on board.

2.3 *Electrified Powertrains*

The vehicle performance features which have been electrified can be used to classify electrified vehicle types (see Fig. 1). Likewise, the vehicle powertrain (engine, motor, and transmission) is also used to further sort electrified vehicle types. The four different traditional electrified powertrain types are shown schematically in (see Fig. 2).

Series EV: The series EV is the simplest electrified vehicle powertrain in terms of quantity of hardware and control strategy required. This powertrain type has all traction power delivered from a single electric power source. Figure 2a shows the electrical power flowing from a DC battery to be converted to AC power before driving the electric motor. The series EV motor can run backwards to become a generator during regenerative braking. The motor/generator converts between electrical and mechanical power in interacting with the vehicle transmission and ultimately wheels. Recently, commercialized vehicles using this powertrain type are the Nissan Leaf, Tesla Model S, and Ford Focus EV.

Series (P)HEV: A series PHEV or HEV powertrain have the same basic layout type as shown in Fig. 2b. Both architectures have the mechanical (engine/generator) and electrical (battery) power sources feed into a common electrical inverter/converter. Once the electrical power has been combined in the inverter/converter, this powertrain behaves similarly to a series EV in having one power/torque flow path to and from the single motor/generator, transmission, and wheels. The series design allows for the efficient distribution of vehicle power demands across the two power sources, particularly advantageous during fluctuating vehicle speed scenarios. A series PHEV would typically be distinguished from a series HEV in having a more powerful battery, engine, and inverter but a less powerful engine. Diesel-electric locomotives and the Chevrolet Volt are commercially available examples of series HEV and series PHEV powertrains, respectively.

Parallel (P)HEV: Parallel PHEV and HEV vehicles require complicated electronic controls and transmission setups to manage two different torque sources (Fig. 2c). By electronically separating the two torque sources, this vehicle transfers the task of managing overall vehicle efficiency to a primarily mechanical operation. By maintaining a mechanical linkage between the engine and transmission,

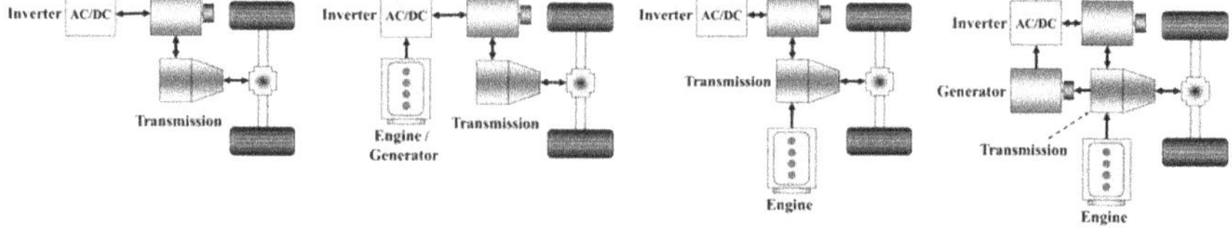

Fig. 2 Electrified vehicle type powertrain **a** Series EV, **b** Series (P)HEV, **c** Parallel (P)HEV, **d** Series/Parallel (P)HEV

this type is most efficient in steady state velocity conditions. However, also due to the mechanical setup, it is not possible to completely turn off the engine while maintaining vehicle functions in a parallel hybrid. This architecture has experienced limited commercially popularity, having been primarily manufactured by Honda (Civic, Insight, Accord) during the 2000s.

Series/Parallel (P)HEV: The series/parallel (P)HEV is the layout which combines the strengths and complexities of the separate series and parallel arrangements. Although series/parallel controls and components (particularly transmissions and quantity of motor/generators) are the most complicated and costly, they enable the greatest flexibility to achieve fuel economy savings. The mechanical and electrical power sources can each separately and in unison provide torque to drive the wheels and meet the driver throttle demand (Fig. 2d). The series/parallel powertrain has been implemented into many vehicles, including Toyota Prius and Ford Fusion.

2.4 Battery Chemistry Applicability

Battery engineers can design packs to meet the range of energy and power requirements that each vehicle feature requires using any battery chemistry. Various battery chemistries (lead–acid, nickel–metal hydride, and lithium-ion) have lent themselves to being commercialized in specific electrified vehicles types based on cost, energy, power, weight, and volume requirements (see Table 1). As a result, various trends for the application of lithium-ion technology have emerged.

Lithium-ion has helped to commercialize the PHEV type as a vehicle class, as the weight of older chemistries (such as nickel–metal hydride or lead–acid) required to get any appreciable electric range along with an engine was impractical. The improved energy gravimetric and volumetric density of lithium-ion has allowed it to completely replace other chemistries for EV applications. Lithium-ion provides several times greater specific energy than either nickel–metal hydride or lead–acid chemistries, without which modern EVs would be unworkable.

The lower energy and power needs of the S/S and HEV types have led to lithium-ion competing with lead–acid and nickel–metal hydride chemistries in the near term. In the future, it is expected that all HEVs will shift to lithium-ion chemistry batteries with the notable exception of Toyota which is expected to

Table 1 Commercialized electrified vehicle type by battery chemistry

Vehicle type	Historical	Recent (2010–2018)	Future (2018)
S/S	PbA	PbA	PbA & LIB
HEV	NiMH	LIB & NiMH	LIB & NiMH (Toyota)
PHEV	N/A	LIB	
EV	PbA & LIB	LIB	

amortize its significant investment in nickel–metal hydride production. While S/S applications will still primarily utilize lead–acid chemistries due to their low cost, it is expected that lithium-ion will begin to gain a foothold in this market as well.

3 Packs

3.1 Battery Pack Design

Electrified vehicle battery packs are required to meet a variety of automotive technical requirements, in addition to meeting the vehicle electrical power and energy demand. The assembly of cells into modules and subsequently packs is what makes the hardware relevant to an automotive designer and user. Mechanically, battery packs are required to be integrated into the existing vehicle crash structure. Packs are also required to manage the electronic control interface with the rest of the vehicle control modules and to maintain their cells within predetermined operating parameters for life and safety. Additionally, battery packs typically have dedicated or vehicle-derived thermal control components, also for performance and safety considerations. The sensitivity of lithium-ion chemistries to mechanical, electrical, and thermal excursions outside design specifications places additional importance on robust battery pack design.

Modules/Packs: In automotive applications, we need to consider not only cell-level figures of merit, but also at the module (a mechanical assembly of cells, often containing electrical/thermal sensing and interfaces) and battery pack level (a mechanical assembly of modules, often containing electrical and thermal control hardware and software). Although module and pack designs can vary substantially, they all add additional weight and volume which effectively de-rates the cell-level performance values.

Individual cell voltages are insufficient to provide kilowatts of power required for electrified vehicles since practical considerations with electric motors, cabling, and power electronics limit current flow to <500 A. A single series string or a series-parallel arrangement of cells is used to electrically and mechanically form a subassembly building block known as a battery module. Battery modules typically contain cell arrangements such that voltage is ≤ 50 V and weight ≤ 22 kg for ease of handling and safety.

Battery modules are combined electrically (most often in series) to provide the full power and energy need for electric vehicles. Depending on the vehicle type and design, the electrochemical cells may account for between 50 and 75% of the pack cost, weight, and volume. Thus, the specific performance of the battery pack system is always less than that of the modules and the modules less than that of the cells. Electrified vehicle battery performance targets are therefore typically set at the pack level to be most relevant to automotive designers.

Mechanical: Battery packs must also be contained physically within the vehicle in such a way as to be safe in the event of crashes and during normal vehicle use and vibrations. A typical vehicle passenger zone is defined as the area between the wheel axles and inside the rocker panels on either side. Placing a battery pack outside this zone is possible, but typically requires significant structural reinforcement to be added to ensure crash integrity, adding significant cost, weight, and volume burdens to the pack. Due to the restrictive nature of volume limitations on the battery pack, the engineering of the battery packaging envelope is always a critical feature in the design of an electric vehicle.

Mechanical packaging location also influences the level of robustness against water and dust intrusion which is required, with pack surfaces external to the vehicle structure requiring the greatest control. Additionally, packing location and battery chemistry type will influence whether a gas vent routing system to the vehicle exterior is necessary in the event of a cell vent during a malfunction (given that lithium-ion cells are sealed unlike nickel–metal hydride types).

Electrical: The electronic controls assist the battery pack in providing the required propulsion power while maintaining the battery pack within the normal operating conditions. A traditional 12 V lead–acid battery is commonly used to power electrified vehicle auxiliary loads (lights, alarm systems, radio, etc.) during engine off conditions to prevent the accidental overdischarge of the high-voltage battery. Upon receiving an engine on signal, the 12 V supply will close several system contactors to electronically connect the high-voltage battery to the vehicle. Packs typically also contain manual service disconnects (MSDs) to electronically disable a pack in the event of a malfunction (i.e., welded contactors) or during maintenance. Low voltage (≤ 50 V) battery packs are typically grounded to the vehicle chassis, whereas high-voltage (≥ 50 V) systems are required to be electronically isolated. For additional safety, packs often contain high voltage interlock (HVIL) circuits that when broken will alert the vehicle to a possible isolation leak.

The pack/module/cell voltages and currents are measured and controlled by the packs battery management system (BMS). The BMS will often reference lookup tables and/or predefined algorithms to set cell voltage/current as well as pack energy/power limits for the battery based on factors such as state of charge (SOC), age, and temperature. The BMS is also responsible for communicating with the vehicle controllers the status of the pack contactors, HVIL, MSD, and temperature information, in addition to moderating the overall pack electrical output and input. For vehicles which are externally chargeable (PHEV and EVs), the BMS is also required to supervise and control the onboard charger and external plug interface. Over time individual cells will have their SOC drift from a common or average value due to imperfections in their assembly, capacity, power, and thermal history. Periodically, the BMS can attempt to perform either passive (i.e., discharge to a resistor) or active (i.e., redistribution of energy among high and low SOC cells) cell balancing events to return the pack subunits to a tighter SOC window to avoid overdischarge or overcharge.

Thermal: Electrified vehicle battery packs commonly employ a thermal management system designed to maintain the cell temperatures within a normal operating range. Lithium-ion battery cell design parameters such as electrolyte composition will strongly influence the preferred operating temperature range. As a practical matter, most chemistries are able to achieve a desirable balance of available energy and power in the range of 10–40 °C. Fortunately for air-cooled systems, this temperature window is similar to that which the vehicle operator will naturally prefer for the cabin with which the pack shares airflow. This results in a small temperature delta between the customer set thermal environment and the battery pack. Unfortunately, for liquid cooling systems, this temperature range is dissimilar from the existing engine/transmission ($\sim$100 °C) and power electronics ($\sim$80 °C) coolant. This difference in temperature windows means that it is not currently possible to commonize existing coolant loops to leverage existing vehicle cooling hardware.

Hybrid electrified vehicles typically have large amounts of waste heat energy from the thermal inefficiencies of the engine to perform heating functions. Since electric vehicles do not have an engine to supply excess heat, they typically employ positive thermal coefficient (PTC) heaters to warm the passenger cabin and potentially the battery pack. These heat sources, the exothermic nature of lithium-ion battery discharge/charge and the joule heating of pack components (on the order of 10% of the pack delivered power) means that most temperature control systems focus on cooling the pack rather than heating.

Cooling systems can vary in complexity and design but typically fall into three categories: passive air, active air, or liquid temperature control. A few vehicles (Nissan Leaf) depend on natural convection; however, most designs incorporate one or multiple fans to provide sufficient heat transfer. A passive air approach receives air from the passenger cabin and relies on the vehicle operator to determine the conditioning strategy (i.e., cooling in summer and heating in winter). This approach has the advantage of simplicity and, as a result, is used in most commercial electrified air-cooled vehicles (Toyota Prius). The active air approach would directly link conditioned air from the vehicles air conditioning system to the battery pack (First-Generation Ford Escape Hybrid, US Model Year 2004–2006) or add a second, dedicated air conditioning system solely for the battery (Lexus LS 600 h, US Model Year 2006–2012). In this way, the active air approach would bypass the operator and cabin temperature preference in favor of the optimum inlet temperature for the battery. Due to the additional air channels involved and/or second cooling system and the resulting associated weight, volume, and cost, the active air approach is much less common than the passive approach.

Liquid cooling typically allows for a greater degree of thermal control than air cooling given the need to use a dedicated heat exchanger and coolant loop. Additionally, liquid cooling is much more volumetrically efficient than air cooling due to much greater specific heat capacity of its cooling medium (typically water/ethylene glycol 50/50% mix) and channels (pipes vs. plenums). Unfortunately, liquid cooling does increase the weight, cost, parts count, and manufacturing complexity when compared to passive air cooling. As a result, only vehicles that require the greater degree of thermal control from liquid cooling (typically large EV or PHEV batteries) implement this solution (Ford Focus EV, Chevrolet Volt PHEV).

The goal of all thermal management systems is to keep the average battery temperatures within the normal operating range and provide for uniform temperatures across the entire battery pack. Uniform battery cell temperatures are needed to minimize operating cell voltage variations caused by differences in resistance, which can have significant temperature dependence. It is equally important to avoid cell-to-cell temperature variations to maintain a uniform state of charge since self-discharge is also significantly temperature dependent.

3.2 *Battery Sizing*

Determining the appropriate battery energy and power to put on board an electrified vehicle requires balancing a variety of competing requirements. Each vehicle class has different needs based on its electrified features. Ultimately, the goal of electrification is to improve the vehicle fuel economy, so an understanding of drive cycles is necessary. After considering the vehicle driving demand, it is possible to size a vehicle's energy and power. After the overall energy and power of the pack is determined, it is necessary to determine the quantity and arrangement of cells to meet these targets. A variety of business factors can determine the ultimate cell selection; we will focus on the technical requirements of pack voltage and capacity.

Drive Cycles: Regulatory fuel economy is confirmed by each government by requiring a vehicle to follow prescribed drive cycles, traces of speed and time. Typically, the ultimate label fuel economy is a convolution of multiple low (urban) and high speed (highway) cycle results. Table 2 and Figs. 3 and 4 illustrate representative drive cycles of the US, Economic Commission for Europe (ECE), and Japan for the purposes of comparison [7]. A visual review of the traces enables a qualitative comparison among the cycles. Among the high-speed cycles, the US06 pattern has the greatest number of speed changes, whereas the EUDC cycle has

Table 2 International drive cycle summary [7]

Name		Time	Length	Speed (kph)		Absolute acceleration (m/sec^2)	
		(sec)	(km)	Avg	Max	Avg	Max
USA	UDDS	1369	11.99	31.51	91.25	0.40	1.48
	HWY	765	16.51	77.58	96.40	0.17	1.48
	US06	596	12.89	77.86	129.23	0.61	3.76
EU	EUC	195	0.99	18.26	50	0.27	1.06
	EUDC	400	62.44	62.44	120	0.19	1.39
Japan	10–15	892	25.58	25.58	70	0.27	0.83

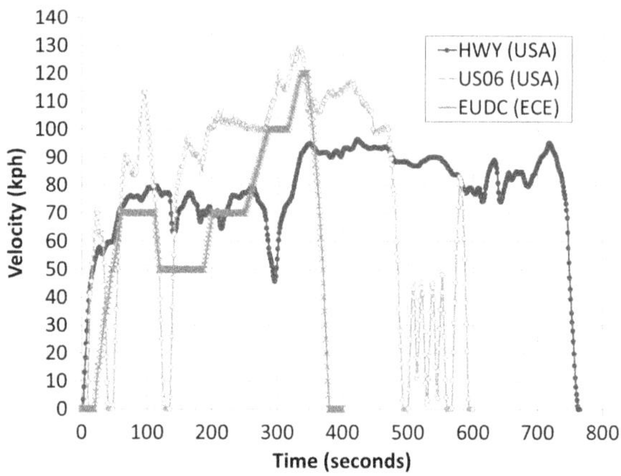

Fig. 3 Speed traces (fast) [7]

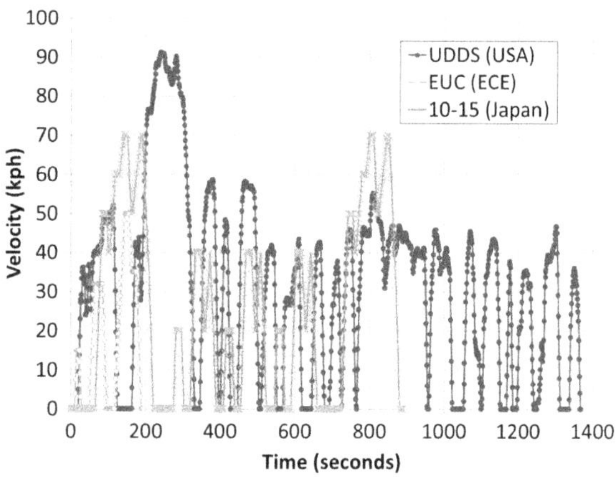

Fig. 4 Speed traces (slow) [7]

several periods of zero acceleration. The low-speed traces show that the UDDS pattern has the largest speed maximum and greatest frequency of speed changes, thereby likely the largest power demand.

By reviewing the average and maximum of each cycles speed (kph) and absolute acceleration (m/sec^2), it is possible to achieve a more detailed cycle comparison. The US06 and HWY cycles have similar average speeds, but very different speed maximums and acceleration values, therefore US06 will be more demanding on a vehicles powertrain. A similar comparison of speed and acceleration in the context of vehicle propulsion needs to be described earlier leads to the following ranking for cycle power demand US06 > HWY > EUDC > UDDS > EUC ~ 10–15.

Power: The battery pack power requirement is driven by several different vehicle functional requirements. For hybrid vehicles, the power required to cold crank an engine is a critical sizing need. Vehicles which need to contribute to the traction power have the battery power demand vary as a function of degree of electrification.

Cold cranking an internal combustion engine is one of the critical functions of the starting, lighting, and ignition (SLI) battery, ubiquitously performed by lead–acid batteries in conventional vehicles. In the case of the low voltage batteries in S/S hybrids or the high-voltage units in HEV and PHEV designs, providing sufficient power to perform this function is an important sizing requirement. The specifics of a cold crank pulse vary from engine to engine, but a representative 12 V S/S specification requires a cycle of 6 kW for 0.5 s and then 4 kW for 4 s, to be repeated 3 times, all at −30 °C [8]. The majority of HEVs and PHEVs need to contain lead–acid batteries on board to drive vehicle auxiliaries during engine off conditions and to open/close the high-voltage contactors between the high-voltage battery pack and the vehicle's electrical system. As a result, the actual engine cranking function are assigned to the high-voltage battery, enabling the size reduction of the supporting lead–acid battery in comparison to traditional vehicles. Although the power demands of this function ($\sim$6 kW) are not large for a typical high-voltage design, the requirement to deliver this power at −30 °C makes this a critical value to battery sizing. Lithium-ion batteries use liquid electrolytes which follow Arrhenius kinetic behavior, leading to significant reductions in available power at low temperatures.

The US06 drive cycle was shown to be one of the most power demanding cycles in Figs. 3 and 4 and Table 2. By performing a battery sizing exercise based on the US06 cycle, it is possible to determine the upper boundary of the traction battery requirement. Figure 4 shows the US06 vehicle speed trace along with the requisite absolute power from the road load required for a typical midsize (C/D class) sedan to match the trace. A review of Fig. 5's power profile shows that in order for a midsize sedan to run the US06 pattern in all electric mode, it would need an electric powertrain that could deliver approximately 83 kW. As a practical matter, electrical losses in the vehicle electrical components mean that a battery pack power of approximately 110 kW is typically needed to deliver a target road load power.

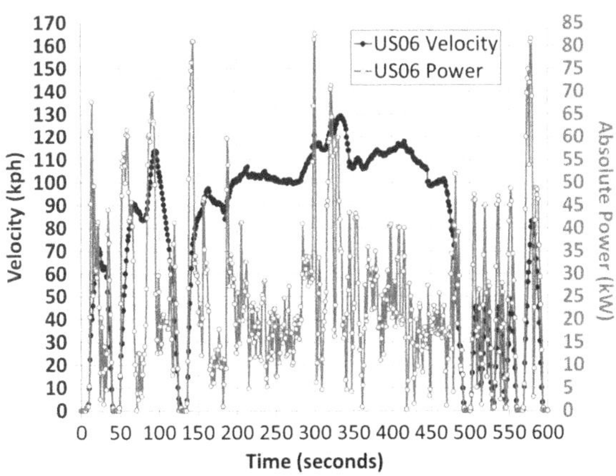

Fig. 5 US06 power profile for C/D sedan

A time histogram of power and velocity of the US06 pattern for a midsize sedan is shown in Fig. 6. A review of this chart shows that the peak power demand identified in Fig. 5 only occurs for a very short portion of the cycle, with powers of 60 and 35 kW sufficient for 95 and 80% of the cycle time, respectively. Using data such as this for a given vehicle design and drive cycle, it is possible to determine the resulting fuel economy savings as a function of battery power. Once this correlation is known the automotive designer must balance the other design constraints (cost, weight, volume, etc.) in determining how much electrical battery power to place on board.

Electric Range: The stress on a battery pack can vary significantly as a function of drive cycle, as was shown above with regards to power. The same is true for energy consumption (Wh/km) which can be defined as varying linearly with the force required to maintain the drive cycle speed trace.

$$\text{Consumption} = \frac{\text{Energy}}{\text{Distance}} = \frac{\text{Power*Time}}{\text{Distance}} = \frac{\text{Power}}{\text{Velocity}} = \frac{\text{Force*Velocity}}{\text{Velocity}} = \text{Force}$$

In the previous sections on propulsion and battery power sizing, it was shown that the vehicle force equation has terms which vary in linear and cubic relations with velocity. A quantitative understanding of consumption can be achieved by again looking at the most power intense cycle, US06. In Fig. 7, the US06 speed trace is compared to the cumulative energy expended throughout the cycle. The potential impact of regenerative braking is seen during the vehicle deceleration events; since this is a plot of road load, the net consumption of 2.54 kWh (for a typical midsize sedan) during US06 should be seen as a lower limit. Battery power must always exceed the required road power to account for inefficiencies such as joule heating of electrical components and electrical losses in the motor and inverter. As a result, this simulation would translate to a consumption of approximately 240 Wh/km when accounting for real-world losses and the challenges in

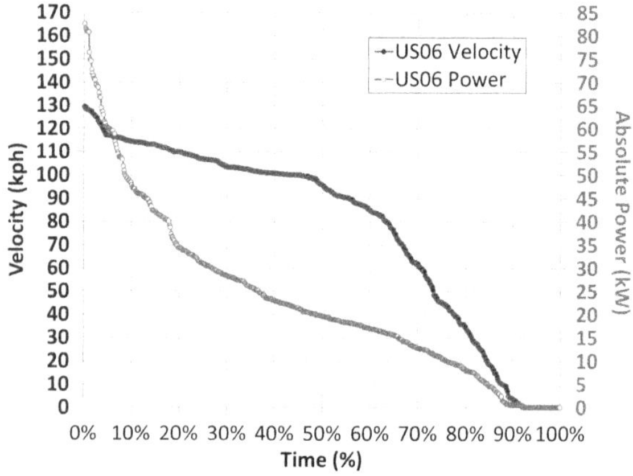

Fig. 6 US06 power histogram for C/D sedan

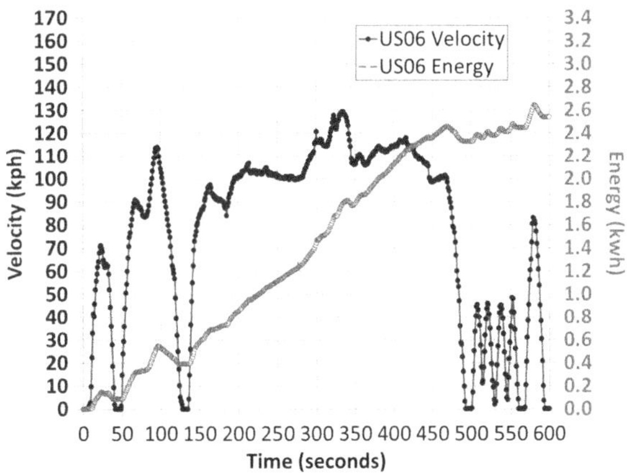

Fig. 7 US06 velocity and energy for C/D sedan

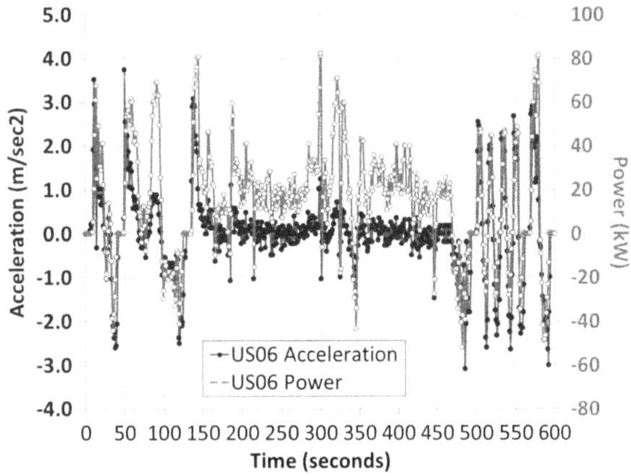

Fig. 8 US06 acceleration and power for C/D sedan

sizing pack power to run all of US06 in electric mode (83 and 52 kW discharge and charge power for a typical midsize sedan, respectively, see Fig. 8).

A similar calculation can be performed with each of the cycles described in Table 2 and Figs. 3 and 4 with consumption values ranging from the high of US06 to an approximate low of 95 Wh/km while running the 10–15 mode cycle. Due to this large impact of driving behavior on energy consumption (a span of $\times$ 2.5), it can be difficult to precisely predict a PHEV or EV's all electric range without knowing the route power needs in advance.

Voltage/Capacity: Sizing the total energy and power requirements of a battery pack are primarily driven by the vehicle road load. Deciding a battery packs number of cells and series-parallel electrical arrangement is mainly determined by available cell technology and vehicle electrical component needs.

A battery pack's desired voltage window is heavily influenced by electrical component (motors, power electronics, wiring, etc.) sizing, efficiency, and cost. S/S hybrids have their voltage requirements very tightly controlled (10.5–15 V) by the existing vehicle electrical components, notably the alternator [8]. Typically, HEVs have pack voltages ranging between 200 and 300 V (with the exception of some mild hybrids with 100–150 V ratings), EVs are 275–375 V and PHEVs are in between the two bands. Owing to the high cell-level voltages of lithium-ion chemistries, it is possible to achieve these S/S and HEV pack voltages using a single string of many cells in series. Some manufacturers have introduced DC/DC boost converters into their battery packs to decouple the battery DC voltage and the corresponding inverter AC voltage.

Adding parallel strings of cells to a battery pack is typically only done when the capacity (Ah) of available cells is not appropriate to deliver the desired pack energy content through a single series string. As a result, parallel strings are only seen in PHEV and EV pack designs due to their higher energy needs. The designed pack capacity is also influenced by the applications typical energy window. Operating within different regions of lithium-ion batteries' state of charge (SOC) window can have a significant impact on the cell's power capability and degradation rate. As a result, each electrified vehicle type has an SOC window in which the competing factors of available energy, power, and degradation rates are optimized.

Figure 9 below shows the typical regions of unused, charge sustaining, and charge-depleting capacity for each vehicle type [9]. Each automaker and cell manufacturer will prescribe different SOC windows for their packs and cells, respectively; the values in Fig. 9 are only for illustration of various concepts. As was stated earlier, the voltage window of a S/S battery is tightly prescribed by the alternator voltages required and so is consequentially its SOC. HEVs are primarily required to deliver high power, low energy pulses, hence they tend to have charge sustaining (CS) regions about 50% SOC to minimize degradation and balance available power between charge and discharge resistance. The charge-depleting (CD) region of a PHEV gives its electric range and distinguishes it from an HEV. However, the lower end of the CD zone is limited to avoid the accelerated

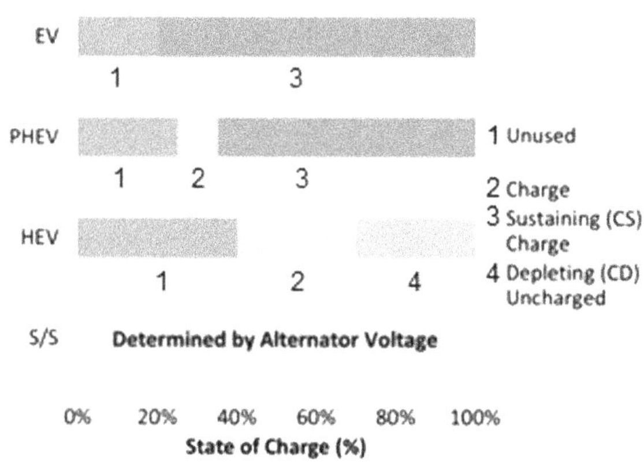

Fig. 9 Representative SOC Windows [9]

degradation of overdischarge (i.e., near 0%) while maintaining sufficient discharge power (>15%) to maintain the CS function. The simplified SOC operation of the EV is shown in its large CD region.

3.3 Pack Performance Targets

Vehicle electrification is a global phenomenon and an appraisal of its development must be cast with a similarly wide viewpoint. Each of the three most mature automotive manufacturing regions (United States, Europe, and Japan) has published EV battery targets. In the US, the United States Advanced Battery Consortium (USABC) is composed of Fiat Chrysler, Ford and General Motors in partnership with the US Department of Energy [10]. The European Council for Automotive Research and Development (EUCAR) is an analogous organization in Europe [11]. In Japan, the Ministry of Economy, Trade and Industry (METI) established an agency to promote the development of new energy technologies called the New Energy and Industrial Technology Development Organization (NEDO) [12]. Each of these respective organizations, USABC, EUCAR, and NEDO has created EV battery targets to guide industry's technology development in 2013, 2009, and 2008, respectively. The most relevant targets for a variety of time periods are shown in Table 3 [13–15]. The USABC goals are for system (pack) and cell-level performance for commercialization in 2020. The EUCAR and NEDO targets span 5-year intervals up to 2030.

A comparison of Table 3 across organizations and time reveals a few consistent trends. A combined appraisal of specific energy implies that a value of at least 200 Wh/kg at the pack level is required for a competitive EV and is predicted in the early 2020's. The EUCAR and NEDO targets for power are noticeably more aggressive than the USABC values. The energy of a battery is an intrinsic property determined by the choice of materials and their electrochemical properties of voltage and capacity. Power on the other hand is an extrinsic property, influenced by the material behavior, but also substantially controlled by specific design choices. Most long-range EVs carry significant battery packs on board which means the relative specific power required is rather low (see Table 16). An appraisal of the life targets shows alignment on a life of the vehicle expectation of 10–15 years.

Table 3 Electric vehicle targets [13–15]

Organization	USABC (2013)		EUCAR (2009)			NEDO (2008)			
Hardware-level	Pack	Cell	Pack			Module			
Timing	2020		2010	2015	2020	2008	2015	2020	2030
Energy (Wh/kg)	235	350	90–100	130–150	180–200	100	150	250	500
Power (W/kg)	470	700	400–750	500–950	600–1250	1000	1200	1500	1000
Calendar life (Y)	15	15	8–10	10	15	5–8	8–10	10–15	10–15
Cost ($/Wh)	0.125	0.10	0.40–0.50	0.30	0.15	1.00–2.00	0.30	0.20	0.10

Table 4 Hybrid electric vehicle targets [14, 18]

Organization	USABC (2002)		NEDO (2008)			
Hardware-Level	Pack		Module			
Timing	Min	Max	2008	2010	2015	2020
Energy (Wh/kg)	7.5	8.3	70	70	100	200
Power (W/kg)	625	667	1800	2000	2000	2500
Calendar life (Y)	15	15	5–8	5–8	8–10	10–15
Cost ($/kW)	20.00	20.00	51.43	28.57	6.00	2.50

It is also possible to compare the high-power (HEV) battery targets of the USABC and NEDO (see Table 4). A clear difference in these targets is seen, primarily owing to the USABC targets focusing on required vehicle useable energy and power and NEDO on nameplate module values. Additionally, it is notable that in a $/100 yen scenario, the USABC and NEDO price targets were scheduled to intersect in approximately 2012. As in the case of EV energy battery targets, there is consensus toward achieving a 15-year battery life expectation for HEV batteries.

4 Current Technology

4.1 Energy and Power

Lithium as an element is a very attractive material to base batteries on owing to its very low electrochemical potential which gives it the promise of a high cell voltage. As a result, lithium-based batteries have been in development for over 40 years since the first intercalation/deintercalation was demonstrated in the 1970s (see Fig. 10) [16]. Twenty plus years of research culminated in the first commercial production of a lithium-ion battery by Sony Energy Device Corporation in 1991 [17]. Sony was able to create a product suitable for the consumer electronics market whose form factor (the 18,650 cylindrical cell or 18 mm diameter and 65 mm height) has since become an industry mainstay and is today produced in the billions annually. Due to its ubiquity and massive production volumes, the 18,650 cell is a useful benchmark for the development of lithium-ion battery technology as a whole. Despite the continuous improvements in commercial lithium-ion technology over the last 20 years, it has been only recently that the technology has been suitable for some automotive applications.

Energy: Module and pack designs can vary substantially; however, they all add additional weight and volume which effectively de-rates the cell-level performance values. In the case of weight, estimates of a 20% loss at the module level and a further 10% loss at the pack level are often made by industry groups [14]. These estimates would yield a battery versus cell pack weight efficiency of 72%, which falls at the high end of practical designs which yield values ranging from 50 to 75%.

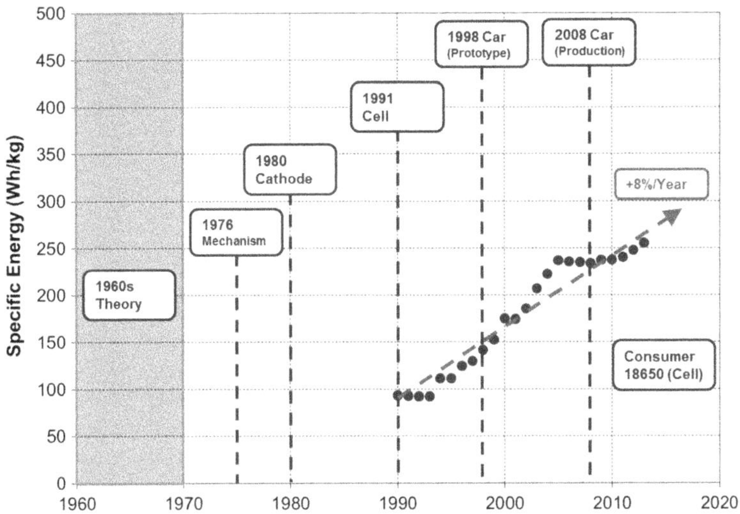

Fig. 10 LIB chemistry and 18650 cell development (Wh/kg)

One of the principle differences between consumer electronics and automotive applications is the duration of use in terms of both calendar and cycle life. Typically, consumer electronics require 500 cycles and 1–2 years of useful life. In automotive applications, the battery is part of a traction system which is expected to last the life of the vehicle, requiring thousands of cycles over 10–15 years [13, 18, 19].

While posing many challenges for automotive applications, the usage of 18,650 cells in automotive applications was first proposed and demonstrated by AC Propulsion of California in the 2006 eBox. The suitability of 18,650 s for automotive aside, this cell format and size is at the forefront of specific energy developments and so considering its rate of improvement can prove instructive relative to the broader class of lithium-ion technology. In Fig. 11, it is shown that when considering the highly engineered and mass produced 18,650 cells, specific energy values have only recently approached the goals outlined by USABC for long-term automotive production.

Although considering the capabilities of 18,650's in abstract designs is a useful calibrating investigation, it is also beneficial to examine existing vehicle designs. Also in Fig. 11, a variety of EV battery specific energies are plotted. Three EVs from Ford show evolution of chemistries through time, the 1998 Lead–Acid Battery Ranger EV, the 1999 Nickel–Metal Hydride Ranger EV and the 2012 Lithium-Ion Battery Focus EV. For comparison, the Nissan Leaf EV, Tesla Roadster, and Model S (using 18,650 cells) are shown as well. It should be noted that the specific energy achieved in vehicles is below the value estimated for the 18,650 cell-based pack specific energy capability available in that year. The specific energy achieved in an actual automotive application is approximately that which was predicted to be possible with an 18,650 cell-based pack about eight to 9 years earlier. This can be

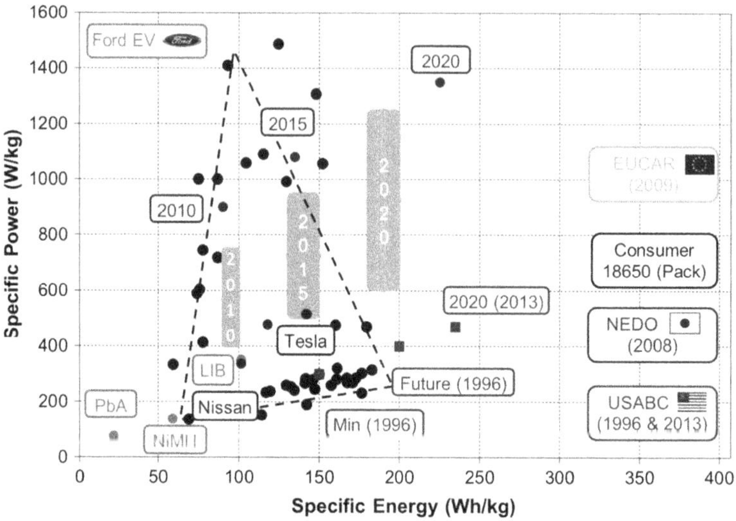

Fig. 11 EV and 18650 energy and power. PbA, NiMH, LIB: developed by Ford; vertical rectangles: EUCAR 2010, 2015, 2020; 1996 and 2013, USABC; 2010, 2015, 2020 (small horizontal rectangles): NEDO; dotted triangle: consumer 18650 (pack) in various applications

attributed to several factors specific to battery designs for automotive applications, but in general communicates the challenges in using cells designed for consumer electronics in an automotive application.

Power: A general principle of battery design is that power and energy can be optimized, but rarely at the same time, and this is shown in the triangle found in Fig. 11, which represents a summary of existing state of the art 18,650 cell-based pack specific power versus energy. These cells are designed for either high energy or high-power applications and the estimated performance envelope of the technology takes the shape of a triangle.

As in the case of energy, it is useful to consider the EV battery power targets of the USABC, EUCAR, and NEDO (as well as the performance of existing vehicles) and those results are also plotted in Fig. 11. It is apparent that the performance of the Ford Focus, Nissan Leaf, and Tesla Model S all fit inside the 18,650 performance triangle. Additionally, the higher power performance targets of EUCAR and NEDO introduced in the methodology are highlighted here when compared to USABC targets.

4.2 Cell Design

Lithium-ion cell designs for automotive have several cell design features that automotive battery engineers need to consider. In the previous battery sizing section, how to determine the energy and power needs of a vehicle were described. In addition to these values, additional cell design considerations are important such as geometry, case, terminals, vents, dimensions, and standardization.

Geometry: The three main geometry options, cylindrical, pouch, and prismatic, are shown in Table 5 [20]. Historically the cylindrical form factor has been used extensively in laptop applications in the 18650 size described previously. Prismatic cells and pouches have been used in cell phones due to their reduced thicknesses and greater packing densities. For automotive applications, a wide variety of cell form factors are used commercially (see Tables 14, 15, and 16). Cylindrical cells have their internal cathode/separator/anode sheets wound on a common mandrel to create what is known as a jelly roll (see Table 5). Prismatic and pouch cells can have their internal electrodes stacked on top of each other or wound into a jellyroll shape which is then compressed flat to fit into their casing.

Case: Common materials used for lithium-ion cell cases are stainless steel and aluminum. For prismatic and cylindrical cells, case thicknesses are typically optimized for weight savings and formability for stamping. Following stamping of the prismatic and cylindrical case, a head assembly containing the terminals, vent port, and electrolyte fill port is typically laser welded to the rest of the case. Pouch cells are typically formed from thin aluminum foil (<50 μm) coated with various external (~15 μm films of polyester, polyimides, etc.) and internal barriers (<100 μm polypropylene) and adhesive layers. Plastic cases used in lead–acid or nickel–metal hydride chemistry cells typically will experience material durability issues in lithium-ion cells.

Table 5 Cell case/jellyroll geometry [20]

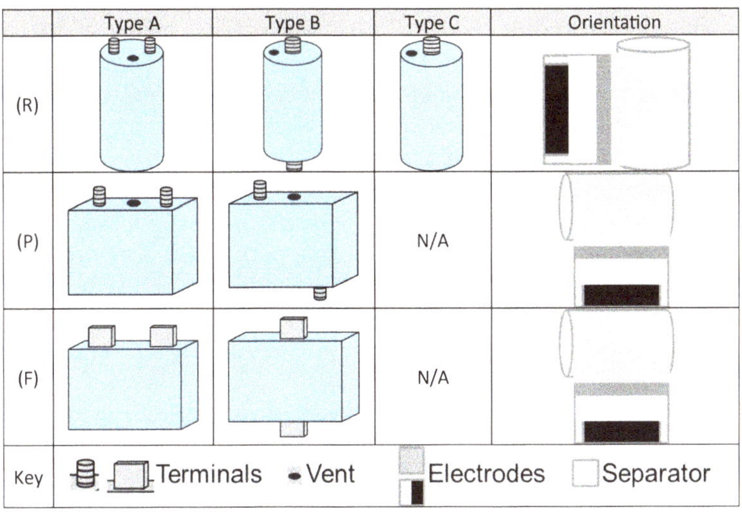

Existing cell cases are not designed to bear structural loads; as a result, automotive designers have to add additional mechanical structures to the pack design to avoid loading the cells themselves. Cell designers must also decide to treat the metallic cases in prismatic and cylindrical cells as either case positive or neutral. Cases which are in electronic contact with their electrodes will develop an electric potential, hence their possible polarity. Cases which have a positive polarity often need external insulation (typically a shrink warp polymer film), whereas those that are case neutral are outfitted with insulation material internally to shield the case from the jellyroll's electrochemical couple.

Terminals: As can be seen in Table 5, three different permutations of terminal positioning are used in lithium-ion cells [20]. The most common approach is to use position both positive and negative terminals on the same face, Type A. This approach allows for one common cell face to contain both terminal interconnects and weld points to the internal jelly roll. By having both terminals on one side, it is possible to simplify the electrical wiring in the resulting pack structure. The Type B terminal layout features terminal on opposing faces and is sometimes used for HEV pouches. The Type B layout typical trades improvements in current and heat distribution for increased packing and manufacturing complexity, hence their use in high-power applications. The Type C design uses the case structure as the negative electrode and therefore only requires one dedicated terminal. Type C layouts are only found in cylindrical cells and are not very common in any industry.

Vents: Lithium-ion cells are designed as sealed systems, unlike many lead–acid and nickel–hydride-based batteries which have reusable pressure valves. A malfunctioning lithium-ion cell will often create gases resulting from the decomposition of its materials. To prevent the buildup of large pressures during such a malfunction, prismatic and cylindrical cells employ one time used vent ports. Concerns surrounding the outgassing byproducts with battery packs mounted near the passenger compartment result in some automakers employing gas routing ducting over each cell's vent port. Due to the nature of their seals, predicting the location of vent in pouch cells can be more difficult. To provide some direction to possible vent events, cell makers can design in weak points into the pouch design to create a region of preferential foil failure.

Dimensioning: In automotive packaging and lithium-ion cell design, not all axes are of equal importance. Typical automotive battery mounting locations are under seats, center columns, or trunks. As a result, the pack height direction is the most important to reduce so as to avoid interfering with passenger and storage spaces. Typically, pack width and length constraints must be determined in context of the specific vehicle and pack mounting and, as a result, neither is as universally important as height.

Table 5 shows the jellyroll axis of wrap relative to the cell shape for cylindrical and the flat wound wraps used for prismatic and pouch cells. During the manufacture of electrodes, one very wide roll is created that is then cut down to the target jellyroll width for each application. As a result of this process, changing the width of the application jellyroll and by extension the electrodes and separators are very costly and involved. Changing the jellyroll thickness and height is comparatively easy to accomplish as different lengths of electrode/separator would be used.

The majority of pack designs feature cells with Type A terminals oriented on the normal axis to the road surface. With this design, decreasing the height axis in a cylindrical cell would require decreasing the electrode width. This contrasts with prismatic and pouch designs where decreasing the height could be accomplished by relatively simple changes of the roll length. This examination of cell and jellyroll dimensioning shows the considerations that cell makers must make before investing in dedicated production lines to serve current and possible future cell sizes.

Standardization: The mass production of automobiles and other high volume goods is made possible by using standardized parts. It is through parts standardization that economies of scale are able to make manufacturing economically competitive. As a result, there have been recent efforts to standardize automotive lithium-ion battery cell sizes. These efforts have met with limited success given the relative immaturity of the technology for the automotive application. Within individual companies, the same cells are used for various vehicles of the same type; however, efforts to communize among companies and countries are more challenging.

In the recent past (2010 onwards), three different countries have indicated national preferences for automotive lithium-ion cell sizes. National standards have been issued by Germany (DIN 91252) and China (QC/T 840) defining 10 and 8 sizes, respectively. The USABC has determined a list of 6 preferred sizes among its members. In this context is not surprising that when ISO attempted to issue an international standard the size list grew to 63 (seeTable 6).

ISO 16898 prescribes a specific naming standard for propulsion intent lithium-ion cells that describes the cell application, shape, terminal type, and dimensions. This naming scheme uses the dimension assignments in Table 7 and the labels in Table 8 to create the name as follows: $A_1A_2A_3A_4N_1/N_2/N_3$. For an example of the ISO 16898 convention, a prismatic (P) lithium-ion (I) cell intended for propulsion purposes (V) with terminals on the same face and a thickness of 26.5 mm, width of 148 mm, and height of 91 mm would be called a VIPA27/148/91 cell.

Table 6 Lithium-ion cell sizes by case and organization [20–22]

Type	DIN 91252	QC/T 840	USABC	ISO 16898
Cylindrical	1	1	0	8
Prismatic	5	7	3	27
Pouch	4	0	3	28
Total	10	8	6	63

Table 7 Cell dimensions in ISO 16898 [20]

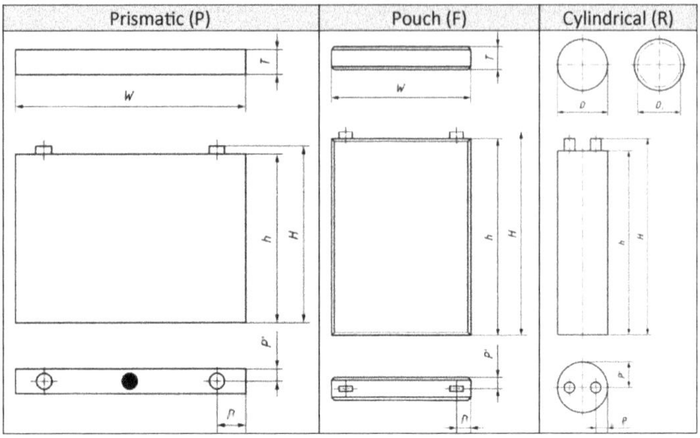

Table 8 ISO 16898 naming scheme [20]

Designation		Description
A1	Application	V = Propulsion
A2	Battery system	I = Lithium-Ion
A3	Shape	R = Cylindrical P = Prismatic F = Pouch Y = Other
A4	Terminal type	A = Terminals same face B = Terminals opposite face C = Case as terminal
N1	Diameter (D) or Thickness (T)	mm rounded up
N2	Width (W)	mm rounded up
N3	Height (H, h)	mm rounded up

4.3 Safety

Achieving appropriate levels of product safety is a prerequisite for designing an automotive systems, battery, or otherwise. A variety of government regulations and industry standards exist which describe mechanical, thermal, and electrical battery abuse tests (see Tables 9, 10, and 12). The response of a battery (cell, module or pack) to these abuse tests can be characterized according to a EUCAR developed rating systems as shown in Table 13.

Mechanical: In the mechanical testing area, there is no one test type that all standards and regulations agree should be performed (Table 9). A common test

Table 9 Mechanical abuse testing standards and regulations [23–31]

Test Type	Industry standard					Government regulations			
	Freedom car	SAE J2929	SAE J2464	ISO 12405-1	ISO 12405-3	UN 38.3	ECE R100	Q/C-T 743	KMVSS 1.48
Mechanical Integrity	•	•	•		•		•	•	
Penetration	•		•					•	
Immersion	•	•	•		•				•
Roll-Over	•		•						
Drop	•	•	•		•	•		•	•
Mechanical Shock	•	•	•	•	•	•	•		
Vibration		•		•	•	•	•	•	

Table 10 Thermal abuse testing standards and regulations [23–31]

Test type	Industry standard					Government regulations			
	Freedom car	SAE J2929	SAE J2464	ISO 12405-1	ISO 12405-3	UN 38.3	ECE R100	Q/C-T 743	KMVSS 1.48
Thermal stability	•								
Fire exposure	•	•	•		•		•		•
High temperature storage	•							•	•
Cycle w/o thermal control	•	•	•		•		•		
Thermal shock	•	•	•	•	•	•	•		
Humidity exposure		•		•	•				
Passive propagation			•						

among all the standards is a mechanical shock test, typically a more aggressive variant of vibration testing for durability, designed to resemble a crash impulse (Fig. 12). Another popular test type found among 7 of the 9 references reviewed is a drop test. The height parameter of this test ranges widely from the 1 m drop to resemble vehicle maintenance conditions of SAE J2929 to the 10 m height of SAE J2464 which was modeled after a similar procedure designed to test the survivability of hydrogen storage tanks for fuel cell vehicles.

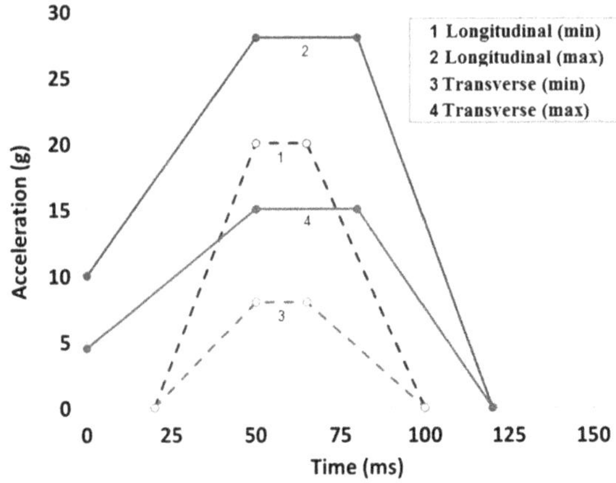

Fig. 12 ECE R100 mechanical shock pulse for M1 (<3500 kg) vehicles [29]

Table 11 ECE R100 fire exposure test [29]

Phase	Time	Setup
A	60 sec	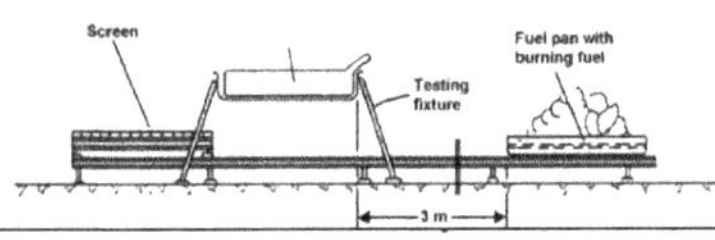
B	70 sec	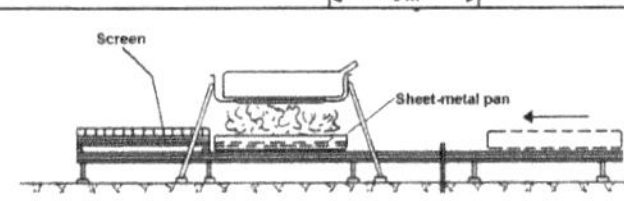
C	60 sec	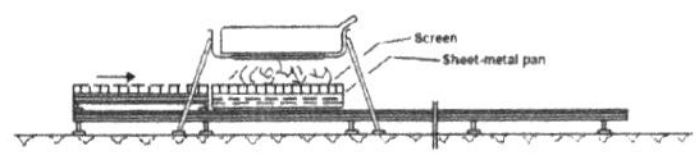
D	3 hour	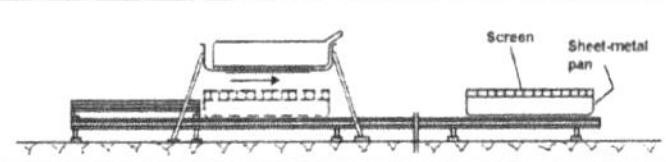

Thermal: The most common thermal-based battery abuse testing involves exposing the test battery to a rapidly increasing thermal environment, either in the thermal shock or fire exposure tests, as seen in Table 10. The thermal shock testing has the battery cycled between high (~60 °C) and low (~−40 °C) storage temperatures in controlled thermal chambers. The fire exposure test will see the battery exposed to actual or simulated heat loads emanating from a fuel fire. The fuel fire test is typically modeled after existing testing performed on plastic fuel tanks (see Table 11 for the 4 phases of a fuel fire exposure test). Given the flammability of liquid electrolyte in lithium-ion batteries, the thermal tests can be particularly informative for the battery designer.

Electrical: The electrical battery abuse testing subcategory is the only one with significant test type agreement among all the standards and regulations referenced in Table 12. Each testing document contains procedures for performing overcharge, overdischarge, and short-circuit testing. It is notable that these three electrical tests are identified as being important for electrical abuse testing by all standards and regulations.

Response Rating: The EUCAR response scoring criteria (see Table 13) is a useful tool to quickly communicate the performance of a battery during an abuse test. Scores of 0 to 2 indicate either no impact or marginal mechanical damage to the device under test (DUT). If the DUT hardware experiences a vent event, it is either scored as a 3 or 4 depending on the quantity of weight loss as a result of the vent. EUCAR scores of 5, 6, and 7 denote major responses to the abuse test in question, demonstrating fire, rupture, and explosion, respectively.

Table 12 Electrical abuse testing standards and regulations [23–31]

Test type	Industry standard					Government regulations			
	Freedom car	SAE J2929	SAE J2464	ISO 12405-1	ISO 12405-3	UN 38.3	ECE R100	Q/C-T 743	KMVSS 1.48
Overcharge	•	•	•	•	•	•	•	•	•
Short circuit	•	•	•	•	•	•	•	•	•
Overdischarge	•	•	•	•	•	•	•	•	•
High-voltage exposure		•							
Partial short circuit	•								
Separator shutdown			•						

Table 13 EUCAR score definitions [23]

Score	Title	Description
0	No effect	No effect. No loss of functionality
1	Passive Protection Activated	No defect; no leakage; no venting, fire or flame; no rupture; no explosion; no exothermic reaction; or thermal runaway. Cell reversibly damaged. Repair of protection device needed.
2	Defect/damage	No leakage; no venting, fire, or flame; no rupture; no explosion; no exothermic reaction; or thermal runaway. Cell reversibly damaged. Repair needed.
3	Leakage (Δ mass < 50%)	No venting, fire or flame; no rupture; no explosion. Weight loss < 50% of electrolyte weight (electrolyte = solvent + salt).
4	Venting (Δ mass $\geq$ 50%)	No fire or flame; no rupture; no explosion. Weight loss of $\geq$ 50% of electrolyte weight (electrolyte = solvent + salt).
5	Fire or Flame	No rupture; no explosion (i.e., no flying parts)
6	Rupture	No explosion, but flying parts of the active mass
7	Explosion	Explosion (i.e., disintegration of the cell).

Most of the industry standards described in Tables 9, 10, and 12 describe a test procedure and then leave it up to the hardware developer to determine what EUCAR response is appropriate. The majority of regulations and SAE J2929 prescribe pass/fail criteria for the DUT abuse response, typically requiring scores of 4 or lower. Given the flammable liquid electrolyte used in the vast majority of lithium-ion batteries, achieving a EUCAR score of 5 or lower and thereby avoiding any fire generation is a desirable engineering goal.

4.4 Vehicles

In recent years, a large variety of automakers have launched lithium-ion electrified vehicles. By reviewing the vehicle battery designs of commercialized products, it is possible to confirm various design trends, for each vehicle type and overall. Tables 14, 15, and 16 lists the cell and pack specifications for a large variety of lithium-ion containing HEVs, PHEVs, and EVs, respectively. S/S hybrid vehicles utilizing lithium-ion chemistries are in development at various automakers, however, they have not yet seen significant commercialization.

HEV: The hybrid electric vehicle (HEV) class is the most technically mature of the electrified vehicle, with some vehicle powertrains on their third or fourth-generation designs. Virtually, all recently launched HEVs utilize lithium-ion with the notable exception of most Toyota hybrids which still use nickel–metal hydride technology. A list of lithium-ion containing HEVs shown in Table 14 highlights common design principles. All HEVs in Table 14 feature one single series string of cells to build up their pack voltage. The average and standard deviations of energy (1.16 $\pm$ 0.46 kWh), power (35 $\pm$ 23 kW), and power/energy (P/E) ratio (29 $\pm$ 8) show some design convergence.

PHEV: Plug-in hybrid electric vehicles have only been enabled by high energy and power density of lithium-ion technology relatively recently. As a result, this vehicle class has the least number of variants in production described in Table 15. Consistent with a new vehicle type, the amount of electric power (80 $\pm$ 45 kW) and energy (9.7 $\pm$ 3.1 kWh) that each automaker places on board also varies. Two groupings of PHEV pack designs emerge, relatively large (>15 kWh and 110 kW) and small ($\sim$6 kWh and 40 kW) pack designs. An examination of pack power/energy (PE) ratio (8.9 $\pm$ 6.8) shows a broad distribution in between HEV and EV values. Although different automakers may size their batteries differently, the underlying cell technology and the P/E ratio that it can achieve while maintaining suitably high energy levels is consistent among the industry.

EV: The commercialization of electric vehicles is currently experiencing a wave of popularity supported by advancements in lithium-ion technology. As a result, there are a large number of vehicle offerings available from automakers, not seen since the lead–acid and nickel–metal hydride powered vehicles of the 1990s (see Table 16). As in the case of PHEVs, there is also a large variety in pack power (109 $\pm$ 59 kW) and energy (35 $\pm$ 22 kWh). However, unlike PHEVs, there is no

Table 14 Lithium-ion containing HEVs [32, 33]

Vehicle	Cell				Pack					
	Supplier	Type	Volt (V)	Cap (Ah)	S/P	Volt (V)	Cap (Ah)	Energy (kWh)	Power (kW)	P/E
Acura NSX	Blue Energy	P	3.6	5.0	72S1P	259.2	5.0	1.30	55	43
Acura RLX	Blue Energy	P	3.6	4.7	72S1P	259.2	4.7	1.22	35	29
Audi Q5	Panasonic	P	3.7	5.0	72S1P	266.4	5.0	1.33	40	30
BMW 5 Series	A123	R	3.3	4.0	96S1P	316.8	4.0	1.27	25	20
Buick Lacrosse/Regal	Hitachi	R	3.6	4.4	32S1P	129.6	4.4	0.57	15	26
Chevrolet Malibu	Hitachi	P	3.6	5.3	80S1P	288	5.3	1.53	60	39
Chevrolet Silverado, GMC Sierra	Hitachi	P	3.6	5.3	24S1P	86.4	5.3	0.46	15	33
Ferrari LaFerrari	Samsung	P	3.6	5.2	120S1P	432	5.2	2.25	120	53
Ford C-Max/ Fusion, Lincoln MKZ	Panasonic	P	3.6	5.0	76S1P	273.6	5.0	1.37	35	26
Honda Accord	Blue Energy	P	3.6	5.0	72S1P	259.2	5.0	1.30	50	39
Honda CR-Z	Blue Energy	P	3.6	4.7	40S1P	144	4.7	0.68	15	22
Hyundai Ioniq	LG Chem	F	3.75	6.5	64S1P	240	6.5	1.56	32	21
Hyundai Sonata, Kia Optima	LG Chem	F	3.75	6.0	72S1P	270	6.0	1.62	38	23
Infiniti Q50/ Q70	AESC	F	3.65	4.1	96S1P	350.4	4.1	1.44	50	35
Infiniti QX60, Nissan Murano	Hitachi	R	3.6	4.4	40S1P	144	4.4	0.63	15	24
Toyota Prius	Primearth	P	3.7	3.6	56S1P	207.2	3.6	0.75	18	24
Volkswagen Jetta	Panasonic	P	3.6	5.0	60S1P	216	5.0	1.08	20	19

clear grouping of design choices but rather a continuum. A comparison of P/E ratios (3.4 ± 1.2) shows much greater consensus of design and technology ratios than absolute power and energy. As was seen in the PHEV designs this reflects similar cell technology power capability at the high levels of energy density required for EV applications.

Overview: A review of Tables 14, 15, and 16 reveals a few overall trends regarding recent lithium-ion pack design. It is notable that there are only a limited number of cell suppliers actually in production relative to the large number of companies in the market. The supplier concentration of lithium cell makers would only increase if one considered the sales volume of each individual vehicle type, as many of the nameplates in these tables are low volume. It is also notable that some

Table 15 Lithium-ion containing PHEVs [32, 33]

Vehicle	Cell				Pack					
	Supplier	Type	Volt (V)	Cap (Ah)	S/P	Volt (V)	Cap (Ah)	Energy (kWh)	Power (kW)	P/E
Audi A3	Panasonic	P	3.6	25	96S1P	345.6	25	8.6	74	8.7
BMW 330e	Samsung	P	3.7	28	74S1P	273.8	28	7.7	65	8.5
BMW 740e/X5	Samsung	P	3.7	26	96S1P	355.2	26	9.2	50	5.4
BMW i8	Samsung	P	3.7	20	96S1P	355.2	20	7.1	96	13.5
Cadillac ELR	LG Chem	F	3.7	15.5	96S3P	355.2	46.5	16.5	110	6.7
Chevrolet Volt	LG Chem	F	3.7	26	96S2P	355.2	52	18.5	120	6.5
Ford C-Max/Fusion	Panasonic	P	3.6	25	84S1P	302.4	25	7.6	43	5.7
Hyundai Sonata, Kia Optima	LG Chem	F	3.7	27.2	96S1P	355.2	27.2	9.7	50	5.2
Mercedes-Benz GLE550e/S550e	A123	F	3.3	22	120S1P	396	22	8.7	85	9.8
Porsche 918	Samsung	P	3.7	5.9	104S3P	384.8	17.7	6.8	230	33.8
Porsche Cayenne	Panasonic	P	3.7	28	104S1P	384.8	28	10.8	70	6.5
Porsche Panamera	Panasonic	P	3.7	24.5	104S1P	384.8	24.5	9.4	70	7.4
Volvo XC90	LG Chem	F	3.7	26	96S1P	355.2	26	9.2	65	7.0

Table 16 Lithium-ion containing EVs [32, 33]

Vehicle	Cell				Pack					
	Supplier	Type	Volt (V)	Cap (Ah)	S/P	Volt (V)	Cap (Ah)	Energy (kWh)	Power (kW)	P/E
BMW i3	Samsung	P	3.7	94	96S1P	347.8	94	32.7	125	3.8
BYD e6	BYD	P	3.2	200	96S1P	307.2	200	61.4	90	1.5
Chevrolet Bolt	LG Chem	F	3.65	59	96S3P	350.4	177	62.0	160	2.6
Chevrolet Spark	LG Chem	F	3.75	27	96S2P	360	54	19.4	120	6.2
Fiat 500e	Samsung	P	3.75	66	97S1P	363.8	66	24.0	83	3.5
Ford Focus	LG Chem	F	3.7	15	86S5P	318.2	75	23.9	110	4.6
Kia Soul	SK	F	3.75	37.5	96S2P	360	75	27.0	90	3.3
Mitsubishi i	LEJ	P	3.7	50	88S1P	325.6	50	16.3	49	3.0
Smart Fortwo	Deutsche ACCUmotive	F	3.6	26	94S2P	338.4	52	17.6	55	3.1
Tesla Model S	Panasonic	R	3.6	3.35	96S74P	345.6	247.9	85.7	270	3.2
Volkswagen e-Golf	Panasonic	P	3.7	25	88S3P	325.6	75	24.4	85	3.5

automakers (General Motors, Ford, Honda, and BMW) are utilizing the same cell and battery pack design across various vehicle nameplates. An examination of nominal cell voltages shows that although NMC (Lithium Nickel Manganese Cobalt Oxide)-based cells are the most popular, there is a wide variety of active materials, such as LMO (Lithium Manganese Oxide), LFP (Lithium Iron Phosphate), and LTO (Lithium Titanium Oxide) being commercialized. Likewise, although prismatic cell (P) geometries are the most popular, all three form factors are being used in all three commercialized applications in the US market.

5 Summary

The recent application of lithium-ion batteries to automobiles has driven significant changes in the design and commercialization of electrified vehicles. The successful application of any battery technology requires a thorough understanding of the landscape of vehicle propulsion and electrified features. Once the requirements of the transportation field are understood, it is possible to appropriately size and design lithium-ion containing battery pack solutions.

The recent resurgence in commercial electric vehicles (EVs) has been enabled by the substantial improvements in electric range made possible by the high energy density of lithium-ion batteries. Future activity in EV battery development will focus on further improvements of lithium-ion gravimetric and volumetric energy densities. The plug-in hybrid electric vehicle (PHEV) has been around as a concept for some time, but the ability of lithium-ion cells to deliver high power while maintaining the majority of their available energy has made them a commercial reality. PHEV lithium-ion batteries that can provide improved electric range (energy density) while maintaining their power and minimizing their cost are the goal of future research and development efforts. Hybrid electric vehicles (HEVs) and stop-start (S/S) hybrids both have a long history of commercialization with competing chemistries such as lead–acid and nickel–metal hydride. For these applications, lithium-ion typically provides an evolutionary improvement in weight and volume for equivalent energy and power requirements. The future success of lithium-ion for these applications will be driven by efforts to reduce cost while maintaining their energy and power performance advantages over competing chemistries.

References

1. H. Ford, *Today and Tomorrow: Commemorative Edition of Ford's 1926 Classic* (Productivity Press, Cambridge, 1988)
2. J. Larminie, J. Lowry, *Electric Vehicle Technology Explained* (Wiley, New York, 2004)
3. D. Rand, R. Woods, R. Dell, *Batteries for Electric Vehicles* (Society of Automotive Engineers, Warrendale, 1998)

4. B. Kong, Top 10 Slowest 0–60 MPH Vehicles Motor Trend Tested in 2012. Motor Trend. [Online] 2012. [Cited: 01 12, 2017.] http://www.motortrend.com/news/top-10-slowest-0-60-mph-vehicles-motor-trend-tested-in-2012-307101
5. A. Masias, A. Kuroda, K. Tojima, *Research and Development of Advanced Automotive Batteries*. Plug-In 2009, Long Beach, 2009
6. P. Savagian, Driving the Volt. SAE Hybrid Technology Conference, San Diego, 2008
7. Dynamometer Drive Schedules. EPA. [Online] [Cited: January 12, 2017.] ttps://www.epa.gov/vehicle-and-fuel-emissions-testing/dynamometer-drive-schedules
8. INL. Battery Test Manual for 12 Volt Start/Stop Vehicles. Idaho Falls: s.n., 2015. INL/EXT-12-26503
9. T. Duong, *Direction for Energy Storage R&D in the Vehicle Technologies Program* (Scalable Energy Storage Beyond Li-Ion, Oak Ridge, 2010)
10. USCAR: U.S. Advanced Battery Consortium LLC. USCAR. [Online] [Cited: January 13, 2017] http://www.uscar.org/guest/teams/12/U-S-Advanced-Battery-Consortium-LLC
11. EUCAR: European Council for Automotive Research. EUCAR. [Online] [Cited: January 13, 2017.] http://www.eucar.be/
12. NEDO. New Energy and Industrial Technology Development Organization. NEDO. [Online] [Cited: January 13, 2017] http://www.nedo.go.jp/english/index.html
13. Battery Test Manual for Plug-In Hybrid Electric Vehicles. Idaho Falls: s.n., 2008. INL/EXT-14-32849
14. NEDO. Secondary Battery Technology Development Roadmap. 2010
15. HELIOS European Project and EUCAR/VDA Research Roadmaps. A. Teyssot, EVS24, Stavanger, 2009
16. S. Whittingham, Science **192**, 1126 (1976)
17. *Linden's Handbook of Batteries*, 4th edn., ed. by T. Reddy. (McGraw-Hill, New York, 2011)
18. INL. FreedomCAR Battery Test Manual for Power-Assist Hybrid Electric Vehicles. Idaho Falls: s.n., 2003. DOE/ID-11069
19. INL. Battery Test Manual for Electric Vehicles. Idaho Falls: s.n., 2015
20. ISO. Electrically Propelled Road Vehicles—Dimensions and Designation of Secondary Lithium-Ion Cells. s.l.: ISO, 2012. ISO/IEC PAS 16898
21. DIN. Electrically Propelled Road Vehicles Battery Systems Dimensions for Lithium-Ion-Cells. 2011. DIN SPEC 91252
22. CATARC. Battery Structure and Size. s.l.: CATARC, 2010. QC/T 840
23. SNL. FreedomCAR Electrical Energy Storage System Abuse Test Manual for Electrical and Hybrid Electric Vehicle Applications. s.l.: SNL, 2006. SAND2005-3123
24. SAE. Electric and Hybrid Vehicle Propulsion Battery System Safety Standard—Lithium-Based Rechargeable Cells. s.l.: SAE J2929, 2011
25. SAE. Electric and Hybrid Electric Vehicle Rechargeable Energy Storage System (RESS). s.l.: SAE J2464, 2009
26. ISO. Electrical Propelled Road Vehicles—Test Specifications for Lithium-Ion Traction Battery Packs and Systems—Part 1: High-Power Applications. s.l.: ISO, 2011. ISO 12405-1
27. ISO. Electrically Propelled Road Vehicles—Test Specification for Lithium-ion Traction Battery Packs and Systems—Part 3: Safety Performance Requirements. s.l.: ISO, 2014. ISO 12405-3
28. UN. Lithium Metal and Lithium Ion Batteries. s.l.: Manual of Tests and Criteria, 5th Revised Edition, 2011. 38.3
29. ECE. Electric Power Trained Vehicles. s.l.: ECE, 2013. ECE R100
30. CATARC. Lithium-Ion Batteries for Electric Vehicles. s.l.: CATARC, 2006. CATARC QC/T 743
31. KMVSS. Traction Battery Safety Test. s.l.: KMVSS, 2009. Test Procedure Annex 1.48
32. INL. Vehicle Testing|Advanced Vehicle Testing Activity. [Online] [Cited: January 13, 2017] https://avt.inl.gov/content/vehicle-testing
33. NFPA. Manufacturers Resources—Electric Vehicle Safety Training. [Online] [Cited: January 13, 2017] http://www.evsafetytraining.org/resources/auto-manufacturer-resources.aspx

The Future of Lithium Availability for Electric Vehicle Batteries

Jamie Speirs and Marcello Contestabile

Abstract Supported by policy, electric vehicles (EVs) powered by lithium batteries are being commercialised in an increasing number of models and their global stock surpassed two million units in 2016. However, there is uncertainty around the future price and availability of lithium, which has consequences on the feasibility of manufacturing lithium batteries at scale. Reaching the EV penetration levels foreseen by governments implies a substantial growth in lithium demand. In this chapter, we review the evidence around future lithium availability for the manufacturing of EV batteries. We examine the methods used to estimate both lithium demand from EVs and lithium supply from brines and ore. The main variables influencing demand are the future size of the EV market, the average battery capacity and the material intensity of the batteries. Supply projections depend on global reserve and resource estimates, forecast production and recyclability. We find that the assumptions made in the literature on the key variables are characterised by significant uncertainty. However based on the available evidence, it appears that lithium production may be on a lower trajectory than demand and would have to rapidly increase in order not to prove a bottleneck to the expansion of the EV market. More research is needed in order to reduce uncertainty on lithium intensity of future EVs and improve understanding of the potential for lithium production expansion and recycling.

J. Speirs (✉)
Sustainable Gas Institute, Earth Science and Engineering,
Imperial College London, London, UK
e-mail: Jamie.speirs@imperial.ac.uk

M. Contestabile
Centre for Environmental Policy, Imperial College London, London, UK
e-mail: Marcello.contestabile@imperial.ac.uk

G. Pistoia and B. Liaw (eds.), *Behaviour of Lithium-Ion Batteries in Electric Vehicles*, Green Energy and Technology,
https://doi.org/10.1007/978-3-319-69950-9_2

1 Introduction

Electric vehicles (EVs) have the potential to contribute to meeting energy and environmental policy objectives of governments worldwide such as reducing dependence on oil-derived fuels and related emissions of greenhouse gases and improving air quality in urban areas. However, the question has been raised whether the availability of critical metals can prove a constraint to the manufacturing of EV batteries at scale [1–3]. Given that lithium batteries are currently the only viable technology for EVs and are likely to remain so for the foreseeable future, it can be expected that large quantities of lithium will be needed to manufacture enough automotive batteries to meet the desired policy objectives [4–6]. This raises questions as to the mining sectors ability to satisfy demand [7, 8]. This chapter addresses the question of lithium availability for the manufacturing of EVs based on research published by the authors in 2014 [9] and updated as appropriate. In Sect. 2, we examine the relevant literature, focusing on the relationship between metal availability and EV manufacturing, highlighting the main variables and assumptions used in previous studies. In Sect. 3, we describe the difficulties in calculating future lithium demand for EV batteries. Section 4 discusses the issues surrounding future supply of lithium. Section 5 compares supply and demand and presents the conclusions that follow from the analysis.

It is important to mention that the term EV can be used to indicate all road vehicle types that have an electric powertrain, including hybrid electric vehicles (HEVs), plug-in hybrid electric vehicles (PHEVs), battery electric vehicles (BEVs) and fuel cell vehicles (FCVs). This classification is broad and does not fully represent all possible powertrain architectures. It is however practical for two reasons: (1) the literature often classifies vehicle technology in a similar way; and (2) these vehicle types all potentially use lithium batteries. In this chapter, we will therefore initially consider all EV types above and we will then restrict our attention to those that use larger lithium batteries and hence account for most of the demand for lithium.

2 The Literature on Availability of Lithium for EV Batteries

A number of authors have explored the relationship between lithium availability and EV uptake over the last 10 years, ranging from pessimistic studies that suggest future EV demand cannot be met by lithium supply [7, 8], to optimistic studies that find no significant constraint to ambitious EV market development projections [10]. These were reviewed by the authors in [9]. Within these studies are a range of different assumptions which lead to a wide range of findings of Table 1.

In particular, the studies we have reviewed generally disagree on the quantity of lithium needed by the future EV market.

Table 1 Comparison of several studies that examine the potential material constraints to lithium battery manufacture

Author	Forecast lithium supply constraint on EV manufacture	EV manufactured (millions per year)	By year	Lithium intensity per vehicle (kg)
Evans [11]	No	5 EV	2015	HEV: 0.23 PHEV: 1.35 BEV: 2.81
Gaines and Nelson [12]	No	~65 PHEV and ~35 BEV	2050	HEV: 0.17–0.64 PHEV: 0.93–5.07 BEV: 3.38–12.68
Gruber et al. [13]	No	>600 EV	2100	HEV: 0.05 PHEV: 1.14 BEV: 3.85
Kushnir and Sandén [14]	No	4500 EV (cumulative)	2100	PHEV: 1.44 BEV: 5.76
Neubauer [15]	No	60 PHEV and 47 BEV	2050	PHEV: 0.6–1.9 BEV: 3.3–7.5
Tahil [8]	Yes	4–8 PHEV	2015–2020	PHEV: 1.5
Yaksic and Tilton [10]	No	3000 EV (cumulative)	2100	EV: 1.27
Martin et al. [16]	No	7.5–9	2020	PHEV: 0.85 BEV: 8.5

Source adapted and updated from [9]

The variables leading to this range of outcomes include:

- the time horizon;
- the number of vehicles manufactured at a given point in the future;
- the size of batteries in different EV types;
- the share of the future EV market taken by different EV types; and
- the quantity of lithium per unit of battery capacity.

Aside from the inherent uncertainty, assumptions tend to vary over time as they are influenced by the development of EV technology as well as its market uptake. The variables above and related assumptions are further discussed in Sect. 3.

As for lithium supply estimates, these are influenced by the following variables:

- lithium reserves;
- future lithium production rates; and
- future recycling rates.

The variables and related assumptions are discussed in detail in Sect. 4.

3 Estimating Lithium Demand from EVs

As mentioned above, the calculation of future lithium demand from EVs involves estimating several variables and hence is subject to significant uncertainty. However, the common elements typically considered are:

- the number of EVs manufactured in the future;
- the size of EV batteries in kWh; and
- the lithium intensity per kWh of battery.

We examine projections of future EV deployment in Sect. 3.1 and the issue of material intensity per vehicle in Sect. 3.2.

3.1 The Future Size of the EV Market

There are several studies presenting a range of different scenarios of the future EV market. To illustrate the range of scenarios found in the literature, in our earlier work [9], we compared high-profile studies [17–21] providing uptake scenarios disaggregated by EV type. The scenarios compared cover a range of timeframes, the earliest beginning in 2008 and the longest projecting to 2050. In the studies reviewed in [9], global PHEV sales estimates in 2050 range from 10 to 79 million vehicles per year, and BEV from 12 to 84 million vehicles per year. It is worth noting that EV uptake scenarios contained in more recent studies [22–26] are generally much more aggressive, with the lower end of the range broadly overlapping with the higher end of the range provided by the older studies.

The scenario selected for the analysis we conducted in [9] was the BLUE Map scenario contained in the 2010 Energy Technology Perspectives (ETPs) report of the International Energy Agency (IEA) [27]. The BLUE scenarios are a set of normative scenarios that allow halving global CO_2 emissions by 2050 compared with 2005 levels at the least cost under different assumptions. The global EV sales by EV type projected in the BLUE Map scenario are summarised in Table 2.

The BLUE scenarios have now been discontinued by the IEA and substituted in more recent editions of the ETP by other normative scenarios that are named according to the level at which they are expected to contain global temperature rise, i.e. the 2 °C scenario (2DS) and the below 2 °C scenario (B2DS). Comparing the EV projections of the 2010 ETP with those of the 2017 ETP [24], it is apparent that the latter is much higher, which reflects the progress made by EV technology and

Table 2 Annual EV sales (millions) in 2030 and 2050 under the IEA 'BLUE Map' and 'BLUE EV shifts' scenarios

	PHEV	BEV
BLUE Map 2030	25	9
BLUE Map 2050	62	47

Source [27]

market uptake over the last few years, and hence, the much bigger role EVs are expected to play in climate change mitigation over the coming decades. The EV uptake levels in the BLUE Map scenario of the 2010 ETP however are broadly comparable with those in the Reference Technology Scenario (RTS) of the 2017 ETP. The RTS is a scenario that depicts a plausible future where today's policy measures are implemented and new measures are introduced based on current policy trends, and where technology develops as is currently expected. For this reason, we decided not to substitute the EV projections of the 2010 BLUE Map scenario with more recent ones, but it is important to stress that the projections should be seen as a lower bound to future EV uptake and that therefore lithium demand from EVs as estimated by our analysis could be much higher if EV uptake proceeded more rapidly.

3.2 *Estimating Lithium Intensity*

Since their re-emergence in the last 10 years, EVs have been relying on lithium-ion (Li-ion) batteries for their propulsion [28]. As discussed below, a number of Li-ion and Li-metal chemistries are currently being developed and it is likely that lithium batteries will continue to dominate the EV market for the foreseeable future. The lithium intensity, i.e. the weight of lithium per vehicle, must be estimated before any estimates of future EV lithium demand can be made based on the EV uptake scenario previously discussed.

Deriving lithium intensity for Li-ion batteries ideally requires knowledge of:

- the nominal voltage of the battery (volts, V);
- the specific capacity of the battery chemistry considered (Ampère-hours per gram, Ah/g); and
- the concentration of lithium in the active materials of the battery when this is assembled (weight per cent, wt%).

While large batteries are required for BEV and PHEV designs, smaller batteries of the order of 1–1.5 kWh are generally sufficient for HEVs and FCVs, where they are only used for storing energy generated on board via regenerative braking and, in the latter case, also shaving the peaks and troughs of fuel cell duty cycles. Since the capacity of PHEV and BEV batteries is likely to be 10–20 times that of HEV and FCV batteries, and since HEVs and FCVs make up a relatively small proportion of the total vehicle market in 2050 based on the scenario presented in Table 2 above, the total lithium demand from HEVs and FCVs is likely to be negligible for the purposes of our study and is hence excluded from our previous analysis [9].

As mentioned above, the amount of lithium contained in an EV battery is a function of its size, particular type of chemistry, construction and rated performance. Hence, it is impossible to define with certainty the amount of lithium that each individual EV battery model will require. Nevertheless, we will discuss each of the main factors influencing the amount of lithium required in an individual EV battery in turn, following the approach used in [9]. On this basis, we identify a range of values for lithium demand per vehicle which is then combined with the global EV demand projections discussed above in order to estimate future global demand for lithium for the EV market.

The calculation of global lithium demand from EVs in year y ($D_{Li,y}$) can be summarised by the following equation:

$$D_{Li,y} = (M \times S \times I)_{\mathrm{BEV}} + (M \times S \times I)_{\mathrm{PHEV}} \tag{1}$$

where M is the market size (annual vehicle sales) of BEVs/PHEVs in year y, S is the average size (kWh) of a BEV/PHEV battery in year y and I is the average intensity (amount of lithium per unit energy capacity (kWh) of a BEV/PHEV battery in year y).

A similar approach has been taken, either implicitly or explicitly, in a number of relevant studies reviewed in [9] (see Table 1). In the remainder of this section, we discuss in turn average battery sizes and average amounts of lithium per unit energy stored. This will allow us to arrive at a plausible lithium demand range, which we will discuss in Sect. 5.

Average battery sizes for BEVs and PHEVs

The rated energy of the battery, expressed in kWh, is one of the main parameters determining the all-electric range (AER) of a BEV or PHEV. The rated energy is declared by the manufacturer and cannot be directly translated into a value for the lithium content of the battery. This is due to a number of reasons, including that the energy stored in an EV battery (and hence its lithium content) is usually significantly higher than its rated energy would suggest, i.e. the battery is over-specified; this is further discussed later on. Let us begin by focusing on the average rated energy of EV batteries, which we will refer to as battery size.

There is no standard battery size for BEVs and PHEVs. It is the prerogative of automotive Original Equipment Manufacturers (OEMs) to manufacture BEVs and PHEVs with different AER capabilities and therefore different battery sizes. Trade-offs exist between AER on the one hand and cost, weight and volume of the battery on the other. This somewhat constrains the extent to which battery size can vary across different models, although for BEVs in particular we are observing that, as the cost of batteries declines [29], OEMs are equipping them with increasingly large batteries in order to meet the range requirements of prospective buyers.

BEV models commercialised over the last 10 years use Li-ion batteries of a wide range of sizes, depending on the size of the car and the desired AER. Battery sizes start at a minimum of around 15 kWh which gives a small city car a range in the order of 130–150 km, up to 100 kWh which gives a large high-performance sedan or crossover a range in the order of 500 km (see Table 3).

Today's PHEVs also use Li-ion batteries, the size of which varies somewhat across vehicle models (see Table 4). Different powertrain architectures are possible for PHEVs, which are suited to different modes of operation and achieve different

Table 3 Key technical attributes of selected BEV models on the market in the UK as of November 2017

BEV model	Battery energy (kWh)	Range (km)	Max speed (km/h)
BMW i3	22–33	130–183	150
BYD e6	75	300	140
Citroen C-Zero	14.5	150	130
Ford Focus Electric	33.5	185	135
Hyundai IONIQ Electric	28	200	165
Kia Soul EV	30.5	150	145
Mercedes-Benz B-class electric drive	36	200	150
Nissan e-NV200	24	170	122
Nissan Leaf	24–30	117–172	150
Peugeot iOn	16	150	130
Renault Fluence Z.E.	22	160	135
Renault Zoe	22–40	240–400	135
Smart for two electric drive	17.6	145	125
Tesla Model S	75–100	370–550	155
Tesla Model X	75–100	320–465	210
Volkswagen e-Golf	35.8	144–201	150
Volkswagen e-Up!	18.7	160	130

Note Ranges are based on a mix of drive cycles and are not directly comparable
Sources CAR Magazine website [30], updated from [9]

Table 4 Key technical attributes of selected PHEV models on the market in the UK as of November 2017

Vehicle model	Battery energy (kWh)	EV range (km)
Audi A3 e-tron	8.8	50
BMW 225xe	7.6	39
BMW 330e	7.6	40
BMW 530e	9.2	47
Hyundai IONIQ PHEV	8.9	50
Kia Optima PHEV	9.8	43
Mercedes-Benz C350e	6.2	31
MINI Countryman PHEV	7.6	40
Mitsubishi Outlander	12	52
Smart for four electric drive	17.6	140
Toyota Prius Plug-in Hybrid	8.8	40
Vauxhall Ampera	18.4	85
Volkswagen Golf GTE	8.7	50
Volkswagen Passat GTE	9.9	50
Volvo V60 PHEV	12	50
Volvo XC90 T8 Twin Engine	9.2	23

Note Ranges are based on a mix of drive cycles and are not directly comparable
Sources Green Car Congress website [31], updated from [9]

AERs, and this largely explains the range of battery sizes observed. In particular, the Toyota Prius Plug-in Hybrid has been designed to have limited all-electric operation capabilities and hence has a relatively small battery pack (8.8 kWh). On the other hand, range-extended electric vehicles such as the Vauxhall Ampera are capable of delivering high performance while operating in EV mode and hence have a significantly larger battery pack (16 kWh).

Lithium content per unit energy stored in BEV and PHEV batteries

Another important parameter for determining the total demand for lithium in EVs is the amount of lithium required per kWh of battery. However, its estimation is far from straightforward, which contributes to the wide range of figures reported in the literature. The studies reviewed in [9] use different methods to derive their estimates, each with its own limitations. Here we discuss the approach we have taken in [9] to estimate the amount of lithium required per kWh of battery, the main factors affecting it and the main differences between our approach and that of others. Our approach shows similarities with those adopted by a number of previous studies [13, 17, 32–34].

As mentioned above, estimating material intensity in batteries ideally requires knowledge of the voltage that the battery is capable of delivering while in operation, its specific capacity and the chemical composition of its active materials. However, this information is only readily available to the battery manufacturers. One method of estimating material intensity (labelled method 'A' in Table 5) is to use industry data where available. This is done in several of the studies cited in Table 5. Alternatively, it is possible to measure voltage and specific capacity of a battery, then disassemble it and analyse its composition in a laboratory. This process (labelled 'B'), sometimes referred to as 'reverse engineering', is often not practical as it is expensive, and results obtained for one particular type of cell would not apply to others. The two remaining options are: to use published data for battery voltage and specific capacity and then make assumptions on their composition (labelled 'C'); or, knowing the particular chemistry of the battery, to estimate the amount of lithium required by starting from the theoretical value required under ideal conditions and then adding to it in order to account for real operating conditions (labelled 'D'). In [9], we have adopted the latter approach. As will become apparent from the discussion, we were not being able to arrive at specific lithium intensity values for EV batteries by simply following this approach. However, it enabled us to derive upper and lower boundaries for the possible range of lithium intensity values and, based on this, assess the validity of some of the figures found in the literature.

As stated in [9], there are three key factors that vary and must be accounted for in a theoretical assessment of lithium intensity in EV batteries:

- Impact of the specific battery chemistry on lithium intensity. A number of lithium battery chemistries are possible, which are characterised by different performance and lithium intensities.

Table 5 Estimates of lithium content per kWh of battery capacity, or lithium intensity, found in the literature

Source	Vehicle application *or* battery chemistry	Material intensity (kg Li/kWh)	Methodology
Chemetall GmbH [36]	BEV (25 kWh)	0.165	A
	PHEV (16 kWh)	0.176	
	HEV (1 kWh)	0.375	
Meridian International Research [7]		0.300	A
Meridian International Research [34]		0.563	D
Kushnir and Sanden [14]	Average for four chemistries	0.160	D
Rade and Andersson [33]	Li-ion (Mn)	0.140	D
	Li-ion (Ni)		
	Li-ion (Co)		
Argonne National Laboratory [12]	HEV4 (1.2 kWh)	0.308	C
	PHEV20 (6 kWh)	0.244	
	PHEV40 (12 kWh)	0.246	
	EV100 (30 kWh)	0.246	
Gruber et al. [13]	Li-ion (Co, Mn, Ni)	0.114	D
Evans [11]		0.113	A
Evans cited by Reuters [28]	Chevrolet Volt (16 kWh)	0.158	A
Engel [35]		0.050	A
Fraunhofer ISI [17]	$LiCoO_2$	0.180	D
	$LiFePO_4$	0.120	
Dundee Capital Markets [18]		0.080	A
National Renewable Energy Laboratory [15]	HEV (1.7 kWh)	0.100	Internal modelling study (C or D)
	PHEV12 (5.6 kWh)	0.108	
	PHEV35 (17.5 kWh)	0.110	
	BEV75 (29.5 kWh)	0.112	
	BEV150 (67 kWh)	0.112	

Source [9]

- Impact of energy losses on lithium intensity. Lithium batteries, like any other battery, when operating only deliver as electrical energy part of the chemical energy stored, while the rest is lost as heat due to internal resistance mechanisms called ‘overpotentials’.
- Impact of over-specification on lithium intensity. The actual capacity of a lithium battery is often much higher than the rated capacity, in order to guarantee durability.

These are dealt with in turn below.

(i) Variation in lithium intensity between different battery chemistries

The amount of lithium used per kWh depends on the stoichiometry of the particular electrochemical reaction that is associated with the battery chemistry considered and on its corresponding electromotive force (E_0). Based on Faraday's laws, the theoretical lithium demand per kWh can be calculated as:

$$I = \frac{m.10^3}{E_0 ac} \tag{2}$$

where I is the lithium intensity in g/kWh, m is the molar mass of lithium in g/mol, E_0 is the electromotive force in volts, a is the fraction of lithium available and c is the charge of 1 mol of lithium ions in Ah/mol.

Using the appropriate values, we get the following:

$$I = \frac{6,941}{E_0 a.\frac{96,485}{3,600}} \tag{3}$$

The conventional Li-ion chemistry (originally commercialised by Sony) is based on the following redox process:

$$6C + LiCoO_2 \leftrightarrow Li_{0.5}C_6 + Li_{0.5}CoO_2 \qquad E_0 \approx 4V$$

where the cathode material $LiCoO_2$ can only exchange roughly half of its lithium content, hence the fraction of lithium available a is 50%. Entering the appropriate values for E_0 and a in formula (3), we estimate the theoretical amount of lithium needed per kWh of a conventional Li-ion battery to be 129.5 g.

Another Li-ion chemistry of practical interest is the one that uses lithium iron phosphate ($LiFePO_4$) at the cathode and lithium titanium oxide ($Li_4Ti_5O_{12}$) at the anode. This chemistry is inherently safer than the conventional one and hence potentially more suited to EVs, particularly PHEVs. The electromotive force E_0 of this system however is substantially lower, at ≈ 2 V. If we assume that 100% of the lithium contained in $LiFePO_4$ and 75% of the lithium contained in $Li_4Ti_5O_{12}$ can be made available, the theoretical amount of lithium needed per kWh will be 172.6 g.

The two examples here provided clearly illustrate that lithium intensity is not the same for different chemistries. Calculating g(Li)/kWh in this way provides a theoretical minimum and not the actual lithium intensity of real EV batteries. However, starting from the theoretical value is useful, not least because it shows that lithium intensity changes from one battery chemistry to another simply as a result of the different electrochemical processes involved. Actual lithium intensity will be higher than the theoretical value for the reasons mentioned above and further discussed below, i.e. the voltage losses that occur during the operation of a battery and the degradation processes that occur during the lifetime of the battery which reduce its capacity, leading manufacturers to over-specify them.

(ii) Impact of energy losses on lithium intensity

The voltage of a lithium-ion battery when operating is significantly lower than its electromotive force E_0, the difference being the result of resistance within the battery. When the cell is operating, its actual voltage, ΔV (the difference in potential between the electrodes), can be expressed as follows:

$$\Delta V = E_0 - (iR_\mathrm{I}) \tag{4}$$

where i is the current being drawn from the cell and R_I is the internal resistance of the cell. R_I is the sum of the ohmic resistance of the electrolyte and electrodes as well as the resistance due to the kinetics of charge transfer at the interface between electrodes and electrolyte. The difference between E_0 and ΔV, usually referred to as overpotential, is a function of both how the cell is operated (i.e. how fast the cell is discharged and its temperature at the time of operation) and how it is constructed (i.e. the chemical composition of the electrodes, their density and thickness, the size of the particles of active material that they contain, the concentration of the lithium salt used as electrolyte and the chemical composition of the solvents used). Hence if we replace E_0 with ΔV in Eq. (3), the lithium demand per kWh will be higher than the theoretical value because ΔV is always smaller than E_0. The difference between E_0 and ΔV depends on a number of complex processes and cannot be estimated theoretically from first principles for any battery chemistry. Its experimental measurement on the other hand is straightforward, though the values obtained for a specific battery model cannot be generalised, not even to other batteries that use the same chemistry.

(iii) Impact of over-specification on lithium intensity

Manufacturers often ‘over-specify’ batteries, typically to reduce the impact of degradation over the lifetime of the battery and hence guarantee the necessary cycle life, which is typically calculated as the number of charge–discharge cycles achievable before the capacity of the battery falls below 80% of its rated value. In many cases, the over-specification of the battery is quite substantial and the battery is operated at reduced depth of charge and discharge. The extent to which the battery is over-specified can vary greatly across manufacturers, chemistries and intended use of the battery. As a result, in the case of heavily over-specified batteries, the actual amount of lithium can be twice the theoretical value.

From the discussion above, it is apparent that directly calculating lithium intensities based on methodology ‘D’ is problematic. Hence, in [9], we examine lithium intensity values for Li-ion batteries available in the literature (see Table 5) and we compare them with the range of values that we have derived by applying methodology ‘D’. The range of lithium intensities in the literature varies widely, between 50 g/kWh and 562 g/kWh, and not all of the estimates listed in Table 5 have the same merit. Firstly, not all methodologies labelled ‘A’ are actual industry data, as many rely on sources quoted in the media or in corporate presentations without reference to either public or proprietary industry data. We discount a number of these estimates on this basis, and we also discount those that provide values that are below the theoretical limits that we have calculated [18, 35]. Methodology 'C' is valid, although the study using it [12] does not disclose references or provide justifications for the assumptions used. Within the studies employing methodology 'D', Tahil [34] and Angerer et al. [17] appear to overstate lithium intensity, while Gruber et al. [13] assume lower values. Finally, a number of studies do not disclose the full details of their assumptions or methods [15, 32]. These observations make it difficult to judge the value of many estimates in Table 5. For this reason in Sect. 5, we use a range (190–380 g/kWh) of lithium intensity which we consider as broadly representative of that found in commercial lithium batteries. The range is based on the lithium intensity estimates we have found in the literature [9], excluding those that appear close to the theoretical minimum [18, 35] as well as the apparent overestimates [34].

Aside from the three factors discussed above that are essential to the theoretical assessment of lithium intensity in EV batteries, it is also worth mentioning that battery research and development may result in a more efficient use of lithium in EV batteries or in its substitution through the development of alternative, non-lithium-based battery chemistries for EVs. Both these developments could in principle impact the demand for lithium from EVs quite substantially, and hence, they are discussed in turn below.

(iv) Potential for lithium weight shedding

The focus of research and development in lithium batteries over the last decades has been on increasing safety, lowering cost, increasing energy density and improving cycle life, with a long-term view towards low environmental impact [37, 38]. Lithium contributes only 1–2% of final battery cost [32]. Accordingly, little discussion about reductions in lithium content can be found in the literature. Rade and Andersson [33] provide one of the few estimates of future lithium intensity of Li-ion batteries based on the improvement of active material utilisation from a current 50% to 60–80% depending on chemistry, leading to intensity reductions of 21–34%. Utilisation of active materials is defined as the share of electrochemically active material contained in the anode and cathode that actually participates in the electrochemical reactions occurring during charge and discharge of the battery. It is unclear whether the lithium intensity reduction suggested in [33] has since been realised or will ever be, so it is not possible to account for it based on the available evidence.

(v) Potential for substitution

Early BEVs such as General Motors' EV1 used lead–acid batteries and more recently the Think City used sodium/nickel chloride (also known as ZEBRA) batteries. However, lithium batteries have significant advantages over these two battery types and it is unlikely that the latter will be used in future BEVs and PHEVs. Since lithium is the lightest metal and has an extremely negative electrode potential, lithium batteries have much higher energy density than lead–acid batteries, allowing EVs to achieve acceptable ranges without imposing a high weight penalty. Moreover, unlike ZEBRA batteries which use molten sodium at 300–350 °C, most lithium battery chemistries operate at room temperature, and because they do not need preheating, they are always available for use, which is a very desirable characteristic for vehicles with no fixed usage patterns such as passenger cars. These favourable characteristics, together with the high power density and long cycle life, explain why lithium batteries are the current technology of choice for BEVs and PHEVs.

Other non-lithium chemistries are being researched that may compete with lithium batteries. However, alternatives to lithium are limited, because prospective systems need to have high energy density and this requires using light metals such as sodium, magnesium and aluminium. Battery systems under investigation include magnesium/sulphur and aluminium/graphite fluoride. However, the practical viability of these systems has not been demonstrated and their future use in EVs depends on significant technological improvement [37]. Metal air chemistries such as sodium air and zinc air are also possible alternatives to lithium air. Sodium air batteries in particular have the potential to mitigate some of the problems of Li-air technology, but technological improvement is needed before this technology becomes practical [39].

4 Lithium Supply

The availability of lithium over the coming decades is contingent upon the rate at which the metal can be produced from natural sources or recycled from spent batteries and brought to market. In order to understand future lithium supply issues, in [9], we explored its geological characteristics and routes to extraction, its existing reserve estimates and current production rate, the extent to which it can be recovered from spent batteries and its forecast production. These are discussed in turn below, based on [9] and with figures updated as necessary.

4.1 Geological Characteristics of Lithium

Due to its reactivity, lithium metal never occurs freely in nature and is instead found in the form of lithium compounds in four main deposit types as follows: minerals, brines, sedimentary rocks and sea water. Minerals and brines constitute the world's main source of lithium today. Lithium-containing minerals are typically coarse-grained intrusive igneous rocks known as pegmatites, such as spodumene, petalite, lepidolite, amblygonite and eucryptite [13]. Brine deposits are currently the largest and cheapest sources of lithium [10] and are mostly found in dry lakes such as the Salar de Atacama in Chile, as well as geothermal deposits and saline aquifers. The third source of lithium is in sedimentary rocks, notably clays such as hectorite and lacustrine evaporates such as the newly discovered jadarite [13, 40]. These sedimentary deposits are currently not commercially recovered. Finally, sea water contains diffuse but very large quantities of lithium. According to Yaksic and Tilton [10], 44.8 billion tonnes of lithium are recoverable from the world's oceans. The economic viability of extracting lithium from sea water is uncertain. The lithium recovered from the above sources is produced in the form of a number of compounds, such as lithium carbonate, lithium hydroxide, lithium chloride and others. Different lithium compounds are used in different applications, lithium carbonate being the one typically used in Li-ion batteries.

4.2 Production and Reserves

Known reserves of lithium exist and are commercially exploited in a number of countries, the relative distribution of which is presented in Fig. 1. The largest share of production is currently in Australia, where lithium is recovered from spodumene deposits. Chile is the second largest producer and has the largest reported reserves of lithium from brine pools located in salt flats throughout the Andes mountain range. The geographical distribution of both reserves and production indicates that lithium supply is unlikely to suffer from the geopolitical supply constraints witnessed for materials with less well geographically distributed resources such as rare earth metals or indium [41–44].

In order to understand the future potential of lithium reserves and production, it is essential to look at their historical development. Figure 2 presents historical world lithium production data by gross weight of the minerals and compounds extracted, published by the USGS [46]. Since 1967, lithium production was reported as 'ore and ore concentrates' from mines and lithium carbonate from brine deposits. Calculating the lithium weight in lithium carbonate is straightforward. However, calculating the lithium content of ore and ore concentrate is problematic

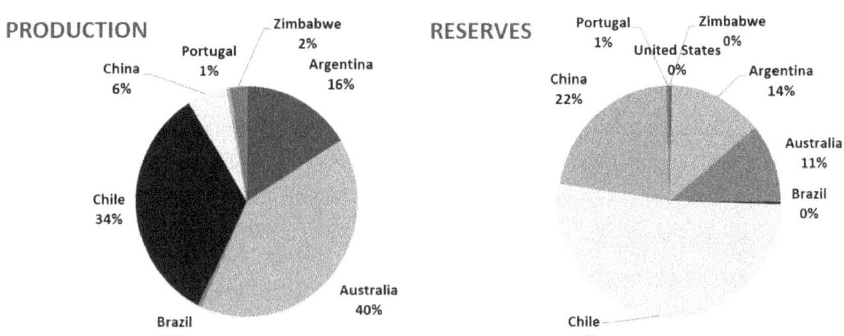

Fig. 1 Distribution of lithium production and reserves in 2016. *Source* USGS [45], updated from [9]

given that the composition of these ores and concentrates is unknown. Despite some inconsistencies in the data, Fig. 2 appears to present a resource which is being exploited through an exponential phase of production.

Similarly, reported reserves are substantially increasing, thanks to continuing exploration [45]. Figure 3 presents different lithium reserve and resource estimates as they developed over time. It is important to note that the estimates are based on different classifications of resources; hence, they are not all directly comparable. The issue is compounded by the fact that explicit descriptions of reserve classifications are not always provided.

In particular, the USGS provides figures for both reserves and reserve base, though reserve base reporting was discontinued in 2010 [45]. Roskill (cited in [36]) provides reserve data for 2009. Garrett [47] provides disaggregate reserve figures for 2004. Tahil [7, 8] provide reserve and reserve base estimates for the years 2005 and 2007. Evans [48, 49] provides reserve and 'in situ' data for year 2008. Finally, Yaksic and Tilton [10] provide estimates of recoverable resources and in situ resources in 2009. The data in Fig. 3 provide a wide range of estimates, with the largest estimate in 2009 over 700% greater than the smallest. This can in part be explained by the different nature of reserve classifications used, but it also reflects the uncertainty around future prospects for lithium production. It is also worth mentioning that the USGS [45] refers to additional 'resources' for several countries, including Bolivia, which as yet has no recorded production or reserves, but the USGS estimates it to have 9 million tonnes of resources. What prevents any of these resources from being reported as reserves by the USGS is unclear. The USGS

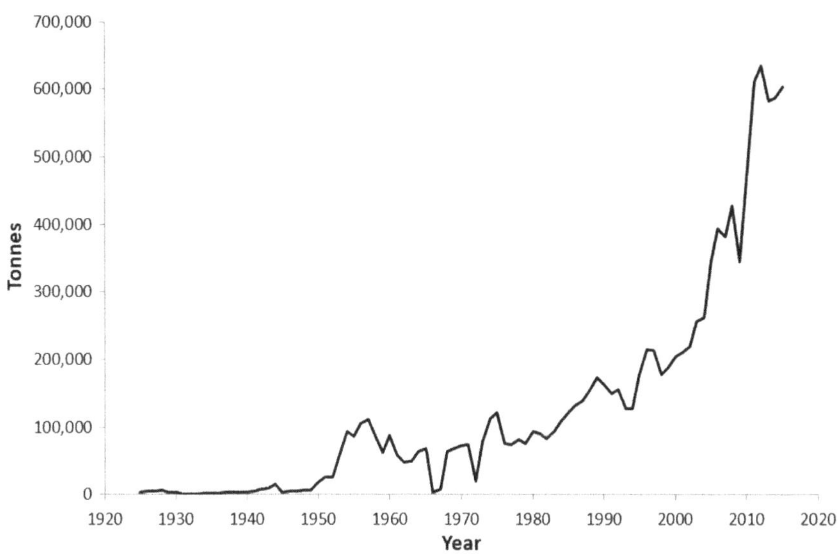

Fig. 2 World annual production by gross weight of lithium ore and ore concentrate from minerals, and lithium carbonate from brines, 1925–2015. *Source* USGS [46], updated from [9]

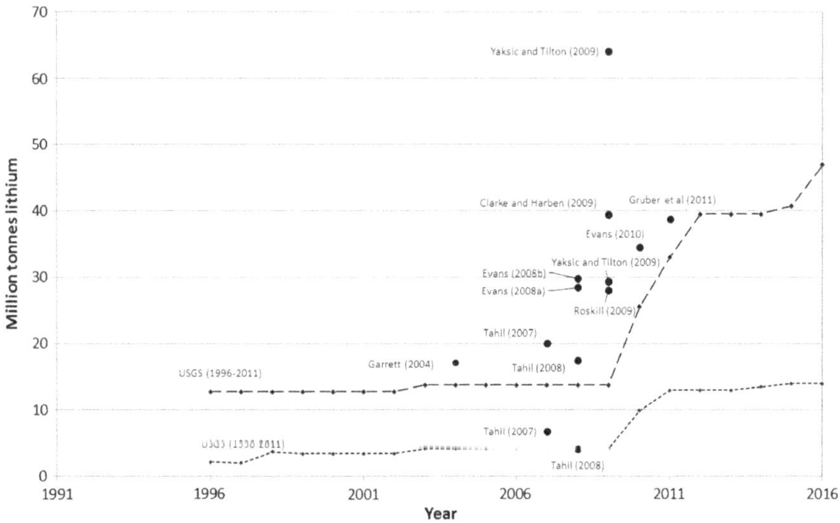

Fig. 3 Reserve and resource estimated from USGS and other studies. *Sources* [7, 8, 10, 11, 36, 45, 47–49], updated from [9]

in 2017 [45] estimates world resources at 47 million tonnes, over three times their reserve estimate but still well below the Yaksic and Tilton [10] estimate.

Given the rapidly growing nature of production (Fig. 2) and the relative increase in reserve estimates over time (Fig. 3), lithium appears to be relatively immature in terms of its exploration and production, with the latter increasing rapidly every year and reserve estimates indicating that new resources are still being discovered.

4.3 Recycling

Historically, only small quantities of lithium have been recycled [45]. The United Nations Environment Programme has estimated lithium end-of-life recycling rates at less than 1% in 2011 [50]. However, more recently we are witnessing an increase in recycling of lithium batteries due to their growing market size which brought about new regulation on their disposal. In Europe, Member States have been obliged to collect 25% of end-of-life batteries by 2012 and 45% by 2016 [51]. This legislation however does not mandate the recycling of lithium. In fact, some Li-ion battery recycling facilities recover cobalt and nickel hydroxides but not lithium [52].

In spite of this, the potential for recycling of lithium from end-of-life batteries is estimated to be significant. Gaines and Nelson [12] estimate that over 40,000 tonnes of lithium could be recycled from batteries in the USA by 2050, assuming 100% recycling rates and a 10-year battery life. Gruber et al. [13] model lithium recycling and estimate that this could satisfy between 50 and 63% of cumulative demand over the 2010–2100 period, assuming recycling rates of 90–100%. Buchert et al. [52], however, note that while the large growth in battery production implies a significant recycling potential, the economic case for recycling lithium remains weak unless its price grows significantly.

A primary issue in recycling lithium from end-of-life batteries is the sorting of collected waste batteries, as even the Li-ion batteries from automotive applications will not all have the same chemistry. In order to develop an efficient recycling process, it is necessary to know the composition of the batteries to be treated [53]. To address this, a number of automatic sorting systems have been developed which use magnetic or electrodynamic sensors, photograph recognition of the label and X-ray imaging, all resulting in varying levels of purity of the separated fractions [54].

For the reasons outlined above, the recovery of lithium from spent batteries remains a niche market [52] and the lithium battery industry has no tradition of using recycled material for manufacturing new batteries [55]. It therefore appears difficult for recycled lithium to contribute half of future supply, as suggested by Gruber et al. [13], unless more targeted legislation is introduced or stronger market incentives develop.

4.4 Estimates of Future Supply

Because the exploitation of lithium as a natural resource is still immature, both production and reserve estimates have been changing over time and several authors have tried to account for these changes within projections of future production or availability. In [9], we had examined a number of estimates of future production to year 2020, which were based on the analysis of lithium reserve exploitation projects that were either in the pipeline or expected, and of future availability, based on assessments of total recoverable resources. These estimates were in the range of 60–110 thousand tonnes per year of lithium metal production in 2020 and ~2 to ~20 million tonnes of lithium metal available in total [7, 8, 10, 13, 18, 33, 56, 57]. In this chapter, we will not discuss the estimates in detail. Instead in Fig. 4, we provide a comparison between the range of production estimates to year 2020 that were made some years ago, with a high-case projection based on [18] and a low-case projection based on [56], and the actual production that has taken place since, as reported by the USGS [45]. As is apparent from Fig. 4, lithium production has substantially underperformed the estimates we had assessed in [9], and extrapolating from the data we can expect lithium production to be at the lower end of the estimated range by year 2020.

As for the assessments of total recoverable resources that we had reviewed in [9], it is worth noting that these tend to be conservative and increase over time. Hence, significantly more lithium may be available if its price increases. Yaksic and Tilton [10] estimate that, at a price of 1.40–2 $ per lb of lithium carbonate,

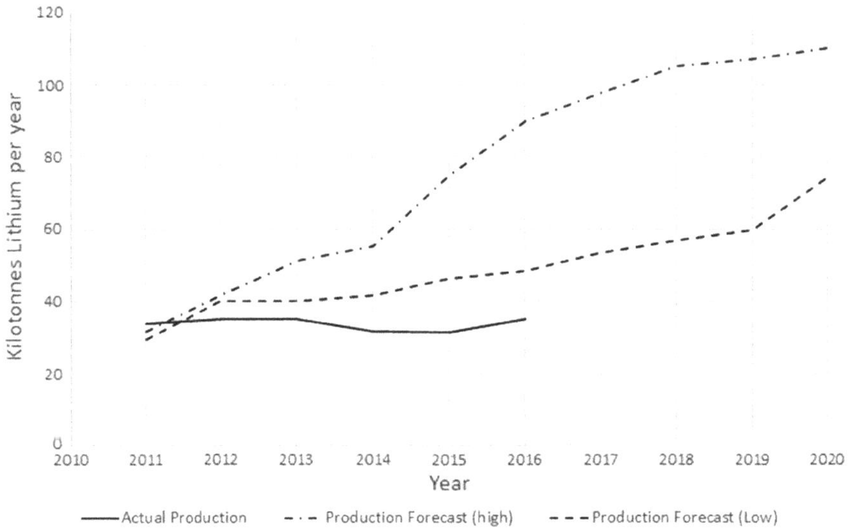

Fig. 4 Comparison between range of estimates of future lithium production from the literature reviewed in [9], compared with actual production. *Sources* [18] for the high production forecast; [56] for the low production forecast; [45] for the actual production

22 million tonnes of lithium are available. However, they also estimate that at a price of 7–10 $ per lb of lithium carbonate, lithium can be extracted from sea water, more than doubling their estimate of available lithium. Hence, it is practical to assume the higher part of the range of estimates reviewed in [9] as a lower bound for future lithium availability.

5 The Balance of Lithium Supply and Demand

In Table 6, we present the range of global lithium demand from BEV and PHEV batteries for the years 2030 and 2050 that we have derived based on the analysis presented in Sect. 3, using Eq. (1) and the values previously discussed for the relevant variables, updated from [9] as necessary. Our estimates of future lithium demand from EV batteries are presented as ‘low’ and ‘high’. Due to the complexity of estimating the lithium content of batteries, we use the range of figures for lithium content per unit energy stored (gLi/kWh), or lithium intensity, that we have discussed in Sect. 3.2 and are the same as in [9]. As for battery sizes, in the ‘low’ case, we use an average of 8 kWh for PHEVs and 30 kWh for BEVs, and in the ‘high’ case, we use 16 kWh for PHEVs and 60 kWh for BEVs. We have revised the figures originally used in [9] in order to account for the trend towards using larger batteries, especially in BEVs (see Tables 3 and 4). Finally, annual EV sales figures are the same as in [9], i.e. based on the IEA’s BLUE Map scenario presented in

Table 6 Global annual lithium demand estimates for BEVs and PHEVs in 2030 and 2050 markets, 'low' and 'high' cases

Variable	Low	High
Battery Size (kWh)—PHEV	4.3	16
Battery Size (kWh)—BEV	16	35
Intensity (gLi/kWh)	190	380
2030 BLUE Map		
Annual sales (million units/yr)—PHEV	25	25
Annual sales (million units/yr)—BEV	9	9
Market share of Li-ion batteries	100%	100%
Range of demand (kilotonnes Li/yr)	89	357
2050 BLUE Map		
Annual sales (million units/yr)—PHEV	62	62
Annual sales (million units/yr)—BEV	47	47
Market share of Li-ion batteries	100%	100%
Range of demand (kilotonnes Li/yr)	362	1449

Source authors' analysis, updated from [9]

Table 2, which we decided to keep because it represents a conservative estimate of future EV market uptake, as discussed in Sect. 3.1. No account of other uses of lithium is included in the demand estimates, and these are purely based on demand for EVs.

As can be seen from Table 6, our estimated lithium demand from EVs increases significantly between 2030 and 2050. This is due to the large growth in annual vehicle sales between the two timeframes, and the changing ratio between PHEV and BEV sales, which have different battery sizes (kWh). The scale of future lithium demand is also very large, with almost 1.5 million tonnes of lithium demand annually in the 2050 market (high case).

In Fig. 5, we combine lithium demand and production estimates, which enables us to derive insights into possible future lithium availability issues. On the left of Fig. 5, we present historical lithium metal production using data from Fig. 2 which shows an approximately geometric growth trend. On the right of Fig. 5 we provide estimates of future lithium supply and demand. The supply estimate is based on modelling by Vikström et al. [58] that extrapolates from historical production; hence, it is on the low end of the production forecast range to year 2020 provided in Fig. 4. The supply projection in [58] stretches to year 2050.

The quantitative impact of recycling on supply is not taken into account, given the ongoing concerns regarding lack of economic incentive and the low recycling rate experienced to date. However, if lithium recycling increased in the future, this would have a positive impact on future lithium availability relative to the estimates presented in Fig. 5.

The range of demand presented in Fig. 5 is large, driven by several factors. First, there is a significant uncertainty regarding the future average battery size and lithium intensity in batteries. There is a paucity of literature discussing the likely development of these factors over time; and the current EV models on the market

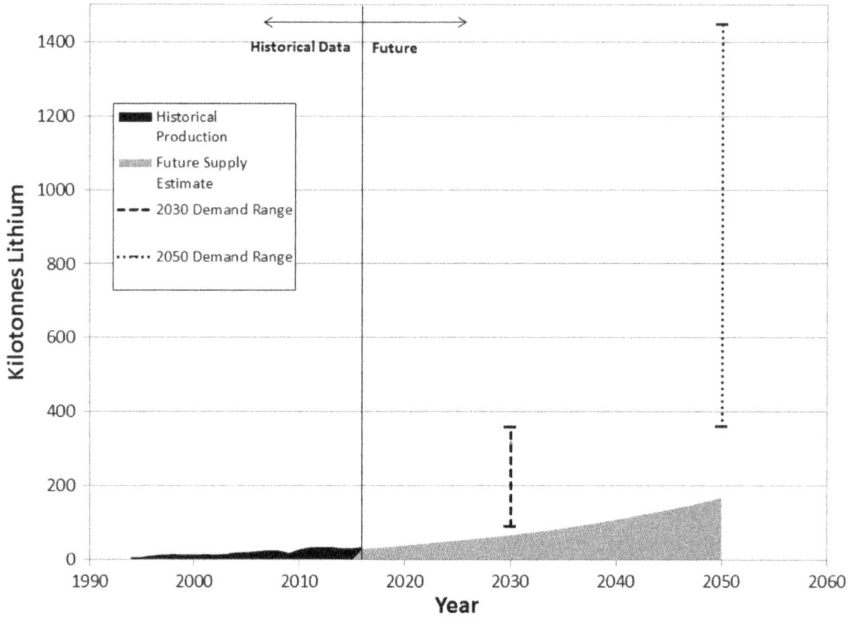

Fig. 5 Comparison of historical production, forecast supply and forecast demand of lithium for EVs *Sources* [45, 46] for historical production, [58] for future projected production, authors' own analysis

and the available studies we have reviewed provide a very wide range of battery sizes and material intensities. Moreover, the future sales of both BEVs and PHEVs are subject of much speculation and a wide range of forecasts is available in the literature. Combining all these factors makes for great uncertainty in future demand for lithium from EV batteries. We can further observe that over the last decade, a clear trend is present where automotive OEMs equip their BEVs with increasingly larger batteries which give their vehicles longer ranges. The increased material intensity that this entails is however likely to be at least in part offset by improvements in battery engineering both at the cell and at the pack levels.

Comparing future supply and demand projections, it is apparent that, if production does not increase rapidly to match the likely rapid growth of demand, lithium availability may become a bottleneck in the manufacturing of EV batteries. Indeed in the last few years, demand has been moderately exceeding supply, which resulted in price increase [45]. The estimated range of future demand is many times greater than current supply. While this is challenging, there is no evidence that future production cannot increase at a sufficient pace. Though long-term exponential growth in lithium production would be unsustainable, if growth could be sustained over the next two decades, meeting future demand may be possible. Supporting this optimism are the significant resource estimates for lithium seen in Fig. 3, though these estimates make no assessment of how easy these resources are to access, and over what timescales they can be produced. In addition to the substantial resources that have been identified, even excluding sea water, end-of-use recycling could contribute to future supply, although it is unclear which lithium price levels will make recycling viable.

Further analysis of material demand for EVs is needed in order to reduce uncertainty concerning the quantity of lithium demanded per battery in the future. Analysis of the production potential of lithium is also needed to better assess which parts of identified lithium resources are economic. While there is evidence in the literature that these steps are being taken, a thorough assessment of the long-term effects of material availability on the deployment of EVs still requires a much improved understanding of the potential for, and the economic implications of, expansion in both lithium production and recycling.

References

1. U. S. Department of Energy (DOE), *Critical Materials Strategy*. Washington D.C. (2010)
2. U. S. Department of Energy (DOE), *Critical Materials Strategy*. Washington, DC. (2011)
3. H. Kara et al., *Lanthanide Resources and Alternatives: A report for Department for Transport and Department for Business, Innovation and Skills* (Oakdene Hollins: Aylesbury, UK, 2010)
4. Climate Change Committee (CCC), *Building a Low Carbon Economy—The UK's contribution to Tackling Climate Change*. London, UK (Dec. 2008)
5. International Energy Agency (IEA), *Energy Technology Perspectives*. Paris, France (2016)
6. UK Energy Research Centre (UKERC), *Energy 2050. Making the Transition to a Secure and Low-Carbon Energy System: Synthesis Report*. London, UK (2009)
7. W. Tahil, *The Trouble with Lithium: Implications of Future PHEV Production for Lithium demand. Meridian International Research*. Martainville, France (2007)
8. W. Tahil, *The Trouble with Lithium 2: Under the Microscope, Meridian International Research*. Martainville, France (2008)
9. J. Speirs et al., Renew. Sust. Energy Rev. **35**, 183 (2014)
10. A. Yaksic, J.E. Tilton, Resour. Policy **34**, 185 (2009)
11. K.R. Evans, The future of electric vehicles: setting the record straight on lithium availability. J. Energ. Security (2009) Available from: http://www.ensec.org/index.php?option=com_content&view=article&id=213
12. L. Gaines, P.A. Nelson, *Lithium-Ion Batteries: Examining Material Demand and Recycling Issues. Argonne National Laboratory*. Argonne, IL, USA (2009)
13. P.W. Gruber et al., J. Ind. Ecol. **15**, 760 (2011)
14. D. Kushnir, B.A. Sandén, Resour. Policy **37**, 93 (2012)
15. J. Neubauer, The Impact of Lithium Availability on Vehicle Electrification, Plug-In 2011 Conference. National Renewable Energy Laboratory. Raleigh, NC, USA
16. G. Martin et al., Energy Storage Mater. **6**, 171 (2017)
17. G. Angerer et al., Raw materials for emerging technologies: A report commissioned by the German Federal Ministry of Economics and Technology (English Summary). Fraunhofer ISI 2009 [31/10/2011]; Available from: http://www.isi.fraunhofer.de/isi-en/service/presseinfos/2009/pri09-02.php
18. Dundee Capital Markets (DCM), *Lithium—Hype or Substance: A look at Lithium Demand and Supply* (Dundee Securities Corporation: Toronto, Canada, 2009)
19. International Energy Agency (IEA), *Technology Roadmap: Electric and Plug-In Electric Hybrid Vehicles*. Paris, France (2011)

20. S. Marcus et al., *A cleantech Resource Crisis: Will Rare Earth and Lithium Availability Thwart Cleantech Growth?* (Cleantech Insight, Cleantech Group LLC, 2010)
21. McKinsey and Company, *Roads Toward a Low-Carbon Future: Reducing CO_2 Emissions from Passenger Vehicles in the Global Road Transportation System.* New York, NY (2009)
22. BP Energy Outlook, 2017 Edition. London, UK
23. International Energy Agency (IEA), *World Energy Outlook.* Paris, France (2016)
24. International Energy Agency (IEA), *Energy Technology Perspectives 2017. Catalysing Energy Technology Transformations.* Paris, France
25. OPEC, *World Oil Outlook* (Vienna, Austria, 2016)
26. Energy Perspectives Statoil, *Long-Term Macro and Market Outlook* (Stavanger, Norway, 2016)
27. International Energy Agency (IEA), Energy Technology Perspectives 2010. Scenarios& Strategies to 2050. Paris, France
28. M. Rosenberg, E. Garcia, Known Lithium Deposits Can Cover Electric Car Boom. Reuters, Feb. 11, 2010 [15/11/2011]
29. B. Nykvist, M. Nilsson, Nat. Clim. Change **5**, 329 (2015)
30. CAR Magazine, 2017. [November 2017]; Available from: http://www.carmagazine.co.uk/
31. Green Car Congress, 2017. [November 2017]; Available from: http://www.greencarcongress.com/
32. D. Kushnir, B.A. Sandén, Resour. Policy **37**, 93 (2012)
33. I. Rade, B.A. Andersson, J. Power Sources **93**, 55 (2001)
34. W. Tahil, *How Much Lithium does a Li-ion EV Battery Really Need? Meridian International Research.* Martainville, France (2010)
35. T. Engel, Lithium bewegt die Welt. Deutsche Gesellschaft für Sonnenenergie e. V. [International Solar Energy Society, German Section], 2007. [15/1/2012]; Available from: http://www.dgs.de/164.0.html?&tx_ttnews%5Btt_news%5D=1330&cHash=5cf989b243
36. M. Engel-Bader, *Chemetall Corporate Presentation: Lithium and the National Economy* (Seminar Sociedad Nacional de Minería, Santiago de Chile, 2010)
37. M. Armand, J.M. Tarascon, Nature **451**, 652 (2008)
38. A.G. Ritchie, J. Power Sources **136**, 285 (2004)
39. E. Peled et al., J. Power Sources **196**, 6835 (2011)
40. G.M. Clarke, P.W. Harben, Lithium Availability Wallmap (LAWM) (2009)
41. C. Candelise, J. Speirs, R. Gross, Renew. Sust. Energy Rev. **15**, 4972 (2011)
42. C. Hurst, *China's Rare Earth Elements Industry: What Can the West Learn? Institute for the Analysis of Global Security (IAGS).* Fort Leavenworth, KS, USA, (2010)
43. J. Speirs et al., Materials Availability: Potential Constraints to the Future Low-Carbon Economy. Working Paper II: Batteries, Magnets and Materials. UK Energy Research Centre. London, UK (2013)
44. J. Speirs et al., Materials Availability: Potential Constraints to the Future Low-Carbon Economy—Working Paper I: A Thin-Film PV Case Study. UK Energy Research Centre. London, UK, 2011
45. U.S. Geological Survey (USGS), Mineral Commodity Summaries 2017. U.S. Department of Interior. Reston, VA, USA
46. T.D. Kelly, J.A. Ober, B.W. Jaskula, Lithium Statistics. U.S. Geological Survey. Reston, VA, USA (2017)
47. D.E. Garrett, *Handbook of Lithium and Natural Calcium Chloride* (Academic Press, Cambridge, MA, 2004)
48. K. Evans, An Abundance of Lithium, 2008. [1/12/2011]; Available from: http://www.worldlithium.com/An_Abundance_of_Lithium_1_files/An%20Abundance%20of%20Lithium.pdf
49. K. Evans, An Abundance of Lithium: Part Two, 2008. [1/12/2011]; Available from: http://www.evworld.com/library/KEvans_LithiumAbunance_pt2.pdf
50. United Nations Environment Programme (UNEP), Recycling Rates of Metals: A Status Report, in Report of the Working Group on the Global Metal Flows to the International Resource Panel, 2011

51. European Parliament, *Directive 2006/66/EC on Batteries and Accumulators and Waste Batteries and Accumulators* (Belgium, Brussels, 2006)
52. M. Buchert, D. Schüler, D. Bleher, Critical Metals for Sustainable Technologies and their Recycling Potential. United Nations Environment Programme (UNEP) and Oko-Institut. Nairobi, Kenya (2009)
53. M. Contestabile, S. Panero, B. Scrosati, J. Power Sources **83**, 75 (1999)
54. A.M. Bernardes, D.C.R. Espinosa, J.A.S. Tenório, J. Power Sources **130**, 291 (2004)
55. K. Kotaich, S.E. Sloop, Recycling: Lithium and Nickel–Metal Hydride Batteries, in Encyclopedia of Electrochemical Power Sources, J. Garche, Ed., 2009, p. 188, Elsevier. Amsterdam, Netherlands
56. E.R. Anderson, Sustainable Lithium Supplies through 2020 in the Face of Sustainable Market Growth, in: 3rd Lithium Supply & Markets Conference. TRUGroup. Toronto, Canada (2011)
57. A. Ebensperger, P. Maxwell, C. Moscoso, Resour. Policy **30**, 218 (2005)
58. H. Vikström, S. Davidsson, M. Höök, App. Energy **110**, 252 (2013)

The Issue of Metal Resources in Li-Ion Batteries for Electric Vehicles

Marcel Weil, Saskia Ziemann and Jens Peters

Abstract The worldwide development and market penetration of electric vehicles (EVs) and hybrid cars has lagged far behind initial expectations and prognoses. However, more recent discussions about petrol and diesel car emissions seem to accelerate the market penetration of battery-based mobility and other alternative options. Many big car manufacturers have announced that they will offer a broad EV fleet by between 2020 and 2024 at the latest, and some even plan to abandon the production of petrol- and diesel-powered cars completely. This might result in a sharp increase in EV market shares and, consequently, in a significant amount of resources needed to produce traction batteries. At present, EVs are produced mainly using different types of Li-ion batteries (LIBs) and only to a lesser extent other battery systems like NiMH. Also in a midterm perspective, LIBs will probably continue to be the preferred energy storage technology for EVs due to their excellent technical performance. This raises the question of whether we will have enough reserves or resources of key metals such as Li, Co, Ni, Cu, Al, Mn or P required for Li-ion traction batteries. In answering this question, a dynamic material flow analysis (dMFA) was conducted to quantify the global demand for these key metals driven by the increasing number of battery vehicles. The calculations also take into account potential recycling of metals from batteries after the use phase, which significantly reduces the pressure on reserves and resources.

M. Weil (✉) · S. Ziemann
Institute for Technology Assessment and Systems Analysis (ITAS), Karlsruhe Institute of Technology (KIT), Karlsruhe, Germany
e-mail: marcel.weil@kit.edu

S. Ziemann
e-mail: saskia.ziemann@kit.edu

M. Weil · J. Peters
Helmholtz Institute Ulm (HIU), KIT, Ulm, Germany
e-mail: jens.peters2@kit.edu

G. Pistoia and B. Liaw (eds.), *Behaviour of Lithium-Ion Batteries in Electric Vehicles*, Green Energy and Technology,
https://doi.org/10.1007/978-3-319-69950-9_3

1 Introduction

Many studies in the past have forecasted a relatively rapid and broad market penetration of full electric vehicles and hybrid electric vehicles (EVs and HEVs) (e.g. [1–3]). However, the reality today is quite different; EVs are in some way still exotic, and Norway is the only country worldwide where a market share above 20% is reached. For 2016, the global stock of EVs and HEVs was estimated to be only about 2 million cars [4]. Nevertheless, the most recent investigations show a high probability that the electric car sales could rise significantly in the near future. The estimated worldwide stock of EVs and HEVs (including plug-in hybrid electric vehicles [PHEVs] and fuel cell vehicles [FCVs]) ranges between 9 and 20 million by 2020 and even between 40 and 70 million by 2025, according to the International Energy Agency (IEA) [4]. The evaluation is based on country targets, original equipment manufacturer (OEM) announcements and scenarios on electric car deployment. However, this optimistic estimation neither considers the recent announcements by Volkswagen and Daimler to invest significantly in electro mobility [5, 6], nor the new car regulation in China which requires car makers to produce a fleet with a total of 10% or more electric vehicles by 2019, and 12% or more by 2020 [7].

Recent discussions about petrol and diesel car emissions seem to further accelerate the introduction of battery-based mobility and other alternative power-train options. The new developments mentioned above may be the reason why other studies (e.g. [8]) potentially underestimated significantly the future energy storage demand for EVs.

In the past, the expensive and resource-intensive NiMH batteries were used for the production of HEVs. However, in the latest Avicenne Energy [3] forecast of future market shares of different battery chemistries, NiMH batteries are (contrary to earlier predictions) not even mentioned anymore, possibly due to the announcement by the HEV market leader Toyota to use LIBs also for HEVs. In the mid- and (possibly) long-term perspective, LIBs will therefore probably be the preferred energy storage technology for EVs and HEVs due to their good technical performance.

If the IEA [4] forecasts and estimations come true, the car mobility sector as a whole will require huge energy storage capacity for future vehicle fleets. The production of the corresponding amounts of batteries for EVs and HEVs will have a significant impact on worldwide reserves and resources (batteries for busses and pedelecs only to a lesser extent; see, e.g. [8]). In fact, the resource requirements and environmental impacts of EV batteries make up a significant share of the total resource requirements of the vehicles over their entire lifetime, with the share being clearly higher than that of comparable combustion engine powertrains [9, 10].

Due to the importance of lithium for LIB systems, there has been a strong research focus on the future availability of lithium as a key raw material [11, 12]. Other metals, despite having much higher mass percentages in the batteries, are often disregarded.

Therefore, the goal of this study is to understand the impact on all major metal resources resulting from the production of traction batteries until 2050. In this study, we will consider different LIB chemistries, including all important raw materials such as Li, Co, Ni, Cu, Al, Mn or P. The raw material requirement until 2050 will be calculated and contrasted with the resource and reserve data for the respective element. Thus, it will be possible to identify and discuss different potential levels of criticalities (regarding resource availability) associated with electric vehicles in the future.

2 Metal Requirements for Different LIB Chemistries

The amount of metals contained in each type of battery is quantified based on existing life cycle assessment studies of LIB. From these, the inventory data (detailed mass balances and required metals) are extracted and unified, assuming the same type of cell container, pack housing and battery management system (BMS) for all battery types. Otherwise, a comparison would give a skewed picture, since different cell containers have different masses and thus significantly affect the material balance of the battery cells. For future application in electric vehicles, pouch cell batteries are assumed that are assembled together with the BMS, wiring, insulation and fastening in a steel box to form a 20-kWh automotive battery pack. Based on the corresponding mass percentage of the battery components and the amounts of metals needed for their manufacturing, the metal requirements of each battery type can be determined on a mass basis (Table 1). The energy density calculated for each of the different battery packs is given in the last line of Table 1, allowing to calculate the amount of metal required to provide a certain energy storage capacity (e.g. amount of metal per kWh). Interestingly, the metal content of the different LIBs ranges only between 43 and 56%, including the contributions of the cell and pack housing and the major metals contained in the BMS. This is because significant proportions of the batteries are organic components, such as solvents for the electrolyte or the separator, carbonaceous materials (graphite and conductive carbons—not considered in this study), binders and plastic, which do not contribute to the metal demand. Apart from that, the cathode-active materials are metal oxides and thus also contain a significant proportion of non-metal elements, basically oxygen. The “passive” battery components such as package and BMS contribute majorly to the demand for aluminium, iron and copper, where they add between 2% (Cu) and 7–8% (Al and Fe) to the total metal requirements per battery pack.

For the resource demand calculation, the data of [13, 14] are used in case of LFP and NCM, respectively (cf. Table 1).

Table 1 Metal requirements (pack level; 20-kWh battery pack; including metals for cell container, pack housing and major metals for BMS) for different LIB chemistries, using unified inventory data [15, 16]. The references below the battery chemistries indicate the source of the original battery modelling data. LTO chemistry is only displayed for comparison

	LTO	LFP	LFP	LMO	NCA	NCM	NCM
Metal	[17]	[13]	[18]	[19]	[17]	[13]	[14]
Li	3.5%	3.7%	1.5%	1.3%	1.6%	1.9%	2.0%
Al	18.0%	12.8%	9.8%	19.3%	13.9%	12.8%	11.8%
Cu	2.0%	12.1%	5.4%	17.8%	13.0%	12.1%	18.5%
Ni	0.4%	0.1%	0.1%	0.1%	10.6%	6.3%	5.5%
Co	0.0%	0.0%	0.0%	0.0%	1.9%	3.1%	5.4%
Mn	0.0%	0.0%	0.0%	9.5%	0.0%	5.7%	5.0%
Fe	13.9%	19.4%	19.7%	8.0%	8.0%	8.0%	8.0%
P	6.9%	9.4%	6.5%	0.0%	0.0%	3.0%	0.0%
Total	44.6%	57.4%	43.0%	56.1%	49.1%	53.0%	56.2%
Wh/kg	52.4	109.6	83.2	114.9	133.6	139.5	131.3

3 Relevance of LIB Chemistries Today and in the Future

The prediction of future trends regarding the relevance of different LIB chemistries is difficult and highly uncertain. Very little information can be found about future trends of different LIB chemistries within the automotive sector. The few studies that are available on the topic do not distinguish between sectors and thus give unrealistic scenarios for automotive applications, where lithium cobalt oxide (LCO) batteries are not an option due to safety concerns. To obtain an idea about possible developments in the sector, the market outlook by Avicenne Energy [3], one of the very few available sources breaking down possible market shares of LIB chemistries to sectors, is used to estimate the relevance of the different LIB chemistries for the automotive sector in the future. The market shares calculated based on [3] for the major LIB types as a basis for the development of scenarios until 2050 are provided in Table 2.

Table 2 Expected market shares of different LIB chemistries for EV, HEV and PHEV applications in 2020 and 2025 versus 2015 [3]

	NMC (%)	NCA (%)	LMO (%)	LFP (%)
2015	25	18	45	12
2020	30	18	26	26
2025	28	23	17	32

4 Modelling of Resource Requirements

The developed dynamic MFA model for calculating the global resource requirements for electric vehicles is schematically represented in Fig. 1. As mentioned above, the recent discoveries regarding the emissions of conventional car engines seem to accelerate the transition towards alternative powertrains. Thus, we assume a strongly increasing EV market penetration and use the optimistic blue map scenario from the underlying publication by IEA [20], including the share development of different vehicle types.

The dynamic MFA model applied in this study is based on the stock dynamics concept developed by Müller (2006) [21]. The core of the model consists of the stock of electric vehicles (EV) that will be needed to fulfil the population's demand for future individual mobility. The EV stock is build up by population (Pop) and car ownership per capita (Car/cap). On this basis, the development of the stock of traction batteries as well as of the respective inflows and outflows (cf. Fig. 1) can be calculated. The demand for new EVs equipped with traction batteries is the primary driver of battery stock, but not the only one. Since battery lifetime is deviating from vehicle lifetime at present, it is assumed that batteries in EVs are replaced after

Fig. 1 MFA model of the EV battery stock

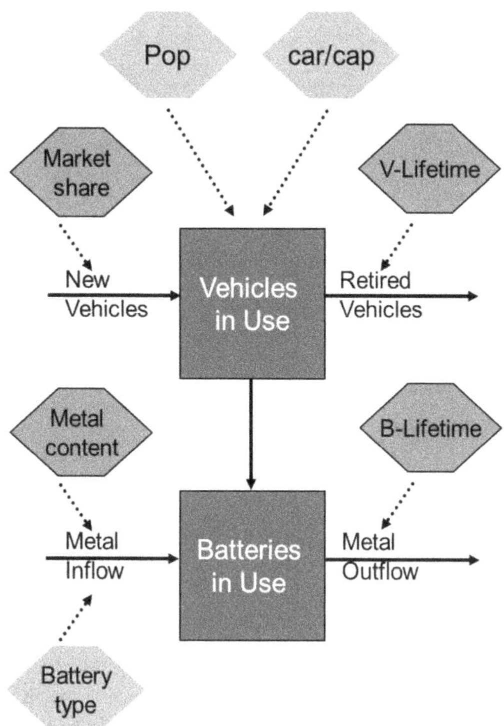

reaching the end of their lifetime (after reaching 80% of the previous capacity). Therefore, battery lifetime is the other important driver of demand for new traction batteries and the materials contained therein.

The system considered in our study includes all relevant processes of the global EV market. The model is applied for a timescale of 35 years, from 2015 to 2050, and all important parameters are displayed in Tables 3 and 4.

On the one hand, the stock of traction batteries in our model can be differentiated by the four vehicle types such as FCV, HEV, PHEV and FEV, which determine the battery size. The estimated battery size for each vehicle class represents average values (of small, compact and premium class cars) and is rather at the lower end. For instance, the first generation of BMWi3 (FEV) was equipped with a 22-kWh battery, which was exchanged for the second generation by a 33-kWh battery [29].

On the other hand, within the battery stock different battery technologies can be further distinguished: lithium metal oxide ($LiCoO_2$ or $LiMn_2O_4$ or LMO), lithium nickel manganese cobalt oxide ($LiNiMnCoO_2$ or NMC), lithium nickel cobalt aluminium oxide ($LiNiCoAlO_2$ or NCA), and lithium iron phosphate ($LiFePO_4$ or LFP), showing individual material compositions (cf. Table 1). We used a percentage distribution between the different battery types for 2015 in the following way: LMO (45%), NMC (25%), NCA (18%) and LFP (12%) (cf. Table 2).

Based on the predictions and trends for different Li-ion chemistries by Avicenne Energy [3] until 2025 (Table 2), we developed three scenarios (low LFP, medium

Table 3 Input parameters for the dynamic MFA model

Input parameters		Data		References
		2015	2050	
Population	Medium case	6.9 bn	8.6 bn	[22]
Car ownership (car/1000 capita)	Medium case	144	320	[23]
Share of EVs in light-duty vehicles [%] by type and battery size [kWh]	FCEV (1.5 kWh)	0	18.6	[20, 24]
	HEV (1.5 kWh)	1.2	7.6	
	PHEV (12 kWh)	0	33.3	
	FEV (25 kWh)	0	25.9	
Vehicle lifetime		15 years		[25, 26]
Battery lifetime		10 years		[27, 28]

Table 4 Considered scenarios for 2050 with different shares in battery technologies

Battery technology	Percentage 2050		
	Low LFP (%)	Medium LFP (%)	High LFP (%)
LFP	25	50	75
LMO	10	10	10
NMC	32	20	7.5
NCA	33	20	7.5

LFP and high LFP) for 2050. The dynamic MFA model calculates linearly from 2015 until 2050. In Table 4, the percentages for the scenarios 2050 are displayed.

To understand to which extent recycling could reduce the potential impact on resource consumption, we consider two different recovery rates (RR): 50 and 90%. These rates are assumed for the whole recycling chain, including collection, dismantling, pre-processing and recovery. The very high recovery rate of 90% is currently reached only for some elements in specific applications, such as Pb in starter batteries. The lead recycling rate for batteries is estimated to be 95% in Europe and the USA and can be considered as the highest recycling rate for metals [30]. But these numbers do not include potential export of cars with lead batteries in Third-World countries (with highly inefficient and contaminating lead acid battery recycling) which reduce the recovery rate noteworthy. In contrast, 50% can be considered as a relatively low recovery rate for many metals. However, for lithium in traction batteries it would already be a great challenge to reach a 50% recovery rate. At present, lithium ends up in the slag of the pyrometallurgical recycling process of batteries and no noteworthy recovery takes place, mainly due to economic reasons [11, 31].

5 Results

In Fig. 2, the calculated raw material requirements until the year 2050 are displayed for all three scenarios. It shows that, in the case of the high LFP scenario, more Li, P and Fe but significantly less Co and Ni are required. In all three scenarios, high amounts of Al, Cu and Fe are necessary, due to their usage also in the BMS and for casing. In the case of Al and Cu, the curves of high LFP scenario and low LFP scenario are quite close to each other, because there are in most cases only minor differences in Cu and Al content between the considered battery compositions. On the contrary, there is a strong deviation of the curves of high LFP scenario and low LFP scenario in case of Ni, due to the significantly different Ni content of LFP, NCA and NCM battery chemistries (cf. Table 1).

In Fig. 3, the cumulative Li demand for the three scenarios until 2050 is displayed. The recycling of metals in batteries reduces the demand for primary resources. To evaluate the recycling effects, also a theoretically recovery rate of 50 and 90% is considered. The calculated Li demand is compared with the known Li reserves and resources in 2016 [32]. The results show that without recycling, the Li demand until 2050 is in all three investigated scenarios higher than the known Li reserves. In this case, even before 2050 the economically less attractive Li resources need to be exploited to cope with the increasing demand for EVs (bear in mind, no other Li-ion battery applications are considered). In contrast, with recycling the demand for primary Li can be covered by the reserves (high LFP scenario, 90% recovery rate) or is lower (medium LFP scenario, 90% recovery rate), or is even noteworthy lower (low LFP scenario, 90% recovery rate).

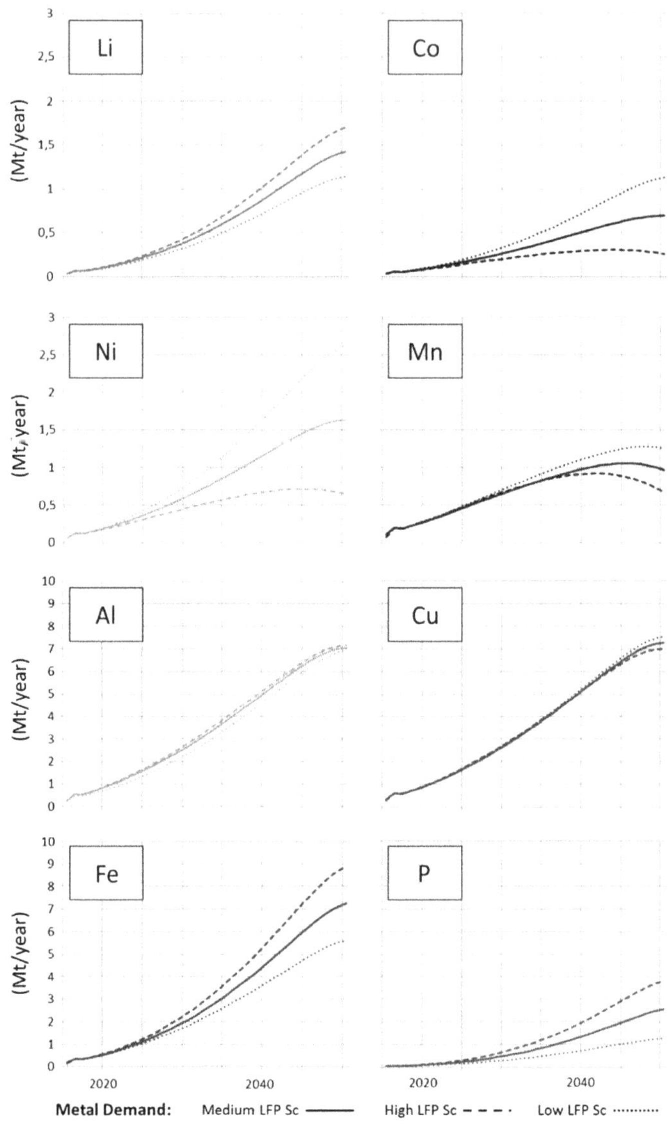

Fig. 2 Calculated raw material requirements until 2050 in Mt/year

The cumulative Co demand for the three scenarios until 2050 is displayed in Fig. 4 and contrasted with the known Li reserves and resources in 2016 [32]. Without recycling, the Co demand until 2050 is in all three investigated scenarios higher than the known Co reserves. In the case of the low LFP scenario, the demand exceeds more than 2.5 times the known reserves and represents approximately 70% of the known resources. Recycling decreases the primary Co demand, but only in

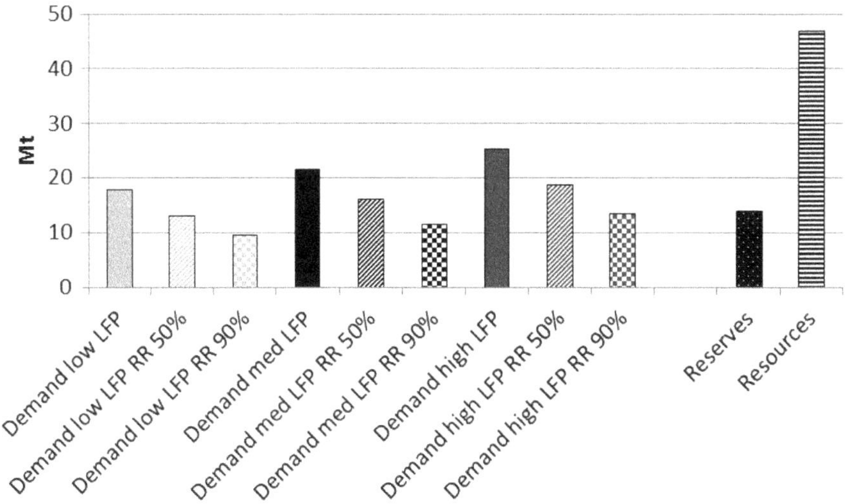

Fig. 3 Calculated cumulative Li demand for the three scenarios until 2050 in Mt, with a theoretical recovery rate of 50 and 90%. Compared with reserve and resource data from [32]

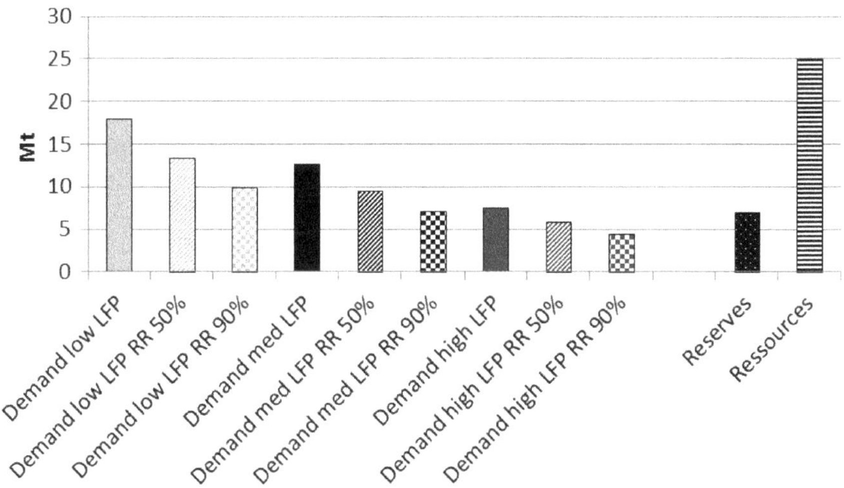

Fig. 4 Calculated cumulative Co demand for the three scenarios until 2050 in Mt, with a theoretical recovery rate of 50 and 90%. Compared with reserve and resource data from [32]

the high LFP scenario the demand is lower than the known reserves. Thus, in comparison with Li, the Co demand in the future can be considered as more critical.

The cumulative Ni demand for the three scenarios until 2050 is displayed in Fig. 5 and contrasted with the known Ni reserves and resources in 2016 [32]. Without recycling, the highest demand corresponds to approximately 50% of

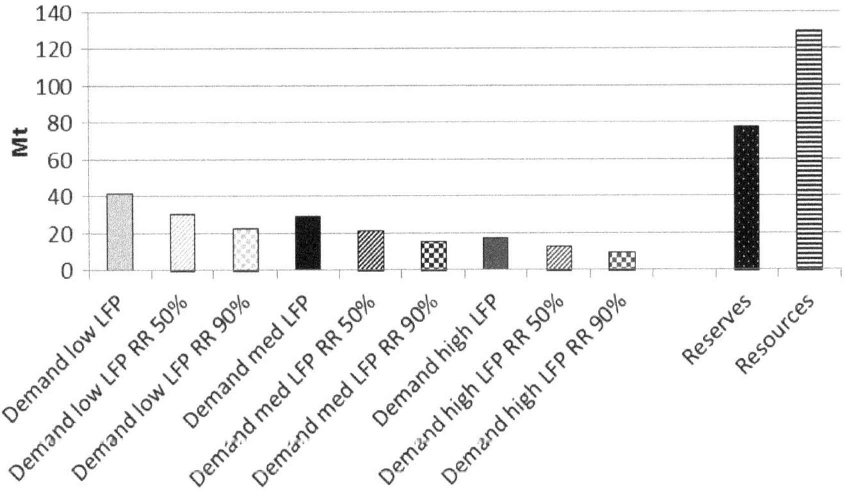

Fig. 5 Calculated cumulative Ni demand for the three scenarios until 2050 in Mt, with a theoretical recovery rate of 50 and 90%. Compared with reserve and resource data from [32]

known Ni reserves until 2050 (low LFP scenario), in all other cases significantly less. In the best case (high LFP scenario, 90% recycling), the primary demand is only 1/8 of the Ni reserves and 1/13 of the known Ni resources.

In Fig. 6, the cumulative Mn demand for the three scenarios until 2050 is displayed and contrasted with the known Mn reserves in 2016 [32]. Without recycling, the highest demand is less than 4% of the known Mn reserves until 2050 (medium LFP scenario), in all other cases significantly less. In the best case (low LFP scenario, 90% recycling), the primary demand is only 1.8% of the Mn reserves.

The cumulative Al demand for the three scenarios until 2050 is displayed in Fig. 7 and contrasted with the known Al reserves in 2016 [32]. There are nearly no differences in demand between the scenarios. Without recycling, the highest demand is only 2.3% of the known Al reserves until 2050 (high LFP scenario), in all other cases slightly less. In the best case (low LFP scenario, 90% recycling), the primary demand is only 1% of the Al reserves and less than 0.5% of Al resources (the latter is not displayed).

The cumulative Cu demand for the three scenarios until 2050 is displayed in Fig. 8 and contrasted with the known Cu reserves in 2016 [32]. There are nearly no differences in demand between the scenarios. Without recycling, the highest demand is approximately 18% of the known Cu reserves until 2050 (low LFP scenario), in all other cases slightly less. In the best case (high LFP scenario, 90% recycling), the primary demand is less than 10% of the Cu reserves and less than 2% of Cu resources (the latter is not displayed).

The cumulative Fe demand for the three scenarios until 2050 is displayed in Fig. 9 and contrasted with the known Fe reserves in 2016 [32]. Between the scenarios, there are nearly no noteworthy differences in demand, which is quite low in

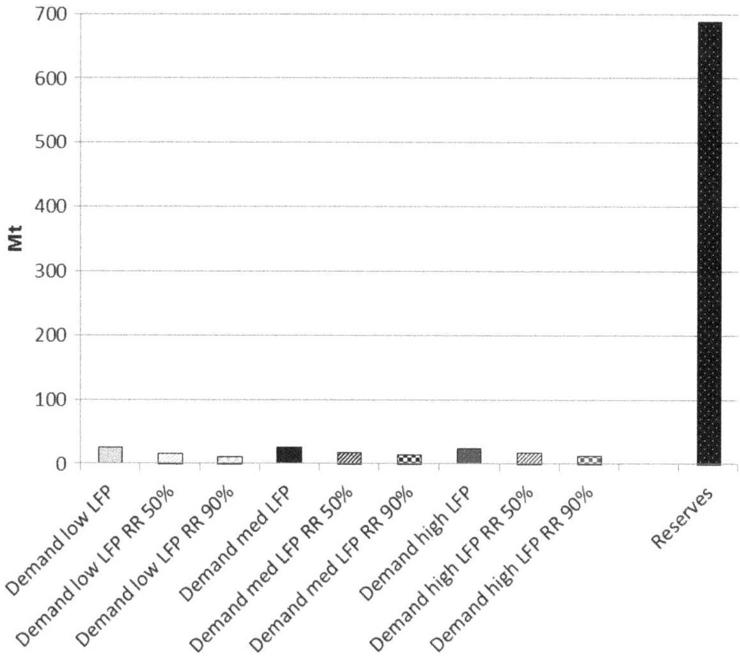

Fig. 6 Calculated cumulative Mn demand for the three scenarios until 2050 in Mt, with a theoretical recovery rate of 50 and 90%. Compared with reserve data from [32]

comparison to the reserve values. Without recycling, the highest demand is less than 0.2% of the known Fe reserves until 2050 (high LFP scenario), in all other cases slightly less. In the best case (low LFP scenario, 90% recycling), the primary demand is less than 0.06% of the Fe reserves and less than 0.02% of Fe resources (the latter is not displayed).

In Fig. 10, the cumulative P demand for the three scenarios until 2050 is displayed and contrasted with the known P reserves in 2016 [32]. Between the scenarios, there are nearly no noteworthy differences in demand, which is in each case quite low in comparison to the reserve values. Without recycling, the highest demand is less than 0.08% of the known P reserves until 2050 (high LFP scenario), in all other cases slightly less. In the best case (low LFP scenario, 90% recycling), the primary demand is less than 0.02% of the P reserves and approximately 0.003% of P resources (the latter is not displayed).

When comparing the results of the demand for the considered elements until 2050 (Figs. 3, 4, 5, 6, 7, 8, 9 and 10) with the known reserves, four groups can be distinguished:

- Very critical (demand potentially higher than known reserves)—Co, Li
- Critical (demand potentially in the double-digit percentage range of known reserves)—Cu, Ni

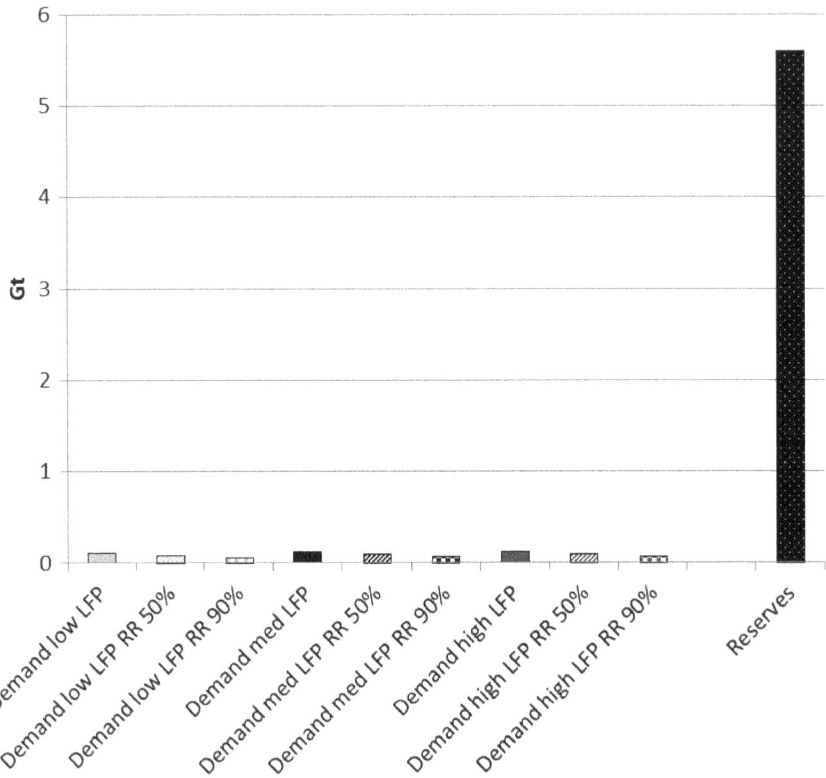

Fig. 7 Calculated cumulative Al demand for the three scenarios until 2050 in Gt, with a theoretical recovery rate of 50 and 90%. Compared with reserve data from [32]

- Less critical (demand potentially in the one-digit percentage range of known reserves)—Mn, Al
- Not critical (demand potentially significantly below 1% of known reserves)—Fe, P.

Thus, the increased use of battery systems like LFP in the future could reduce the impact on cobalt reserves, but not on lithium. A more promising strategy would be to focus on systems that need no or only little Co and Li. In this respect, sodium-ion battery systems [33] (or Mg-based systems [34]) would be more favourable and would even allow the use of Al instead of Cu (as a critical element) as conductor foil, in case of sodium-ion battery systems.

The presented results strongly depend on the parameters used in the model. If, for instance, the car per capita ratio were to remain constant or even decrease in the future (e.g. due to the broad introduction of autonomous driving), the resource issue would be less critical than indicated here. On the other side, if the average battery size of the different vehicle types were to increase, the resource problem would be even more serious in the future. Predictions of resource requirements until 2050 are

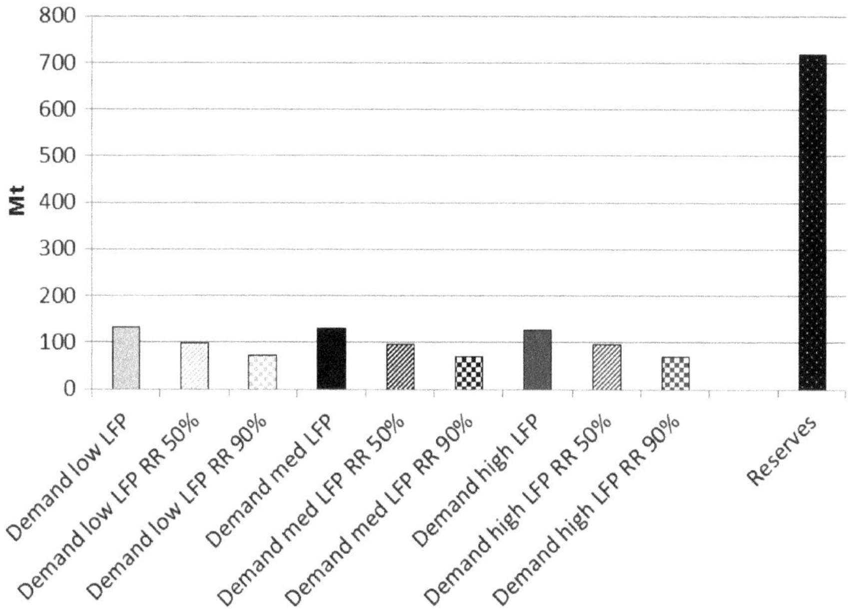

Fig. 8 Calculated cumulative Cu demand for the three scenarios until 2050 in Mt, with a theoretically recovery rate of 50 and 90%. Compared with reserve data from [32]

clearly subject to high uncertainties and also depend on developments in other sectors. Nevertheless, the modelling of realistic scenarios helps to identify potential resource bottlenecks, which should be considered when developing strategies for future battery research.

6 Conclusions

This study investigated the potential raw material requirements for Li-ion batteries for electric vehicles and hybrid cars. The developed dynamic MFA model allows the estimation of the demand for primary elements until 2050, with and without recycling. The results are contrasted with the known reserves (and resources, if applicable) of the respective elements. The findings show that Li and especially Co can be considered as very critical elements, because the calculated demand until 2050 is potentially noteworthy higher than known reserves. In the case of Ni, and especially of Cu, a significant share of the reserves would be depleted until 2050. Therefore, future research should also focus on the substitution of these metals in order to decrease the impact on these resources. In contrast, Mn and Al seem to be less critical from a battery perspective, and Fe and P are uncritical. For the more

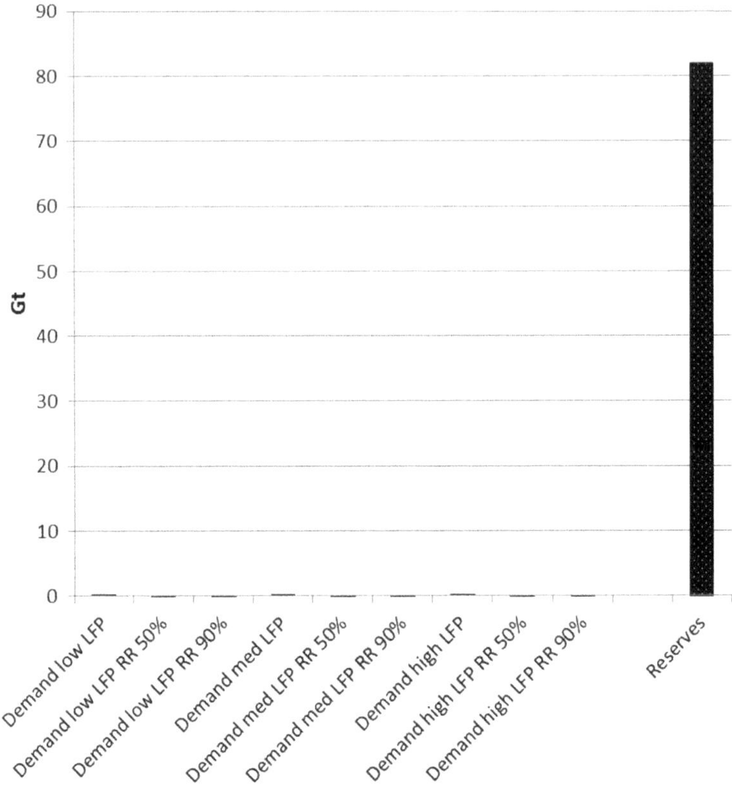

Fig. 9 Calculated cumulative Fe demand for the three scenarios until 2050 in Gt, with a theoretical recovery rate of 50 and 90%. Compared with reserve data from [32]

critical elements, high recovery recycling can reduce the pressure on reserves and resources noteworthy, but cannot solve serious resource problems.

However, one should bear in mind that the present study only focuses on potential resource requirements for batteries in the automotive sector. While this sector will most probably cause the highest demand for battery capacity in the future, other sectors such as portable electronics, pedelecs, power tools and stationary batteries for households, industry or electric grid will also have a noteworthy impact on battery capacity demand. When also taking into account these applications for energy storage (besides the basic demand in other sectors), the pressure on reserves and resources increases significantly. Hence, for several types of lithium-ion batteries, shortages in key raw materials can be expected. Apart from resource availability aspects, also political factors can increase the criticality of such elements, which are often sourced from politically unstable or unreliable countries. Thus, from a responsible research and innovation (RRI) perspective, the development of batteries based on abundant elements like Fe, P or Na, Ca and Mg should be targeted by future battery research.

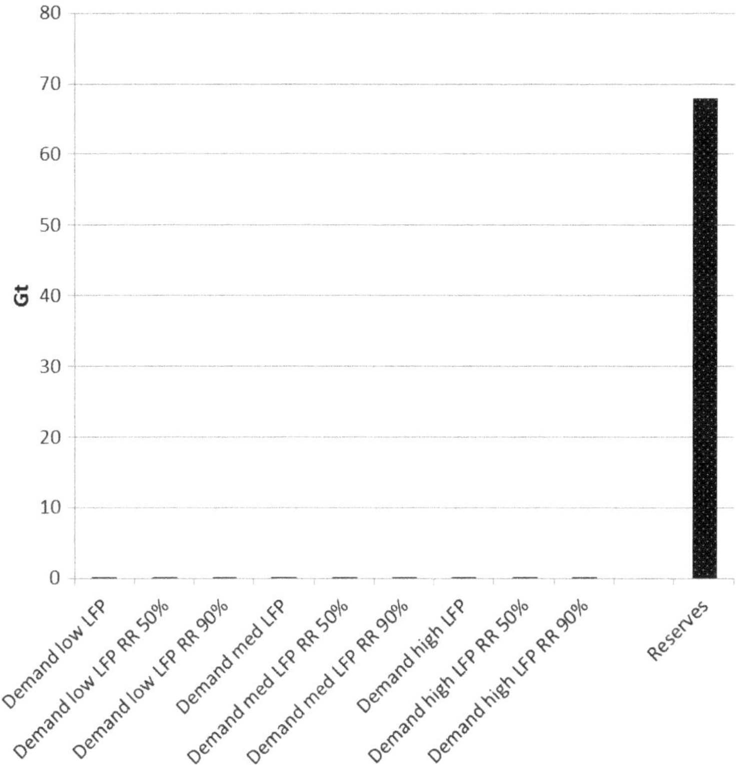

Fig. 10 Calculated cumulative P demand for the three scenarios until 2050 in Gt, with a theoretical recovery rate of 50 and 90%. Compared with reserve data from [32]

References

1. E. Heymann, O. Koppel, T. Puls, *Electromobility—Falling Costs Are a Must* (Deutsche Bank AG, DB Research, Frankfurt am Main, Germany, 2011)
2. H.K. Ng, A.D. Vyas, D.J. Santini, *The Prospects for Hybrid Electric Vehicles, 2005–2020: Results of a Delphi Study* [ANL Report 1999-01-2942]. Argonne National Laboratories, Argonne, USA (1999)
3. *Avicenne, Worldwide Rechargeable Battery Market 2015–2025* (Avicenne Energy, Paris, France, 2016)
4. *IEA, Global EV Outlook 2017—Two Million and Counting* (International Energy Agency, Paris, France, 2017)
5. V.W. AG, The Volkswagen group launches the most comprehensive electrification initiative in the automotive industry with 'Roadmap E, Volkswagen AG-News, 11-Sep-2017. [Online]. Available: https://www.volkswagenag.com/en/news/2017/09/Roadmap_E.html. Accessed: 06 Oct 2017
6. S. Koeller, Daimler beschleunigt seinen Elektrifizierungs-Fahrplan, electrive.net - Branchendienst für Elektromobilität, Berlin, Germany, 11 Sep 2017
7. K. Lant, *China Pushes Electric Vehicles and Makes Producing Fossil Fuel Vehicles Tougher, Futurism—Earth & Energy*, 30 Sep 2017

8. Öko-Institut, Strategien für die nachhaltige Rohstoffversorgung der Elektromobilität. Synthesepapier zum Rohstoffbedarf für Batterien und Brennstoffzellen. Studie im Auftrag von Agora Verkehrswende, Darmstadt, Germany, 04–2017–DE, 2017
9. C. Bauer, J. Hofer, H.-J. Althaus, A. Del Duce, A. Simons, App. Energy **157**, 871 (2015)
10. M. Weil, J.F. Peters, M.J. Baumann, H. Dura, B.M. Zimmermann, Elektrochemische Energiespeicher für mobile Anwendungen im Fokus der Systemanalyse. TATuP **24**, 20 (2015)
11. M. Weil, S. Ziemann, in *Lithium-Ion Batteries: Advances and Applications*, 1st edn., ed. by G. Pistoia (Elsevier, Amsterdam, 2014), pp. 509–528
12. B. Simon, S. Ziemann, M. Weil, Resour. Conserv. Recy. **104**, 300 (2015)
13. G. Majeau-Bettez, T.R. Hawkins, A.H. Strømman, Environ. Sci. Technol. **45**, 4548 (2011)
14. L.A.-W. Ellingsen, G. Majeau-Bettez, B. Singh, A.K. Srivastava, L.O. Valøen, A.H. Strømman, J. Ind. Ecol. **18**, 113 (2014)
15. J.F. Peters, B. Simon, G. Rodriguez-Garcia, M. Weil, *Building a Common Base for LCA Benchmarking of Li-Ion Batteries, 26th SETAC Ann* (Meeting, Nantes, France, 2016)
16. J.F. Peters, M. Weil, J. Clean. Prod. Available online: 7 Oct 2017
17. C. Bauer, *Ökobilanz von Lithium Ionen Batterien* (Paul Scherrer Institut, Labor für Energiesystem-Analysen (LEA), Villigen, Switzerland, 2010)
18. M. Zackrisson, L. Avellán, J. Orlenius, J. Clean. Prod. **18**, 1519 (2010)
19. D.A. Notter et al., Environ. Sci. Technol. **44**, 6550 (2010)
20. *IEA, Energy Technology Transitions for Industry—Strategies for the Next Industrial Revolution*, International Energy Agency, Paris, France, 2009
21. D.B. Müller, Ecol. Econ. **59**, 142 (2006)
22. *UN, World Population in 2300*, Department of Economic and Social Affairs, United Nations (UN), New York, US, ST/ESA/SER.A/236, 2004
23. R. Modaresi, D.B. Müller, Environ. Sci. Technol. **46**, 8587 (2012)
24. M. Weil, S. Ziemann, B. Zimmermann, *New Materials and Components for Future Batteries —A Resource Perspective, World of Energy Solutions Battery & storage—Materials and Components* (Stuttgart, Germany, 2013)
25. R. Ball, N. Keers, A. Marcus, E. Bower, *Mobile Energy Resources for Grids of Electricity (MERGE)—Deliverable 2.1: Modelling Electric Storage Devices for Electric Vehicles, Project Deliverable*, 2011
26. D.B. Müller, J. Cao, E. Kongar, M. Altonji, P.-H. Weiner, T.E. Graedel, *Service lifetimes of mineral end uses, U.S. Geological Survey (USGS)* [Final Report, Minerals Resources External Research Program, Award Number: 06HQGR0174] (2007)
27. D. Kushnir, B.A. Sandén, J. Clean. Prod. **19**, 1405 (2011)
28. P. Keil, A. Jossen, *Aging of Lithium-Ion Batteries in Electric Vehicles: Impact of Regenerative Braking, EVS28* (Kintex, Corea, 2015)
29. Green Car Congress, BMW boosts battery capacity of MY2017 i3 to 33 kWh with higher energy density Li-ion cells; up to 114 miles combined cycle range, Green Car Congress, 02 May 2017
30. *ILA, Environmental and Social Responsibility for the 21st Century* (International Lead Association (ILA), London, United Kingdom, 2014)
31. M. Romare, L. Dahllöf, The Life Cycle Energy Consumption and Greenhouse Gas Emissions from Lithium Ion Batteries. A Study with Focus on Current Technology and Batteries for Light-Duty Vehicles, IVL Swedish Environmental Research Institute, Stockholm, Sweden, ivl Study C 243, 2017
32. US Geological Survey, Mineral Commodity Summaries 2016. US Geological Survey
33. M. Keller, C. Vaalma, D. Buchholz, S. Passerini, ChemElectroChem **3**, 1124 (2016)
34. Z. Zhao-Karger et al., J. Power Sources **323**, 213 (2016)

Will Current Electric Vehicle Policy Lead to Cost-Effective Electrification of Passenger Car Transport?

Marcello Contestabile and Mohammed Alajaji

Abstract Encouraged by the falling cost of batteries, electric vehicle (EV) policy today focuses on accelerating electrification of passenger cars, paying comparatively little attention to the cost of the particular type of EVs and charging infrastructure deployed. This chapter first discusses the strong influence that EV policy design has on the development of particular EV types. It then illustrates recent research conducted by the authors, showing that EV policy with a strong bias towards long-range battery electric vehicles (BEVs) risks leading to higher overall costs in the medium term. The costs could possibly exceed the ability of governments to sustain the necessary incentives and of automotive original equipment manufacturers to internally subsidise EVs until battery cost drops sufficiently. While the research does not fully explore the latter issue and its potential to stall the EV transition, it does show that the incremental cost of different EV and infrastructure mixes over the whole passenger car fleet can differ quite substantially and that promoting a balanced mix of BEVs and plug-in hybrid electric vehicles (PHEVs) may set the electrification of passenger cars on a lower-risk, lower-cost path. Examining EV policy in the UK and in California, we find that it is generally not incompatible with achieving balanced mixes of BEVs and PHEVs; however, it could be better designed if it paid more attention to cost and technology development risk.

Keywords Electric vehicle policy · California · United Kingdom
Case study · PHEV · BEV

M. Contestabile (✉) · M. Alajaji
King Abdullah Petroleum Studies and Research Center (KAPSARC),
P.O. Box 88550, Riyadh 11672, Saudi Arabia
e-mail: marcello.contestabile@gmail.com

M. Alajaji
e-mail: mjalajaji@gmail.com

M. Contestabile
Centre for Environmental Policy, Imperial College London,
Exhibition Road, London SW7 2AZ, UK

G. Pistoia and B. Liaw (eds.), *Behaviour of Lithium-Ion Batteries in Electric Vehicles*, Green Energy and Technology,
https://doi.org/10.1007/978-3-319-69950-9_4

1 Government Support to Electric Vehicles

Road transport accounted for 21% of global energy consumption and 17% of global CO_2 emissions in 2013 [1]. CO_2 emissions from road transport have been growing steadily and will continue to do so if road transport is not progressively decoupled from fossil fuels [2]. Stabilising global temperature increase to <2 °C relative to pre-industrial levels will require a combination of improved fuel efficiency and deployment of alternative fuels in road transport, particularly advanced biofuels, electricity and hydrogen [3, 4]. Scenarios may differ as a multitude of energy technology mixes are possible [5]; however, it is generally accepted that electric vehicles (EVs) will have a major role to play, especially in large markets such as the USA, Europe, China and India. Electrification of passenger car transport also has the added benefit of reducing emissions of local air pollutants in urban areas, the impact of which on public health is of growing concern in both developed and developing countries [6].

For the reasons noted, electrification of passenger car transport is receiving strong support from several national governments worldwide which seek not only to meet their environmental protection goals but also to develop national value chains in this emerging industry [7]. Alongside aspirational targets set by several governments, electrification is increasingly being driven by regulation. Most notably, the California's zero-emission vehicle (ZEV) mandate sets mandatory targets for EV sales; this type of regulation is being adopted across the USA and Canada. In the European Union, the directive on the deployment of alternative fuel infrastructure [8] mandates that Member States develop national policy frameworks for future EV charging infrastructure roll-out.

In order to achieve their targets, national and local governments are introducing sets of incentives to EV adoption, particularly purchase grants, tax exemptions, non-monetary incentives such as free parking and access to restricted lanes, and financial support for the development of extensive charging infrastructure [7, 9]. Incentives are necessary to overcome the substantial cost gap currently existing between EVs and conventional internal combustion engine vehicles (ICEVs) and the first mover disadvantage that characterises the development of alternative fuel infrastructures [10]. For their part, automotive OEMs are producing an increasingly diverse range of EV models in order to comply with mandates and standards while developing market and technology knowledge. EV fleet penetration on a global level is still low; however, their market share is growing fast [11]. In some countries, such as Norway and the Netherlands, EV market shares have reached substantial levels, while the USA, Japan and China lead the way in terms of the absolute size of the EV stock, and several new markets are starting to develop [11].

Despite some early success stories and the growing momentum behind the EV transition, substantial hurdles remain on the path to a high level of EV penetration worldwide, because of strong economic, institutional and behavioural barriers, combined with the slow rate of turnover of passenger car stocks [10, 12, 13]. For this reason, in today's policy discourse much emphasis is placed on identifying those mixes of policy instruments that are most effective at accelerating the deployment of EVs and related charging infrastructure [7]. Comparatively, little attention is devoted to clearly articulating a vision of future self-sustained electrification of passenger car transport that does not solely rely on the cost of EV batteries rapidly falling. However, considering that the current high levels of government incentives cannot be sustained indefinitely and that there is a limit to the extent to which automotive original equipment manufacturers (OEMs) can internally subsidise EV technology, we argue that policy should also be designed taking account of the need to guide the EV transition towards low-cost and low-technology risk pathways.

The research presented in this chapter is based on Ref. [14] and addresses the question whether today's EV policy is conducive to a future cost-effective use of this technology, considering the policy objectives it aims to achieve, particularly carbon emission reduction. The research has explored the incremental costs of future mixes of EVs and charging infrastructures that are broadly compatible with today's policy and market trends, and that provide similar carbon emission reductions at fleet level. The results of the cost analysis conducted provide a basis for discussing the key features and possible implications of current EV policy, and to identify opportunities for making it more robust under uncertainty.

The contents and structure of this chapter follow closely Ref. [14], adapted by the authors as deemed appropriate.[1] Section 2 discusses the effect that policy-driven EV deployment has on EV innovation pathways, including the possibility of technological lock-ins. Section 3 presents the methods used in the study and their limitations. Section 4 describes the current policy framework and deployment targets for the UK and California, the two case studies we have chosen. Section 5 discusses the fleet structure and driving patterns for the UK and California, two important inputs to the case studies. Section 6 presents the results of the case studies conducted. Section 7 discusses their policy implications and concludes the chapter.

2 EV Deployment Policy and Its Effect on Innovation

Different approaches have so far been used to EV policy support across different markets, which can be explained by their specific characteristics, the widely differing underlying taxation of conventional vehicles and fuels, and the lack of generally accepted best practices. As a result, different patterns of deployment of EVs and charging infrastructure have begun to emerge in the most active countries and regions, i.e. China, Europe, Japan and the USA [7, 9, 11]. In particular, different ratios of pure battery electric vehicles (BEVs) and plug-in hybrid electric vehicles (PHEVs), and of rapid charging and slow charging infrastructure can be observed across leading markets [9, 11]. BEVs are those EVs that operate solely on electricity, while PHEVs can operate on both battery power and an internal combustion engine, especially once the battery is depleted. In PHEVs, the internal combustion engine and electric components of the powertrain can be arranged either in parallel or in series; the latter is also referred to as range-extended electric vehicles (RE-EVs). In this chapter, we will use the term PHEV for both types, unless otherwise specified. The term slow chargers is here used to indicate charging points of 3–7 kW power; rapid chargers supply power of the order of 40–50 kW. Figures 1 and 2 provide illustration of the different patterns of EV and charging infrastructure deployment observed today [15].

Evidence shows that incentives strongly influence the overall rate of EV uptake and the relative market shares of BEVs and PHEVs [16]. In Norway, for example, BEVs receive generous support, whereas PHEVs have only recently become eligible for some, hence the rapid rate of uptake of BEVs. In the Netherlands, incentives for BEVs and PHEVs have been similar, hence the dominance of PHEVs that offer better functionality. In California, where BEVs qualify for higher incentives than PHEVs, their market shares are comparable [17]. Hence, government incentives to EV purchase, combined with the underlying taxation of conventional fuels and vehicles, determine the type of EVs that are most competitive and also the market segments in which the value they offer relative to ICEVs is highest. This in turn influences the EV types and models that automotive OEMs will commercialise in order to achieve highest possible sales.

Moreover, public charging infrastructure is a strong enabler of BEV adoption [18] and it is for this reason that some countries are building extensive networks of public chargers, be they rapid or slow, even if the current level of EV penetration does not justify their presence [10, 17]. The aim is to anticipate future user needs, and therefore, the particular type, density and location of charging points are intended to reduce range anxiety and increase the perceived utility of BEVs for prospective adopters to a level comparable to ICEVs. However, it is difficult to predict how well this will work in practice and the extent to which the infrastructure will actually be utilised [10, 17].

The trends shown in Figs. 1 and 2 can change in future as policy support measures are periodically adjusted by governments in response to market developments. In particular, also encouraged by recent evidence showing a rapid rate of decrease of EV battery cost [19], a growing number of countries are currently increasing their support for BEVs relative to PHEVs, which some consider as a transitional technology. However to the best of our knowledge, the relevant policy documents do not explicitly discuss the overall cost of the particular EV and charging infrastructure mixes they seek to promote. Because the emphasis is on rapidly electrifying passenger car transport, it is therefore possible that the EV and charging infrastructure mixes that will be deployed in the short and medium term will not provide the most practical and cost-effective way of achieving the intended goals.

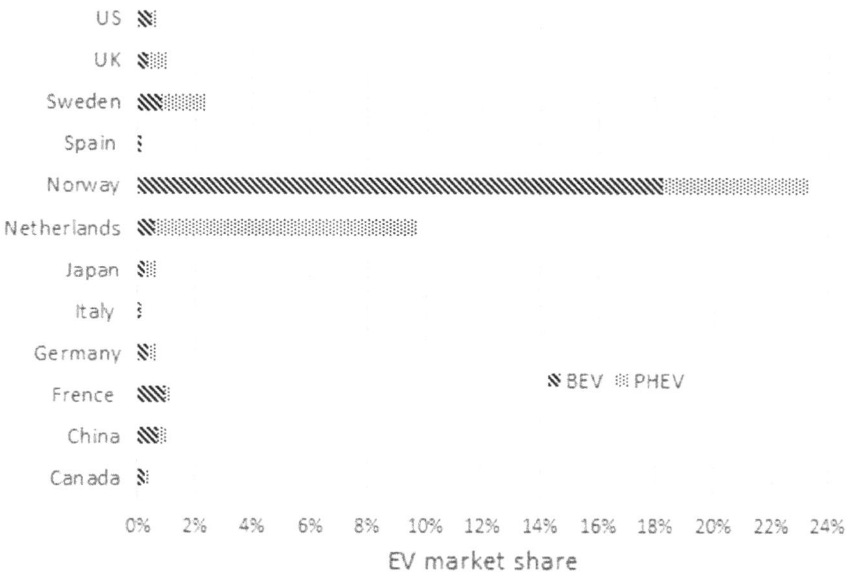

Fig. 1 Market share of EVs in selected countries in 2015, broken down by BEVs and PHEVs. *Source* Adapted from [15]

Fig. 2 Charging point/EV ratio in selected countries in 2015, respectively, for fast (top) and slow chargers (bottom). *Source* Adapted from [15]

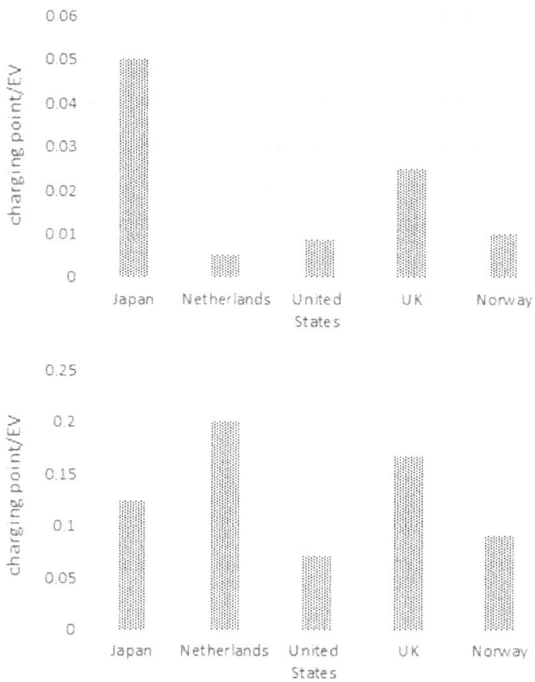

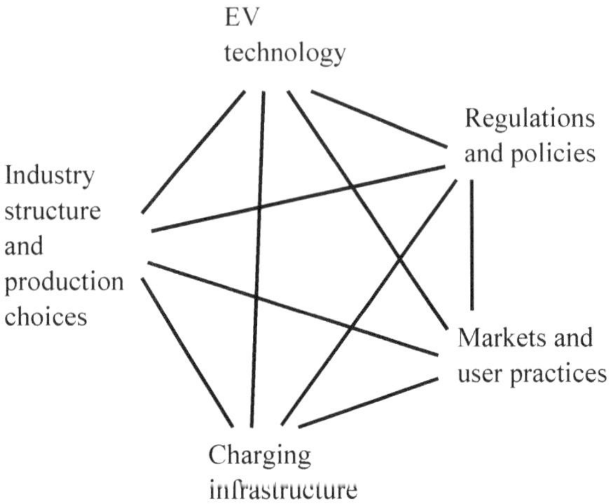

Fig. 3 Schematic representation of the complex interlinkages among the multiple dimensions of the EV socio-technical system. *Source* Adapted from [22]

This is problematic because the process of development and adoption of new technology such as EVs exhibits path dependence and is prone to lock-in effects [20]. In other words, the type of EVs and infrastructure initially deployed will influence the behaviour and preferences of adopters and the development of related institutions and hence will contribute to pushing future EV technology and infrastructure development down a certain path. This will in turn further influence consumer adoption of new EV models and the development of policy and regulation, in a process that in technology studies is generally referred to as co-evolution [21, 22]. This is schematically illustrated in Fig. 3.

As policy and regulation co-evolve with the new technology and the preferences of its users become entrenched, electrification of passenger cars will become increasingly locked into certain mixes of EV and charging infrastructure types. In the early phases of the EV transition, these mixes are made competitive by the policy incentives that support the electrification process. However as higher levels of adoption are reached and policy support measures are withdrawn, costs will increasingly be borne by the automotive OEMs and passed on to EV users. The actual cost of electrification will then become very important in determining whether or not the EV transition will be able to sustain itself. Switching to more cost-effective electrification paths later on would still be possible but challenging. Meanwhile, the whole EV transition could risk stalling. In the light of this, posing the question whether today's EV and infrastructure policy is conducive to cost-effective electrification of passenger cars becomes more important.

3 Assessing the Incremental Cost of Different EV Mixes

The future cost and emissions of different types of EVs has been studied extensively. Numerous studies can be found in the literature that cover the whole spectrum of economic assessments of EVs, from detailed powertrain cost and performance modelling, aimed at guiding the design of systems or components, to studies comparing the lifetime cost and emissions of different EV types in order to inform policy-making. Common to most of these studies is the use of the Relative Cost of Ownership (RCO) metrics that allows to estimate the cost of individual EVs over their lifetime and to compare them across different powertrain types. As the name suggests, RCO does not consider all costs but only those that are relevant to the comparison being made. For a critical review of the relevant studies in this area, we refer the reader to [14]. These studies have generated a large amount of knowledge on the economics of EVs; however, they tend to focus on single vehicles as opposed to whole fleets and charging infrastructures.

In order to address this gap, the authors of this chapter developed a model that calculates the incremental cost and emission savings of future EV and charging infrastructure mixes at the level of a whole fleet of passenger cars. The model performs RCO calculations for single vehicles and integrates them over the fleet by including all the key factors with the minimum possible level of detailed complexity. The model relies on inputs from a number of specific studies and technical modelling activities. An overview of the model structure is provided in Fig. 4.

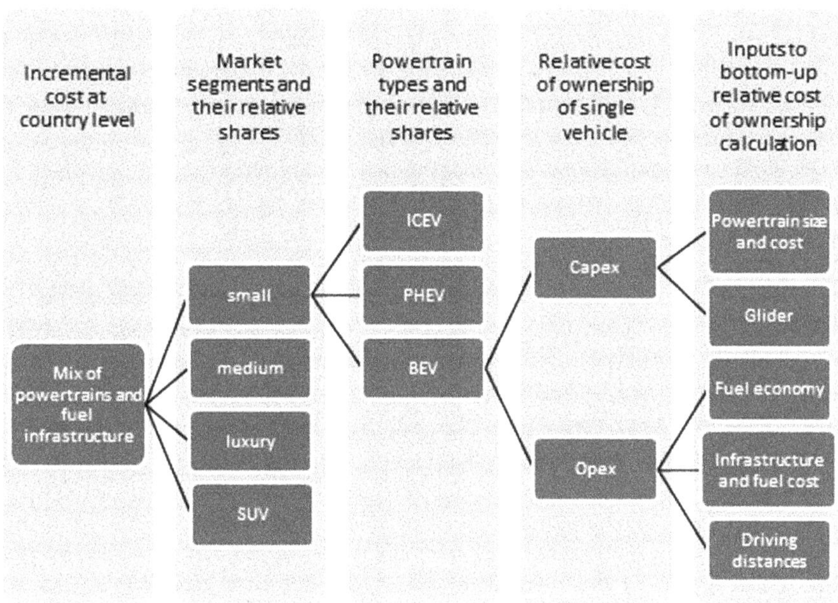

Fig. 4 Schematic representation of the model developed by the authors to estimate the incremental cost of passenger car fleet electrification for different EV and infrastructure types. *Source* Adapted from [14]

It is worth mentioning that the model does not include all those cost elements that are common to both EVs and ICEVs, nor does it include vehicle and fuel taxation. The latter corresponds to implicitly assuming that EVs will be taxed the same as ICEVs. This is a fair assumption for all taxes except those on emissions of CO_2 and other pollutants, which we expect to be proportional to the environmental performance of the vehicles. However, to ensure our analysis is meaningful, we compare only EV mixes that are characterised by similar average tailpipe CO_2 emissions across the whole fleet. In this way, our results are not influenced by assumptions on the price of CO_2 emissions. Furthermore, to ensure that the utility of the BEVs and PHEVs modelled is acceptable, we set the range of BEVs and the availability of charging infrastructure so as to satisfy either the stated requirements of passenger car users or at least their observed driving needs. For a full description of the model, we refer the reader to [14].

Moreover, considering that the cost of electrification depends in part on the technology inputs and in part on the specific vehicle market examined, in [14] we have taken a case study approach. We have selected the UK and California because both are aggressively pursuing electrification of passenger cars and their markets are illustrative of larger regional markets, such as Western Europe and North America. Another reason is the availability of data and documents in English. Finally, by comparing and contrasting the two cases, we have tested the extent to which general lessons can be learned about the cost-effectiveness of policy-driven electrification. We have based our analysis around the year 2030, because: (a) current policy targets tend to refer to the 2025–2030 timeframe; (b) the level of adoption foreseen is such that lock-in effects may begin to occur; and (c) technology projections become very uncertain beyond 2030.

As part of the analysis in [14], we developed a set of electrification scenarios for the UK and California around the year 2030 that are based on narratives broadly consistent with current policy and market trends. It is worth stressing though that the scenarios are not intended as predictions of the composition of EV and infrastructure mixes in 2030 based on the current choices being made by policymakers. In fact, policy-making in the UK, California and elsewhere is flexible enough to allow shifting direction should it be required, and anticipating future decisions of policymakers and their effects is beyond the scope of the research we have conducted. However, due to the path-dependent nature of the innovation processes discussed, major changes in direction will no doubt involve time lags and additional costs.

The main limitations of the analysis presented in this chapter are that: (a) it is of a snapshot type, i.e. it assesses the incremental cost of electrification for the year 2030 only instead of the cumulative cost until that year; and (b) it does not apportion the incremental cost of electrification among government, car manufacturers and passenger car buyers. The implications of these limitations on our policy insights are discussed in Sect. 7.

4 EVs in the UK and California: Current Policy and Future Deployment

4.1 The UK

The UK is subject to EU transport and environmental policy (although this will change as a result of the UK's referendum vote to leave the EU). In particular, the post-2020 EU fleet average CO_2 emission standards for passenger cars and the alternative fuel infrastructure directive will provide the strongest drivers for electrification at European level. In addition to that, the UK has set itself the legally binding target of reducing total GHG emissions, to which transport is a major contributor, by 80% relative to 1990 levels in 2050. In order to fulfil its domestic and European obligations, the UK is committed to supporting the development and deployment of ultra-low emission vehicles (ULEVs), particularly EVs, which the government also sees as an opportunity to revive the country's automotive industry [23]. The UK government aspires to achieve near-complete decarbonisation of passenger car transport by 2050 and supports electrification through financial and non-financial incentives, which are periodically revised based on observation of market and technology trends. The UK government's approach to ULEVs is in general technology neutral, although the recently revised EV grant now has different levels for long and short All-Electric Range (AER) EVs, which is intended to increase BEV over PHEV sales. Moreover, the maximum number of grants available has been capped at the same level for BEVs and PHEVs, which also suggests a desire to balance the sales of either EV type. Table 1, taken from [14], provides an overview of EV and infrastructure deployment levels, government incentives supporting it and future estimated EV and infrastructure levels required in order to support the UK achieve its climate change policy goals [12, 17, 24–26].

The evidence gathered from EV users in the UK so far shows that adoption of EVs is mainly by affluent, multi-vehicle households in urban areas. EVs are typically used as the main car, relied upon for the majority of daily trips and driven on annual mileages comparable to those of ICEVs, while the latter are used more for infrequent, longer journeys [27]. EV consumer research suggests that key barriers to EV adoption remain price and for BEVs also range, with users expressing desire for longer-range vehicles that can also be used conveniently for infrequent, longer trips [17]. EV owners strongly prefer charging overnight at home. This is due to convenience and not much influenced by availability of public infrastructure [27]. However, a fully developed charging infrastructure, particularly rapid, is also perceived as necessary for further BEV market expansion [17]. Analysis conducted by [25] suggests that, to complement private residential charging, the most valuable option is rapid charging infrastructure. However, the business case for it is still challenging, due to the expected low utilisation rate; continued government support will therefore be required in order for the rapid charging infrastructure to develop in the UK [25].

Table 1 Deployment, policy support measures and future estimated EV and charging infrastructure needed in order to meet the UK's GHG emission reduction targets

	Deployment level	Government incentives	Government targets/ deployment requirements
EVs	1% market share in 2015 (1/3 BEVs, 2/3 PHEVs)	"Plug-in Car Grant" amounts to up to 35% of the vehicle's retail price, for a maximum of £4500 for EVs with AER of at least 110 km (currently BEVs) and £2500 below 110 km (currently PHEVs). The grant originally offered a maximum of £5000 per EV, irrespective of AER, and was amended in March 2016 Exemption from road user charges, notably London's congestion charge	Aspirational target of 100% ULEV new car registrations in 2040. No mandated EV targets The committee on climate change estimates that meeting UK's GHG emission targets requires between 4 and 8 m EVs on the road in 2030
Charging infrastructure	Ca. 30,000 home charging points, 7000 workplace (open access) and 8000 public charging points (7100 slow and 900 fast) as of Feb 2015	Grant for home chargers covering 75% of cost up to £700 Government match funding for private and public entities that deploy chargers in selected locations ("Plugged-in Places") Highways Agency committed to investing £15 M in order to add 1000s of new charging points on the strategic road network. The aim is that motorists will be no more than 30 km from a charge point 95% of the time	EU regulation requires the UK to develop a roll-out plan for charging infrastructure. The directive indicates a target density of 0.1 chargers/vehicle, depending on the type of EVs and chargers deployed It is estimated that a network of 2100 rapid charging sites (10 charging points per site) could provide UK-wide coverage Around 70% of UK households have access to private parking; however, this is as low as 10% in certain urban areas

Sources [12, 17, 24–26] in [14]

4.2 California

Like the UK, California has set itself the target of achieving an 80% reduction of GHG emissions by 2050 relative to 1990 levels [28], with an interim target of 40% reduction by 2030. This complements strong air quality policy, including the low-emission vehicle standards of the California Air Resources Board. In order to facilitate the achievement of the intended reduction in emissions of GHGs and air pollutants from road transport while supporting the development of a clean car industry in California, in 2012 Governor Brown issued an executive order aimed at facilitating the rapid commercialisation of zero-emission vehicles (ZEVs) [29]. The executive order sets specific EV deployment targets, the strategy for achieving which is set out in the 2013 ZEV Action Plan of February 2013 [30], updated in 2015 based on a review of the progress achieved until then [31]. The strategy includes providing incentives to EV adoption and infrastructure deployment as well as studying future infrastructure needs. An infrastructure study was conducted by the National Renewable Energy Laboratory (NREL) in 2014. The targets and key elements of the strategy are summarised in Table 2 [30–35] taken from [14].

It is also worth mentioning that the 2015 ZEV Action Plan explicitly states that incentives should be cost-effective and withdrawn as early as possible: "Financial incentives continue to play a critical role in making ZEVs cost competitive with conventional vehicles during the early phases of their deployment, until economies of scale lead to cost reductions and a fully self-sustaining market" [31].

The executive order targets are broadly in line with the EV penetration levels required by the ZEV mandate, although the exact EV numbers required by the latter will depend on the particular compliance strategies chosen by the OEMs. In particular, the ZEV mandate sets a minimum number of credits that large and intermediate volume manufacturers have to earn or purchase to comply and avoid fines. The credits are earned through manufacturing pure ZEVs (i.e. BEVs, a newly introduced category of range-extended EVs called BEVx, and fuel cell electric vehicles (FCEVs); the latter are not discussed in this chapter) and other ULEVs (such as PHEVs, also referred to in the regulation as transitional ZEVs or TZEVs). BEVx are defined as full BEVs also equipped with a small ICE auxiliary power unit enabling them to operate at reduced power when the AER is exhausted, and with a non-electric range not exceeding their AER. Hence, the structure of the powertrain of a BEVx is similar to that of a RE-EV, but the components are sized differently and the utility of the vehicle is substantially lower. The number of credits for each EV is awarded proportional to its AER, based on different formulas for ZEVs and ULEVs. Although the regulation allows OEMs a certain degree of flexibility in the way they meet their credit obligations, a minimum ZEV credit floor applies. A synthesis of the ZEV mandate credit mechanism is provided in Table 2; we refer the reader to the relevant regulation for full details [35, 36]. However, it is important to note that the ZEV mandate will play a strong role in defining the future split between BEVs and PHEVs in California, ensuring that BEVs (either pure BEV or BEVx) retain a substantial share. Moreover, the credit mechanism for ZEVs is a

Table 2 Deployment, policy support measures, future EV mandates and charging infrastructure needs in California

	Deployment level	Government mandates/ incentives	Government targets/ deployment requirements
EVs	3.2% market share in 2015 (1/2 BEVs, 1/2 PHEVs) 120,000 EVs on the road in Jan 2015	ZEV mandate currently forces the commercialisation of BEVs and PHEVs in sufficient numbers for individual car manufacturers to generate the necessary number of credits Federal tax rebate of up to $7500 (proportional to EV battery size). California clean vehicle rebate, a state rebate of $2500 for BEVs and $1500 for PHEVs Non-financial incentives such as access to high occupancy vehicle (HOV) lanes and parking benefits	Executive order sets a target of 1 m ZEVs on the road by 2020 and 1.5 m ZEVs by 2025, and for new vehicle purchases in light-duty fleets of government agencies to reach 10% ZEVs by 2015 and 25% by 2020 Post-2018, ZEV credits are earned by BEVs and BEVx with AER >80 km proportional to their AER (e.g. 160-km AER = 1.5 credits; 480-km AER = 3.5 credits). PHEVs with AER comprised between 16 and 120 km also earn credits proportional to the AER (0.4–1.10 credits, respectively) Large-volume car manufacturers have to earn the majority of their credits from pure ZEVs (i.e. BEVs, BEVx and FCEVs)
Charging infrastructure	3224 public charging stations for a total of 9577 public charging points in California as of March 2016	The California energy commission administers a number of programmes providing funding for new charging infrastructure. It also conducts and commissions studies on the future need for charging infrastructure across the state The California building code requires all recently constructed parking lots or housing to put electrical capacity in place to easily install EV chargers	Executive order mandates the roll-out of the necessary charging infrastructure to support the ZEV targets NREL study estimates that, to support the 1-m EVs by 2020 target, between 20 and 50 k public chargers will be needed. It suggests two alternative options: "Home dominant": 100 k workplace and 22,250 public chargers (of which 550 rapid); "Public access": 167 k workplace and 48,600 public chargers (of which 1550 rapid)

Sources [30–35] in [14]

contributing factor to the emergence of the long-AER BEVs manufactured by Tesla Motors Inc. and increasingly other OEMs as well and will most likely continue to influence future OEM decisions about the AER of their EVs.

As in the UK, EV owners in California are predominantly affluent, highly educated, multi-vehicle households and use their EVs as the main car for frequent, shorter journeys, with similar annual mileages to ICEVs [37]. BEV users in California report that for full satisfaction their vehicles would need to have a range of more than 250 km [37]. In California, charging of EVs happens mainly at home, as also in the UK. PHEV users' charging behaviour is currently being investigated, but early results suggest that long-AER PHEVs are used on electricity as much as possible. EV users generally were not entirely satisfied with public charging infrastructure, although this is improving as infrastructure coverage increases [17, 37].

5 Fleet Structure and Driving Patterns in the UK and California

Two very important differences between the UK and California for the purpose of our study are the structure of the fleet and the vehicle usage patterns. In [14], we modelled the structure of the 2025–2030 fleet in a simplified way: we assumed that the overall size of the fleet will stay the same as today and we only considered the four largest market segments, accounting for the majority of the fleet. We sized the segments in the year 2025–2030 based on new passenger car sales for a reference year, and we modelled them based on the characteristics of the best-selling cars for that same year [32, 38]. Modelling the fleet as we did has its limitations, and a stock model would provide more accurate projections of future fleet composition. However, given the nature of our study, the simplified approach we have taken is acceptable. Moreover, as previous studies have shown [39], even a relatively simple segmentation approach like ours can provide substantial additional insight compared with treating the whole passenger car market as homogenous. The details of how the future fleets in the UK and California are modelled in [14] are provided in Tables 3 and 4 [32, 38, 40–44]. In the tables, the vehicle segments are named as is most common in the relevant markets. Note that in our analysis the reference vehicle weight is reduced relative to today's based on future scenarios on the use of lightweight materials [45], and the powertrain size is downscaled accordingly [46, 47], both of which reduce the cost difference between ICEVs and EVs, particularly BEVs.

As for the vehicle usage patterns, in [14] we analysed data from the UK National Travel Survey [40] and the US National Household Travel Survey [44], respectively, and we derived the frequency distributions of daily distances driven shown in Figs. 5 and 6. These distributions are used when calculating utility factors of PHEVs and relative shares of home versus public charging for BEVs.

Table 3 Structure of the UK passenger car market in 2030 as modelled in [14], based on [38, 40, 41]

Segment	Reference model	Weight (kg)	Power (kW)	Annual mileage (km)	Fleet share (%)
Mini/Supermini (A/B)	Ford Fiesta	1050	64	12,950	40.6
Medium (C/D)	Volkswagen Golf	1300	92	14,950	40.8
Executive/Luxury (E/F)	Mercedes C-class	1550	135	17,450	4.8
Dual purpose/MPV (H/I)	Vauxhall Zafira	1550	105	22,200	11.3

Table 4 Structure of the Californian passenger car market in 2025 as modelled in [14], based on [32, 42–44]

Segment	Reference model	Weight (kg)	Power (kW)	Annual mileage (km)	Fleet share (%)
Small	Toyota Corolla	1270	98	19,850	26.8
Medium	Honda Accord	1475	140	18,900	31.1
Luxury	Mercedes E-class	1735	224	20,600	10.7
SUV	Ford Explorer	2010	216	21,000	20.1

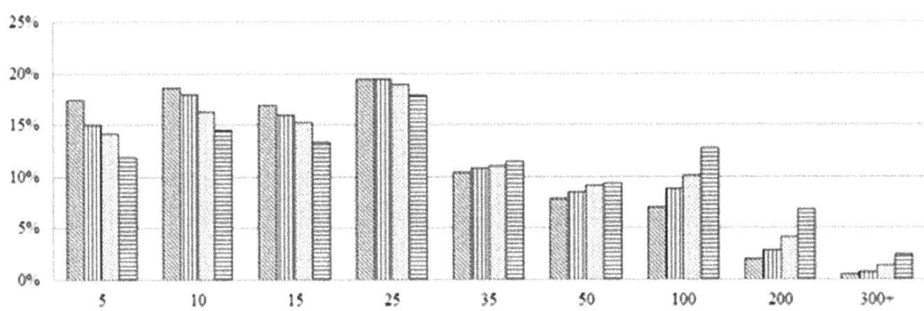

Fig. 5 Frequency distribution of daily distances driven in the UK (in km), by vehicle segment. *Source* Authors' analysis of UK National Travel Survey data [40] in [14]. In each group, from left to right: A/B, C/D, E/F, H/I

As can be observed from the tables and figures, the UK and California markets differ in terms of the attributes and shares of their main segments, and the associated driving patterns. In particular, the structure of the fleet in California is slanted towards larger and heavier vehicles. This is generally the case comparing North America with Europe. Moreover, in the UK larger vehicles are on average driven more frequently on longer distances and have higher annual mileages than smaller

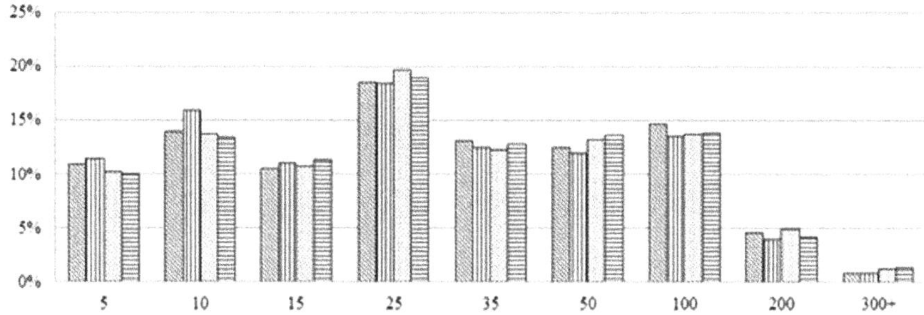

Fig. 6 Frequency distribution of daily distances driven in California (in km), by vehicle segment. *Source* Authors' analysis of US National Household Travel Survey data [44] in [14]. In each group, from left to right: small, medium, luxury, SUV

vehicles, whereas in California all segments are on average driven similarly and have comparable annual mileages. In [14], we have made the simplifying assumption that this will not change until 2030.

6 UK and California Scenario Analysis

6.1 *The UK*

Based on the current status and future targets for electrification discussed in Sect. 4.1, a set of key scenarios for the year 2030 were built in [14] and their incremental cost was estimated relative to a base case where the whole passenger car fleet is composed only of ICEVs.

All scenarios are consistent with the UK aspiration of 60% EV market share, or 8 m EV fleet, by 2030 but differ in terms of the EV types and related infrastructure deployed. It is also worth noting that, despite the difference in EV types across scenarios, the average fleet tailpipe CO_2 emissions are comparable. The key elements of each of the four scenarios modelled are listed below and further illustrated in Table 5.

- Scenarios 1 and 2 are based on the current trend of seeking to balance the relative shares of BEVs and PHEVs through incentives. Hence, we assume a 40/60 split between BEVs and PHEVs and the country-wide charging infrastructure needed for rapid adoption of BEVs.
- In Scenario 1, rapid charging infrastructure is modelled based on the analysis by [25]; slow charging infrastructure is based on the indicative target of the European Commission Directive [8], i.e. the equivalent of at least 0.1 public charging points per EV.

Table 5 UK 2030 EV and infrastructure scenarios modelled in [14]

Scenario	BEV share (%)	BEV segments	PHEV share (%)	PHEV segments	BEV AER (km)	PHEV AER (km)	# Slow chargers	# Fast chargers
1	40	C/D, E/F, (H/I)	60	C/D, E/F, H/I	250	50	300,000	20,000
2	40	C/D, E/F, (H/I)	60	C/D, E/F, H/I	250	50	0	20,000
3	0	–	100	C/D, E/F, H/I	–	100	–	–
4	20	A/B	80	C/D, E/F, H/I	150	100	–	5000

- In Scenario 2, only rapid charging infrastructure is present, because this is seen by users as most valuable, and hence, the public charging point per EV ratio becomes 0.01.
- Taken together, Scenarios 1 and 2 represent possible upper and lower bounds for a country-wide charging infrastructure in the UK that is capable of supporting a fleet consisting of a mix of 250-km-range BEVs and 50-km-AER PHEVs. A 250-km range may be the least required for BEVs to offer similar functionality to ICEVs. Finally, we assume that BEVs will penetrate the market across all segments, with the exception of the small car segment (A/B) where fuel-efficient ICEVs currently benefit from a relatively low level of taxation and where a long-range BEV will be both expensive and not required. PHEVs are also present in all segments except A/B, due to the same reasons of cost-competitiveness with ICEVs.
- Scenario 3 meets 60% EV penetration by 2030 with the least amount of battery capacity and infrastructure installed. This means only using 100-km-AER PHEVs that do not require public charging infrastructure at all and use the same type of batteries as BEVs, sharing with them all other components of the electric powertrain. Thus, they could generate the necessary scale economies that would also be needed for BEVs to become competitive. However by not developing the charging infrastructure and user preferences for BEVs, the latter could become locked out, potentially delaying the achievement of full electrification of passenger cars post-2030. A PHEV-only scenario is also clearly not consistent with current policy.
- Finally, Scenario 4 offers a compromise where 100-km-AER PHEVs dominate the market, with the exception of the A/B segment where only 150-km-range BEVs are present. This is in principle compatible with the current structure of the Plug-In Vehicle Grant that does not favour longer-AER BEVs and potentially rewards long-AER PHEVs in the same way as BEVs. In this way, the BEV option remains open but, by targeting the smaller vehicles typically used for shorter distances in urban areas, expensive long-AER BEVs and a country-wide infrastructure are no longer needed. Adoption of BEVs as urban vehicles can be further encouraged by developing charging infrastructures accordingly.

Figure 7 shows the incremental cost of the scenarios as calculated in [14]. The error bar indicates the full range of uncertainty associated with future battery technology development. In particular, the highest cost corresponds to today's battery technology cost ($300/kWh) and energy density (100 Wh/kg) as reported by the leading industry players, while the lowest cost corresponds to battery technology meeting its long-term cost reduction target ($100/kWh) and doubling its energy density. The midpoint case falls exactly in between with respect to both battery cost and energy density; as far as cost is concerned, $200/kWh is considered as a plausible scenario for 2025–2030 based on recent projections [19]. The same logics apply to PHEV batteries, with cost comprised between $350 and $135/kWh, and energy density between 70 and 150 Wh/kg. The battery cost and energy density scenarios in [14] are based on [15, 19, 48, 49]. While it is unlikely that battery technology will not improve at all by 2025–2030, using today's state of the art as worst-case scenario also gives a sense of the extent to which different EV mixes will require policy support while battery technology develops, and hence of the cumulative transition cost and technology risk associated with each particular scenario.

It is worth noting that the absolute value of the incremental cost of EV scenarios as shown in Fig. 7 is influenced by the relative cost of gasoline and electricity as well as other variables and hence should be regarded as indicative only. For this reason, the main insight that can be gleaned from the analysis here illustrated is the relative cost of the different EV scenarios. This is particularly sensitive to battery technology development, less so to other parameters. Accordingly, only the effect of the former is discussed. However, when examining the results obtained, it is also important to note that they are based on assumptions that particularly favour BEVs over PHEVs. Specifically, in [14] we have assumed EV batteries to last the whole lifetime of the vehicle; due to the larger size of the BEV battery pack, having to

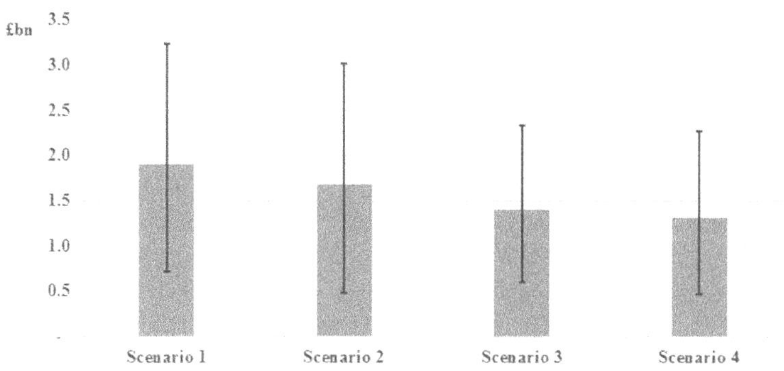

Fig. 7 Incremental annual user cost in British pounds of the UK EV and infrastructure scenarios modelled in [14]. The error bars indicate uncertainty associated with future battery technology development

replace it would incur a much higher-cost penalty than in the case of PHEVs. Moreover, possible grid reinforcement costs associated with public charging infrastructure are excluded, which also favours BEVs over PHEVs. So in effect the risk associated with large numbers of long-range BEVs could be much higher than the results presented suggest.

As shown in Fig. 7, Scenarios 1 and 2 show the greatest cost sensitivity to future battery development, around 40% higher than Scenarios 3 and 4. This means that initially such an EV mix would have to be subsidised substantially more than one dominated by long-AER PHEVs. Even at a BEV battery cost of $200/kWh, Scenarios 1 and 2 would cost around £400–600 M a year more than Scenario 4. Only with batteries that cost in the order of $100/kWh and have double the energy density of today's best in class would the cost of all scenarios converge. By comparing Scenarios 1 and 2, we can also see that the effect of reducing infrastructure coverage to only rapid chargers is relatively minor, of the order of £200 M a year. However, the model in [14] shows that even in Scenario 2 the utilisation level of the charging infrastructure is likely to be low: assuming that BEVs are charged at home overnight and their AER is used in full, there would be on average less than two rapid charging events a day per charger. This makes the business case for building an extensive rapid charging infrastructure problematic. Scenario 4 is cheapest and allows to keep the BEV option open while not being more sensitive to battery technology development than Scenario 3. Moreover by strategically siting the rapid charging infrastructure in and around urban areas, better utilisation levels could be achieved at around four charges a day on average. It therefore follows from our analysis that pursuing an EV and charging infrastructure mix of the kind of Scenario 4 would provide a relatively low-cost, low-risk electrification path for the UK.

6.2 California

Based on the discussion of current state and future EV targets in California provided in Sect. 4.2, a set of key scenarios for the year 2025 was built in [14] and their incremental costs were estimated, following the same logics as for the UK.

All scenarios are consistent with the target of 1.5 m EVs on the roads in 2025 set by the executive order of the Governor of the State of California. Although they differ widely in terms of the types of EV and infrastructure deployed, all scenarios are characterised by comparable average fleet tailpipe CO_2 emissions. The key elements of each of the five scenarios modelled are listed below and further illustrated in Table 6; their respective incremental cost is shown in Fig. 8. It is worth noting that, with the exception of Scenario 1, all other scenarios mirror those chosen for the UK, in order to allow for easier comparison between the two case studies.

Table 6 California 2025 EV and infrastructure scenarios modelled in [14]

Scenario	BEV share (%)	BEV segments	PHEV share (%)	PHEV segments	BEV range (km)	PHEV range (km)	# Slow chargers	# Fast chargers
1	40	All	60	All	300	50	75,000	2250
2	40	All	60	All	250	50	75,000	2250
3	40	All	60	All	250	50	30,000	750
4	0	–	100	All	–	100	–	–
5	20	Small	80	Medium, luxury, SUV	150	100	35,000	1000

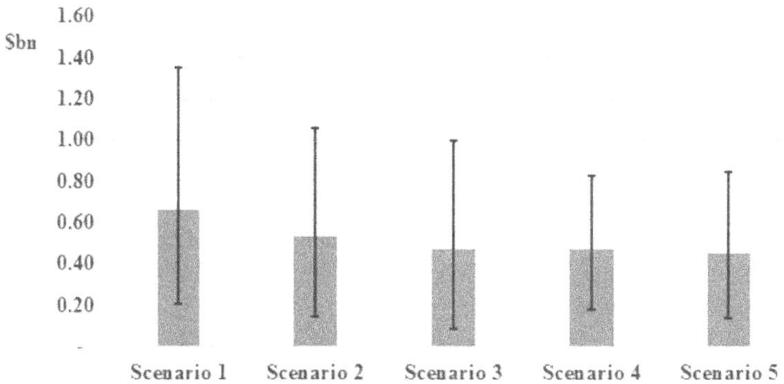

Fig. 8 Incremental annual user cost of California EV and infrastructure scenarios modelled in [14]. The error bars indicate uncertainty associated with future battery technology development

- Scenarios 1, 2 and 3, in addition to meeting the Governor's target, also broadly fulfil the requirements of the ZEV mandate in terms of number of credits and ZEV floor.
- We assume that the ZEV floor is met using BEVs only. We do not model BEVx due to their reduced utility; however, we will later discuss their possible role from a qualitative standpoint, and we do not consider the effect of possible deployment of FCEVs.
- We assume the BEV/PHEV ratio to be 40/60, although in reality this will vary somewhat depending on the compliance strategy chosen by the OEMs. In particular, longer-AER EVs qualify for more credits; hence, fewer of them would be required. Given that the difference between the particular scenario we have chosen and possible other compliance scenarios is in the order of few percentage points, we will ignore it for the sake of simplicity.
- BEVs and PHEVs feature in all segments of the passenger car market, which is not incompatible with today's rapidly growing offer of new EV models.
- The only difference between Scenarios 1 and 2 is the range of the BEVs, which is 300 and 250 km, respectively. A longer-range BEV earns more credits,

so it could provide OEMs with a cheaper way of complying with the ZEV mandate, while better matching the stated preferences of Californian BEV users. A shorter-range BEV that does not fully satisfy the desire of the users in terms of range however allows higher utilisation of the battery installed and hence is more economical. We also assume that the remaining ZEV credits are earned with 50-km-AER PHEVs, which qualify for ca. 0.8 ZEV credits each and enable meeting the overall 1.5 m EV target.
- In both cases, we assume that public infrastructure is provided based on the 2020 "public access" scenario of the NREL study [34], scaled up to 2025 as appropriate.
- Comparing Scenarios 1 and 2 allows us to test the effect of BEV range on the incremental cost of electrification.
- Scenario 3 is the same as Scenario 2, except that the public infrastructure provision is reduced based on the "home-dominant" scenario of the NREL study. Comparing Scenarios 2 and 3 allows to assess the impact of different types of public charging infrastructure in California on incremental costs.
- Scenarios 4 and 5 meet the Governor's target but do not comply with the ZEV mandate.
- Scenario 4 is based on the same logic as the corresponding one for the UK, i.e. to achieve electrification with the least deployment of battery capacity and charging infrastructure. This means using 100-km-AER PHEVs and no public charging infrastructure at all.
- Scenario 5 also mirrors the corresponding one for the UK, with 150-km-range BEVs adopted in the small vehicle segment, complemented by 100-km-AER PHEVs in all other segments.

Comparing Fig. 8 (California) with Fig. 7 (UK), we observe a similar trend across scenarios. In the case of California, however, we have added a scenario (Scenario 1) with longer-range BEVs. Comparing Scenarios 1 and 2 shows that further increasing BEV range has a strong impact on battery technology risk. We also notice that the economics of long-AER PHEVs in California are overall worse than in the UK, mainly due to the effect of the extra weight of the series PHEV powertrain in large, powerful cars. This explains why, as battery technology improves, the incremental cost of Scenarios 2 and 3 converge with those of Scenarios 4 and 5 more rapidly than the corresponding scenarios for the UK. Comparing Scenarios 2 and 3 shows that the effect of reducing public infrastructure provision is small, albeit not negligible. The cost of Scenario 3 in particular could start to converge to those of Scenarios 4 and 5 already at a battery cost of $200/kWh, assuming that battery lifetime was not an issue. Whether rapid uptake of BEVs with 250-km range can be achieved without extensive public charging infrastructure though is not known. Scenario 4 shows that, if not complying with the ZEV mandate was an option, the same level of tailpipe emissions could be achieved at lower risk using only PHEVs with 100-km AER; based on the model results, these have a utility factor of around 85% if fully charged at home every day and without using public infrastructure and require, at least initially, substantially less support than the other scenarios discussed so far. Finally, Scenario 5 which combines 100-km-AER PHEVs with 150-km-range BEVs may be preferable as it is only marginally riskier than Scenario 4 while probably sufficient to continue promoting BEV innovation.

7 Policy Analysis and Recommendations

Despite the substantial differences in terms of passenger car market structure and usage patterns in the UK and California, in [14] we find that the incremental cost of the different EV and charging infrastructure mixes we explored follows qualitatively a similar pattern. In particular, the analysis suggests that lower-cost, lower-risk electrification of passenger cars in the 2030 timeframe can be achieved through a balanced mix of relatively short-range BEVs and long-AER PHEVs. This can be broadly extrapolated to the North American and European markets in general. While it is apparent that once BEV batteries achieve their cost reduction target and increase their energy density substantially, long-range BEVs have the potential to outstrip PHEVs on a cost basis, to rely on this happening rapidly is potentially risky. Hence, the main implication of our findings is that, by designing policies primarily aimed at accelerating complete electrification of passenger car transport through supporting the rapid roll-out of long-AER BEVs and extensive charging infrastructures, the EV transition may be set on a higher-cost, higher-risk path which could eventually result in its losing momentum and possibly stalling.

When discussing the policy implications of the analysis in [14], however, the limitations of the methods used should not be overlooked. As already mentioned in Sect. 3, the analysis performed is of the snapshot type, i.e. it assesses the incremental cost of electrification for the year 2030 only. For a full picture of the technology risk associated with the different electrification pathways, their cumulative cost until the year 2030 should be estimated instead. Moreover, the analysis does not apportion the incremental cost of electrification among government, car manufacturers and passenger car buyers. If this was done, further policy insight could be gained. In fact, it is possible that the car manufacturers will prefer, for technical or strategic reasons, to pursue a technology pathway that we have identified as higher risk, and that they will be prepared to internally subsidise it as needed. However, we do not want to speculate on this in the absence of sufficient evidence. Similarly, we do not want to speculate on the preferences of mainstream EV buyers in the year 2030. Should they be prepared to pay a significant premium for long-AER BEVs over PHEVs, though, this would somewhat alter the findings of our research.

We now proceed to further examine the UK and California EV policy based on the findings of [14]. In general, EV policy in both the UK and California today shows, to different degrees, signs of favouring the rapid development of the BEV market alongside that of PHEVs. The UK approach is generally cautious, and no commitment has so far been made for the long term, particularly on EV incentives which are reviewed periodically. On the infrastructure side, however, the development of a country-wide network of chargers may soon be underway. Hence, we argue that, due to the path-dependent nature of EV innovation processes, even a relatively cautious approach based on monitoring market and technology development and periodically revising support measures may unintentionally lead to higher-cost, higher-risk pathways. The probability of this happening is higher though in the case of California, where the unique technology-forcing approach of the ZEV mandate has already had a strong effect on EV innovation and will continue to do so in the foreseeable future. The analysis conducted also allows us to infer that the specific design of both EV and infrastructure incentives and mandates can potentially have a strong impact on the cost of future EV pathways and hence it should be more carefully considered. Specific aspects of the UK and California policy are here briefly discussed in turn.

In the case of the UK, in the absence of a ZEV mandate that influences the direction of EV innovation, the type of EV and infrastructure deployed in the UK under the current policy framework will largely depend on the combined effect of the incentives provided, EV models offered and users' needs and preferences. In this context, the recently introduced two-tier incentive system for EVs, with a step in the value at the 110-km-AER mark, arguably favours shorter-AER over longer-AER PHEVs. This could be rebalanced by either moving the Plug-in Vehicle Grant step to an AER of 80–100 km so that long-AER PHEVs could also benefit from the higher grant available to BEVs, or by making the value of the grant for PHEVs proportional to their AER. As for BEVs, the flat rate of the grant currently provided is in principle favourable to short-range BEVs in the city car segment, although these face tough competition from small, fuel-efficient ICEVs that benefit from low CO_2-based taxation. In all other segments, based on current driving patterns, it is plausible that BEV users will require their vehicles to have relatively long ranges if they are to penetrate the market rapidly, for which a higher grant may be initially required. The other important influence on BEV adoption in the UK is the EV charging infrastructure strategy, which is currently under development. Focusing on providing extensive urban and sub-urban charging infrastructure, particularly of the rapid type, could further support the uptake of short-AER BEVs in the small car segment. On the contrary, a country-wide infrastructure may indirectly encourage adoption of long-AER BEVs in larger car segments and probably also result in low infrastructure utilisation levels.

In California, the ZEV mandate shapes EV innovation and forces the deployment of substantial numbers of these vehicles. In particular, the current structure of the ZEV credits strongly supports longer-range BEVs and is likely to have played an important role in the development of such vehicles, initially by Tesla Motors Inc. and increasingly also by other OEMs. Post-2018 the ZEV credit structure will change and the support for longer-range BEVs will weaken but continue to exist. The effect that this will have on the compliance strategies of the OEMs remains to be seen, though it is plausible that they will introduce more BEVs with sufficiently long AER that appeal to customers while earning the manufacturer more ZEV certificates per vehicle. On the other hand, short-range BEVs are not a natural fit in California, not even in the small vehicle segment as these vehicles tend to be driven for similar distances to larger vehicles. As for PHEVs, the current structure of ZEV credits and EV incentives favours longer-AER PHEVs. However, the ZEV floor present in the mandate limits the contribution these will likely make to the EV fleet. Despite the increased flexibility granted by the recently introduced category of BEVx, which is allowed to generate up to half of the credits needed to meet the ZEV floor, the tension between supporting strong BEV innovation and achieving the necessary level of CO_2 emission reduction at a low cost remains. In conclusion, the evolution of the California ZEV mandate suggests that costs are increasingly being taken into consideration. However, given the strong influence that this has on EV innovation in the USA and beyond, we argue that more could be done to guide the EV transition towards a path that is robust under uncertainty. An option could be to calculate credits based on actual electric miles driven as opposed to range. Our analysis suggests that a balanced mix of EVs in California in 2025–2030 should include long-AER PHEVs and possibly also relatively short-range BEVs in specific segments where they can be competitive such as shared urban car fleets. BEVx, if adequately supported, could facilitate the subsequent transition from long-AER PHEVs to BEVs.

Finally, it is worth mentioning that in our analysis we have not taken into account new technology paradigms such as autonomous vehicles, shared ownership and mobility services. In the long term, these could have a profound effect on the passenger car market structure and use patterns. However, our modelling approach lends itself to this type of analysis and we recommend that these effects are accounted for especially if extending the timeframe of the analysis beyond 2030.

References

1. International Energy Agency (IEA), in *World Energy Outlook* (2015)
2. U.S. Energy Information Adminstration (EIA), *International Energy Outlook* (2014)
3. IEA, *Energy and Climate Change: World Energy Outlook Special Report* (2015)
4. S.K. Ribeiro et al., Chapter 9: Energy End-Use: Transport, in Global Energy Assessment—Toward a Sustainable Future. 2012: Cambridge University Press, Cambridge, UK and New York, USA and the International Institute for Applied Systems Analysis, Laxenburg, Austria
5. IPCC, Climate Change 2014: Synthesis Report. Contribution of Working Groups I, II and III to the Fifth Assessment Report of the Intergovernmental Panel on Climate Change [Core Writing Team, R.K. Pachauri and L.A. Meyer (Eds.)]. 2014, IPCC, Geneva, Switzerland
6. OECD, in *The Cost of Air Pollution: Health Impacts of Road Transport.* (OECD Publishing, 2014)
7. N. Lutsey, in *Transition to a Global Zero-Emission Vehicle Fleet: A Collaborative Agenda for Governments. 2015, The International Council on Clean Transportation (ICCT).* 1225 I Street NW Suite 900, Washington DC 20005
8. European Union, Directive 2014/94/EU of the European Parliament and of the Council of 22 October 2014 on the Deployment of Alternative Fuels Infrastructure (2014)
9. IEA, Global EV, in *Outlook—Understanding the Electric Vehicle Landscape to 2020. 2013, International Energy Agency, Electric Vehicles Initiative and Clean Energy Ministerial*
10. NRC, in *Overcoming Barriers to Deployment of Plug-In Electric Vehicles.* 2015, committee on overcoming barriers to electric-vehicle deployment, board on energy and environmental

systems, division on engineering and physical sciences and transportation research board. National Research Council of the National Academies (The National Academies Press, 500 Fifth Street, Washington D.C)
11. IEA, Global EV, in *Outlook 2015. Electric Vehicles Initiative and Clean Energy Ministerial*
12. Element Energy, Ecolane, and University of Aberdeen, Pathways to High Penetration of Electric Vehicles. Final Report for the Committee on Climate Change. (Element Energy Ltd. 20 Station Road, Cambridge CB1 2JD, 2013)
13. J. Struben, J.D. Sterman, Environ. Plann. B-Plann. Des. **35**, 1070 (2008)
14. M. Contestabile, M. Alajaji, B. Almubarak, Energy Policy **110**, 20 (2017)
15. IEA, Global EV, in *Outlook 2016. Beyond One Million Electric Cars. International Energy Agency, Electric Vehicles Initiative and Clean Energy Ministerial*
16. P. Mock, Z. Yang, Driving Electrification—A Global Comparison of Fiscal Incentive Policy for Electric Vehicles. 2014, The International Council on Clean Transportation (ICCT). 1225 I Street NW Suite 900, Washington DC 20005
17. Brook Lyndhurst, Uptake of ultra low emission vehicles in the UK—a rapid evidence assessment for the department for transport (Office for Low Emission Vehicles, 2015)
18. W. Sierzchula et al., Energy Policy **68**, 183 (2014)
19. B. Nykvist, M. Nilsson, Nature. Clim. Change **5**, 329 (2015)
20. M. Åhman, L.J. Nilsson, Util. Policy **16**, 80 (2008)
21. M. Dijk, M. Yarime, Technol. Forecast. Soc. Change **77**, 1371 (2010)
22. F.W. Geels, J. Transp. Geogr. **24**, 471 (2012)
23. A. Chase, P. Wells, G. Alberts, Investing in the Low Carbon Journey: Lessons from the First Decade of UK Policy on the Road to 2050. Prepared for the LowCVP by E4tech with Cardiff Business School. 2014
24. OLEV, Plug-in Car and Van grants eligibility. 28 March 2016; Available from: https://www.gov.uk/plug-in-car-van-grants/eligibility
25. Element Energy, *Transport Energy Infrastructure Roadmap to 2050. Electricity Roadmap* [2015, Element Energy Report for the Low Carbon Vehicle Partnership]
26. DfT, Road Investment Strategy: for the 2015/16–2019/20 Road Period, Department for Transport, Editor, 2015
27. R. Hutchins, et al., *Assessing the role of the plug-in car grant and plugged-in places scheme in electric vehicle take-up. 2013: transport research laboratory (TRL).* Report for the Department for Transport (DfT), Social Research and Evaluation Division
28. California, Governor's Office, Executive Order B-30-15. 2015
29. California, Governor's Office, Executive Order B-16-2012. 2012
30. California, Governor's Office, 2013 ZEV Action Plan. A Roadmap Toward 1.5 Million Zero-Emission Vehicles on California Roadways by 2025, G.s.I.W.G.o.Z.-e.V.G.E.G.B.J.F. 2013, Editor. 2013
31. California, Governor's Office, 2015 ZEV Action Plan. An Updated Roadmap Toward 1.5 Million Zero-Emission Vehicles on California Roadways by 2025, G.s.I.W.G.o.Z.-E.V.G.E. G.B.J.A. 2015, Editor. 2015
32. California Auto Outlook, Comprehensive Information on the California Vehicle Market, vol 12, No. 1, Released Feb 2016, Covering Fourth Quarter 2015. 2016, California New Car Dealers Association
33. AFDC (Alternative Fuel Data Center), Electric Vehicle Charging Station Locations. 2016 [cited 2016, May 11]; Available from: http://www.afdc.energy.gov/fuels/electricity_locations.html
34. NREL, California Statewide Plug-in Electric Vehicle Infrastructure Assessment—Final Project Report. 2014, National Renewable Energy Laboratory, 15013 Denver West Parkway, Golden, CO 80401: Prepared for California Energy Commission
35. California Secretary of State, Zero-Emission Vehicle Standards for 2018 and Subsequent Model Year Passenger Cars, Light-Duty Trucks, and Medium-Duty Vehicles, in *California Code of Regulations Section 1962*:2, ed. by California Air Resources Board (2013)

36. California Secretary of State, Zero-Emission Vehicle Standards for 2009 through 2017 Model Year Passenger Cars, Light-Duty Trucks, and Medium Duty Vehicles, in *California Code of Regulations Section 1962*:1, ed. by California Air Resource Board (2013)
37. Center for Sustainable Energy, California Plug-in Electric Vehicle Driver Survey Results—May 2013. Clean Vehicle Rebate Project, California Center for Sustainable Energy, 9325 Sky Park Court, Suite 100, San Diego, CA 92123
38. SMMT, Motor Industry Facts 2013. The Society of Motor Manufacturers and Traders, 71 Great Peter Street, London SW1P 2BN UK
39. G.J. Offer et al., Techno-economic and behavioural analysis of battery electric, hydrogen fuel cell and hybrid vehicles in a future sustainable road transport system in the UK. Energy Policy **39**, 1939 (2011)
40. DfT, National Travel Survey, 2002–2006 [computer file]. 2nd Edition., Department for Transport Editor. 2008, Data Archive [distributor], Aug 2008. SN: 5340: Colchester, Essex: UK
41. Whatcar.com. 2016; Available from: http://www.whatcar.com/
42. U.S. News Best Cars. (no date); Available from: http://usnews.rankingsandreviews.com/cars-trucks
43. Cars.com. 2016 [Aug 2016]; Available from: https://www.cars.com/
44. DoT, National Household Travel Survey 2009, F. H. A. U.S. Department of Transportation, Office of Highway Policy Information, Editor. 2009
45. Lotus Engineering Inc., An Assessment of Mass Reduction Opportunities for a 2017–2020 Model Year Vehicle Program. 2010, Submitted to: The International Council on Clean Transportation
46. A. Brooker, J. Ward, L. Wang, Lightweighting Impacts on Fuel Economy, Cost, and Component Losses, in SAE 2013 World Congress & Exhibiton. 2013
47. S. Pagerit, P. Sharer, A. Rousseau, Fuel Economy Sensitivity to Vehicle Mass for Advanced Vehicle Powertrains, in SAE Paper 2006-01-0665, SAE World Congress. 2006: Detroit, April 2006
48. Element Energy, Cost and Performance of EV batteries—Final Report for the Committee on Climate Change. 2012, Element Energy Limited: 20 Station Road, Cambridge CB1 2JD
49. A. Moawad et al., *Assessment of Vehicle Sizing, Energy Consumption, and Cost Through Large-Scale Simulation of Advanced Vehicle Technologies* (Argonne National Laboratory, Energy Systems Division, 2016)

Conventional, Battery-Powered, and Other Alternative Fuel Vehicles: Sustainability Assessment

Lambros K. Mitropoulos and Panos D. Prevedouros

Abstract The substantial impacts of transportation on environment, society, and economy strongly urge the incorporation of sustainability into transportation planning. Major developments that enhance transportation sustainability include alternative fuels, electric drive and other novel technologies for vehicle propulsion. This chapter presents a sustainability framework that enables the assessment of transportation vehicle characteristics. Identified indicators are grouped into five sustainability dimensions (environment, technology, energy, economy, and users). The method joins life cycle impacts and a set of quantified indicators to assess the sustainability performance of seven popular light-duty vehicles and two types of transit buses. The hybrid diesel electric bus received the highest sustainability index and the internal combustion engine vehicle the lowest. Fuel cell and hybrid electric vehicles were found to have the highest sustainability index among all passenger vehicles. The sustainability performance of some new technologies currently suffers from limitations in engine and battery performance, comfort and convenience, and availability of charging stations.

1 Introduction

Electric vehicles (EVs) date back to the late 1800s. In the early 1900s, they were competing well with internal combustion engine vehicles (ICEVs). This excerpt from The History of the Electric Vehicle [1] captures that era well:

> …Ferdinand Porsche, founder of the sports car company by the same name, developed an electric car called the P1 in 1898. Around the same time, he created the world's first hybrid electric car – a vehicle that is powered by electricity and a gas engine. Thomas Edison, one

L. K. Mitropoulos · P. D. Prevedouros (✉)
Department of Civil and Environmental Engineering, University of Hawaii at Manoa, 2540 Dole Street, Holmes Hall 383, Honolulu HI 96822, USA
e-mail: pdp@hawaii.edu

L. K. Mitropoulos
e-mail: lampros@hawaii.edu

G. Pistoia and B. Liaw (eds.), *Behaviour of Lithium-Ion Batteries in Electric Vehicles*, Green Energy and Technology,
https://doi.org/10.1007/978-3-319-69950-9_5

of the world's most prolific inventors, thought EVs were the superior technology and worked to build a better EV battery. Even Henry Ford, who was friends with Edison, partnered with Edison to explore options for a low-cost electric car in 1914.

Yet, it was Henry Ford's mass-produced Model T that dealt a blow to the electric car. Introduced in 1908, the Model T made gasoline-powered cars widely available and affordable. By 1912, the gasoline car cost only $650, while an electric roadster sold for $1,750. That same year, Charles Kettering introduced the electric starter, eliminating the need for the hand crank and giving rise to more gasoline-powered vehicle sales.

Fast forward to the 1990 Clean Air Act Amendment and the 1992 Energy Policy Act, both of which re-energized EV development in the USA, including the memorable but failed launch of the 80-mile-range EV1 by General Motors (GM) in 1996 with production ending in 1999 followed by the infamous recall of all 2234 EV1s in 2002. In response to these regulations, Toyota's compromise in the 1997 launch of the Prius hybrid drive was a market success. It took more than nine years to reach one million Prius sold in May 2008, but only a little over two more years to reach two million Prius sold in September 2010, aided by the Great Recession and the simultaneous increase of the price of oil at well over $80 per barrel in 2008 which represented new historical high prices, with proportional impacts on the price of gasoline and diesel fuels.

One of the unintended but beneficial consequences of the Great Recession (the economic recession from 2007 to 2009) in combination with oil prices and concerns about climate change due to anthropogenic pollution gave rise to alternative technologies ranging from cleaner fossil fuels such as compressed natural gas (CNG) and liquefied natural gas (LNG) to high-performing pure EVs such as the Fisker Karma which was produced in 2011 and 2012 only. Some of these alternatives enjoyed major governmental support (e.g., $7500 federal tax credit—and additional state tax credits—for the purchase of an EV in the USA) or other subsidies.

CNG has been in use for light-duty vehicles for decades in Europe, Asia, and Australia. LNG provides a much longer range, but storage and dispensation have much more complex and expensive requirements. This is reflected in their relative availability in the USA: There are over 1600 CNG but only 123 LNG fueling stations. In the UA, CNG has had more success in heavy vehicle fleets; for example, Los Angeles, New York City, Phoenix, Fort Worth, Dallas, San Francisco, San Diego, Denver, Atlanta, Washington, D.C., have natural gas transit bus fleets. However, in 2016 all types of natural gas vehicles were only 0.06% of the total fleet population; even in Italy, which has over one million of natural gas-powered vehicles, their portion in the county's fleet is under 2% [2]. Thus, this technology was not included in our assessment.

Although Fisker Karma did not succeed, the similarly priced Tesla Model S, built in California at an abandoned and subsequently refurbished Toyota plant with help from government subsidies, became substantially successful in the luxury market upon its launch in 2012. The Model S replaced thousands of internal combustion engine-powered cars by other manufacturers. Within a few years, other manufacturers responded with hybrid versions (e.g., BMW i8 in 2014, Mercedes S500 Plug-in Hybrid in 2015).

Table 1 shows that it only took five years past the end of the Great Recession for EV and plug-in hybrid sales to surge past 100,000 with five models (out of 22 available in the US market) accounting for 74% of the sales in 2014 [3]: Nissan Leaf (30,200), Chevrolet Volt (18,805), Tesla S (16,689), Toyota Prius PHV (13,264), and Ford Fusion Energy (11,550).

Fuel cell-powered vehicles are part of the "hydrogen economy" concept that was initially supported during the early years of the President Obama administration. However, focus quickly shifted to bailouts of the auto industry (e.g., cash for clunkers) and support for battery development. A fuel cell vehicle (FCV) uses hydrogen as the fuel which is processed by the fuel cell to produce electricity that drives an electric motor like a regular EV.

Unlike a battery-powered EV, a FCV has an exhaust and dispels water vapor. California, Germany, and Japan have specific initiatives favoring FCV and expect substantial growth in vehicle availability and adoption past 2020 [4–7]. FCVs were included in our analysis because of their growth potential and the fact that they contain a battery and are driven by one or more electric motors. Table 2 shows that FCVs are competitive in all but two categories: price and number of refueling stations.

The technologies assessed herein with our sustainability assessment tool include the following: ICEV, hybrid electric vehicle (HEV), FCV, EV, plug-in hybrid electric vehicle (PHEV), diesel bus (DB), and hybrid diesel electric bus (HDEB).

ICEV is the baseline light-duty vehicle in these comparisons given that they comprise over 90% of all the light-duty vehicle fleets in the USA. The diesel-fueled internal combustion engine for light-duty vehicles was omitted because the performance and pollution outcomes reported by several manufactures were false and became the subject of international re-tests, investigations, and lawsuits between 2014 and 2016 affecting millions of popular light-duty vehicles. The reader can search for the diesel emissions scandal (i.e., dieselgate).

In parallel to the technological and fuel diversification of the vehicle fleet, interest in sustainable development and sustainable transportation has grown rapidly in the first ten years of the millennium due to the environmental, social, and economic impacts that these sectors have on communities, regions, and Earth. Initially, sustainability was used to depict concerns mostly associated with

Table 1 Light-duty EVs sold

Year	US sales	% of world
2010 and 2011	17,425	–
2012	52,607	–
2013	97,507	–
2014	122,438	38
2015	116,099	21
2016	158,614	20
2017[a]	182,253	22

Note ([a]) Based on first eight months of 2017

Table 2 Comparable US market Toyota cars in 2017

Type	ICEV	HEV	PHEV	FCV
Model	Corolla LE	Prius II	Prius Prime	Mirai
MSRP[a]	\$19,000	\$25,000	\$28,000	\$57,500
Tax credit	n.a.	n.a.	\$7500	\$7500
MPG/MPGe[b]	32	54	54	67
Range (miles)	~410	~600	~640	~310
0–60 mph (s)	9.8	10.5	10.9	9.0
Fuel stations	167,000	167,000	15,500	33

Notes [a]Manufacturer's suggested retail price
[b]Miles per gallon equivalent (MPGe) is used to represent the fuel efficiency of the FCV

environmental issues; however, it has been expanded to include energy, economy, and social issues. The energy aspects are of major interest to the analysis of transportation modes because they require large quantities of energy not only for their operation but also for their manufacture (e.g., vehicles) and construction (e.g., highways). Construction, manufacture, and operation of transportation modes generate pollutants and waste. The impacts of the transportation sector on the environment and economy, and the social effects of transportation on communities are so extensive that necessitate the incorporation of sustainability into the transportation planning process.

Attempts to incorporate sustainability into transportation planning have resulted in research on the development of variables defined as measures, indicators, or indices representing elements of sustainability. However, indicators that measure impacts on mobility, safety, and environmental effects are applied mainly at aggregated level for all vehicles on a road network by considering only the operational stage of the transportation system. The aggregation of transportation performance measures limits one of the sustainability's principal roles in transportation planning, which is to assist agencies in evaluating new transportation modes that are proposed for introduction in a network. For example, today's common planning tools do not provide justice to the assessment of bus and BRT routes with electric or fuel cell buses. Similarly, the impacts of electric, fuel cell, and hybrid cars on high-occupancy vehicle (HOV) and high-occupancy toll (HOT) lanes require extensive customization in order to produce representative assessments of energy and pollution impacts.

When evaluating transportation modes, sustainability factors should be integrated with technological and user factors into a framework that will assist decision makers to (a) understand impacts during vehicles' lifetime and (b) justify selection of a transportation mode that fulfills sustainability requirements and public's need for safe and efficient transportation. Thus, vehicles with alternative propulsion systems require the breakdown of vehicle components and characteristics for the proper understanding of their performance and impact over their life cycle. Disaggregation of modes by vehicle type and the sustainability assessment enable effective planning and policy making. Disaggregation by vehicle type in this chapter is the consideration of different technological, operational, and functional characteristics of vehicles.

This chapter extends a previous study in sustainable transportation planning [8] which uses a framework and a set of indicators, to assess seven vehicle types and propulsion options, including ICEV, hybrid electric vehicle (HEV), FCV, EV, PHEV, DB, and HDEB. The sustainability framework was the methodological guide for developing a sustainability assessment tool. The estimates are both technology- and policy-sensitive and thus useful for short- and long-term planning. Indicators are aggregated to a single sustainability index (SI) for each dimension and vehicle type. Indices enable decision makers and policy analysts to explore variations in sustainability performance for alternative scenarios or vehicle mixes (e.g., more hybrid and less regular bus service).

2 Sustainability Assessment of Urban Vehicles

2.1 Sustainability Dimensions

The sustainability framework consists of five dimensions that are captured by goals that govern transportation systems: environment, technology performance, energy, economy, and users. The goals of the framework are to help a community meet its needs by

- Minimizing environmental impact and energy consumption and
- Maximizing its economy, user and community satisfaction, and technology performance. (Technology refers to the features of modes that support community livability, enhance public health, safety and comfort for all their users.)

Environment. The environmental impact of transportation vehicles begins when materials to manufacture components are gathered and ends when the vehicle is disposed or recycled. Fuel efficiency and alternative fuels are two primary areas among several options that vehicle manufacturers pursue in order to reduce emissions and minimize overall environmental impact of vehicle manufacture and usage.

Technology performance refers to all components of systems made by humans to meet their needs. It is one of the most rapidly developing and resource consuming sectors. Sustainable technology helps people meet their mobility needs with safety and comfort while minimizing the consumption of non-renewable energy sources, and maximizing the reuse and recycling of materials. Weight reduction, high-strength materials, engine and aerodynamic enhancements, and alternative propulsion systems are some of the technologies that contribute to sustainable transportation.

Energy is a major component of transportation and is directly connected with the environment and economy. Energy availability, demand, price, and actual consumption have short- and long-term impacts. Consumption of non-renewable energy sources generates emissions that are harmful to humans in the short term and deprives energy from future generations in the long term. Globally, vast amounts of energy are needed for transportation infrastructure development, vehicle manufacture, and transportation operations.

Economy. The creation of a sustainable economy requires the disciplined use of energy and technology. An unsustainable economy contributes to the destruction of the environment, has a multitude of social impacts and results in suboptimal transportation services. In this context, a sustainable economy facilitates a lower cost for urban mobility by assessing vehicle costs including purchase, registration, insurance, operation, parking and fuel costs, and promoting vehicle types and technologies that minimize total cost.

Users represent a large set of stakeholders including individuals (e.g., residents or travelers), groups (e.g., schoolchildren), private companies (e.g., taxis, fleet operators), and public agencies (e.g., regulatory, operation and maintenance agencies). Transportation mode outputs, including traffic delay, reliability, safety, comfort, and convenience, determine user choice as to when, how, and how much they choose to use a transportation mode. User perceptions and preferences vary; hence, vehicle and mode choices also vary. Vehicles with performance deficiencies are less attractive to users and become unsustainable in the long term.

Utilization of the technology performance and users dimensions in the sustainability framework instead of the traditional sustainability dimension of "society" allows for an assessment of technology's capabilities and limitations and takes explicitly into account the preferences and restrictions of users and other stakeholders.

2.2 *Sustainability Indicators*

A set of indicators is used for assessing the sustainability performance of seven urban transportation vehicles based on the framework's five dimensions. Performance measures used for evaluating the sustainability dimensions are shown in Table 3. The indicators in Table 3 address objectives by identifying individual vehicle features that contribute toward maximization of sustainability. When the impacts (i.e., positive or negative) of those features of sustainability are aggregated for all vehicles on the network, their value determines goal achievement and ways to move a transportation system toward sustainability.

Indicators are specifically modified to apply to vehicles based on factory specifications. For example, the indicator comfort used in the literature to assess a transportation system is divided into indicators including passenger space, cargo space, and legroom. Some of the identified indicators in Table 3 are directly adopted from the literature, such as emissions, energy, and cost. However, indicators including emissions, energy, trip cost, fuel cost, or trip time that usually apply to vehicle operation stage only are generalized over a vehicle's life cycle.

Table 3 Sustainable transportation indicators [8]

	Goal	Indicator	Indicator description and reference
Environment	Minimize environmental impact	Carbon dioxide (CO_2), methane (CH_4), nitrous oxide (N_2O), GHG, volatile organic compound (VOC), carbon monoxide (CO), nitrogen oxides (NO_x), particle matter (PM_{10}), sulfur oxides (SO_x)	Emissions are an outcome of all life cycle stages of a vehicle including manufacture, fueling (primary energy production including raw material recovery and storage; fuel production including transportation, storage and distribution), operation (running: driving, start-up, tires, brakes, evaporative, idling; and processes that support the lawful usage of vehicles: insurance, registration, license and taxes), maintenance, and disposal
		Noise	Noise is representative of average urban speeds at a distance of 50 ft
Technology performance	Maximize technology performance	Fueling frequency	Time required to fuel a vehicle; significant for short ranging modes
		Maintenance frequency	Number of times a vehicle has to replace parts and fluids to maintain a safe service to its users during its lifetime
		Vehicle storage	Physical characteristics of a vehicle
		Engine power	Maximization of vehicle power
Energy	Minimize energy consumption	Manufacturing energy	Energy is required for all vehicle life cycle stages; as described for the emission indicators in environment
		Fueling energy	
		Operation energy	
		Maintenance energy	
Economy	Maximize and support a vibrant economy	Life cycle cost	Includes the cost for buying, operating (fueling/charging or using the vehicle); insurance, license, registration, taxes, and maintenance. Indicator values can be replaced with local data
		Parking price	Monthly expenses for parking the vehicle (national average) downtown. Indicator values can be replaced with local data
		Subsidies	Portion of costs covered by taxpayers. Indicator values can be replaced with local data

(continued)

Table 3 (continued)

	Goal	Indicator	Indicator description and reference
Users	Maximize user satisfaction	Mobility[a]	Number of pass./veh./hour that choose to or desire to utilize the subject mode
		Demand	Provision of social and economic opportunities by the transportation network
		Delay	Real travel time minus the travel time of a vehicle when it travels at 30 mi/h
		Safety	Number of accidents or fatalities that have been recorded for a specific vehicle type
		Global availability	Time during which a vehicle is not available to its users during a day. Indicator can be changed based on regional-local specific requirements (it is expressed as an annual percentage)
		Reasonable availability	Time during which a vehicle is not available to its potential users during the 19 h of the day (5–12 a.m.) when 98.8% of total trips occur [9]. Indicator values can be replaced with local data (it is expressed as an annual percentage)
		Legroom, cargo space	Physical vehicle characteristics which maximize user comfort and convenience
		Fueling opportunities	Available locations for fueling or charging a vehicle (regional planning). Indicator values can be replaced with local data. Indicator is not applicable to public transit modes

Notes [a]Indicators in hatched cells apply to specific projects and take values based on regional data; thus, they are not quantified in the generic version presented herein

The indicator cost includes purchase, fuel, insurance, registration, taxes, and maintenance cost over lifetime.

2.3 Vehicle Types

Vehicle type refers to vehicle propulsion technology (e.g., internal combustion engine, electric motor, or hybrid) and basic functionality (e.g., car/van, light truck, bus). All vehicles were assumed to use the same highway infrastructure; roads and related traffic infrastructure were not part of the assessment. The analysis is conducted for seven vehicle types that entered the market in 2011. The seven urban vehicles examined are as follows:

- Internal combustion engine vehicle (ICEV)
- Hybrid electric vehicle (HEV)
- Fuel cell vehicle (FCV)
- Electric vehicle (EV)
- Plug-in hybrid electric vehicle (PHEV)
- Diesel bus (DB)
- Hybrid diesel electric bus (HDEB).

Vehicle specifications were necessary for estimating their impact on the five sustainability dimensions (i.e., environment, technology performance, energy, economy, and users) that capture the goals of a transportation system. For vehicle types, such as the ICEV, the HEV, the FCV, the EV, and the PHEV, the most representative vehicles were selected based on their sales volume [10]. A summary of vehicle specifications and assumptions is shown in Table 4. For consistency in calculating annual fuel consumption and life cycle costs, all vehicles were assumed to operate in an urban environment. Annual distance traveled and lifetime are the same per vehicle type. The average age of passenger vehicles is 10.6 years [11].

While the sustainability assessment is based on the 2011 figures in Table 4, a larger variety of vehicles with non-ICE powertrains were developed and many of them became very successful, such as the Ford C-MAX HEV, while others were quickly discontinued, such as the Honda Fit EV which had a three-year production run from 2012 to 2014. A sample of modern EV, HEV, FCV, and PHEV is shown in Table 5 along with several of their characteristics such a price, weight, battery capacity, range, acceleration, and the EPA miles per gallon (MPG) or miles per gallon equivalent (MPGe) for vehicles with electric, fuel cell and hybrid propulsion.

The comparison between the 2012 and 2017 versions of the base Tesla Model S is enlightening: While price and weight increased by 7 and 12%, respectively, other attributes improved far more substantially: Battery capacity increased by 88%, range increased by 66%, and acceleration improved by 9%. The EPA MPGe decreased by 10% indicating an improved efficiency per mile.

Table 4 Vehicle characteristics in the sustainability assessment [10–15]

		ICEV	HEV	FCV	EV	PHEV	DB	HDEB
		Toyota Camry LE	Toyota Prius III	Honda Clarity	Nissan Leaf	GM Volt	New Flyer 40′	
Weight	lbs	3307	3042	3582	3500	3781	26,000	28,500
Average occupancy	Passengers	1.20	1.20	1.20	1.20	1.20	10.50	10.50
Average lifetime	Years	10.6	10.6	10.6	10.6	10.6	12	12
Average annual miles	Miles	11,300	11,300	11,300	11,300	11,300	41,667	41,667
Lifetime miles	Miles	119,780	119,780	119,780	119,780	119,780	500,000[a]	500,000
Cost to buy (MSRP)	$US dollars	$22,055	$26,550	$39,075	$35,200	$39,145	$319,709	$531,605
Fuel economy	MPG/MPGe[b] (urban)	22.0	48.0	72.0	94.0	35.0	3.9	6.2
Fuel price ($2011)	$US dollars	$3.48	$3.48	$8.00[c]	$0.11[d]	$3.48 $0.11	$3.84	$3.84

Notes [a]For maintaining bus fleets in a good state of repair, US agencies need to replace buses that have exceeded the industry standard retirement age of twelve years/500,000 miles

[b]Miles per gallon equivalent (MPGe) is used to describe the fuel efficiency of the EV and FCV

[c]When all subsidies are excluded, the price of hydrogen is estimated to be $8.00 per kg including $1.14 for taxes, based on the assumption that the delivered hydrogen cost is $42.46 per gallon gasoline equivalent in 2004, and that the FCV is 2.7 times as efficient as the ICEV under average driving conditions [16]

[d]Electricity cost is given in $ per kWh

Table 5 Sample light-duty vehicles with electric, fuel cell, and hybrid power plants

	Year	Brand	Model	Type	Approx. price in US$	Curb weight (lb)	Battery type	Battery (kWh)	City (MPG/MPGe)	Electric-only range (miles)	0–60 mph (s)
1	2017	BMW	i3	EV	$44,450	2635	Li-ion	33	na/137	97	6.5
2	2017	BMW	i8	PHEV	$143,400	3394	Li-ion	7.1	28/76	15	4.2
3	2017	BMW	330e	PHEV	$44,100	3900	Li-ion	7	30/71	14	5.9
4	2016	Chevrolet	Malibu	HEV	$28,750	3388	Li-ion	1.5	49/na	na	7.4
5	2017	Chevrolet	Bolt	EV	$37,495	3580	Li-ion	60	na/128	238	6.3
6	2010	Chevrolet	Volt	EV	$34,095	3543	Li-ion	18.4	42/106	420	8.8
7	2017	Ford	C-MAX SE	HEV	$24,175	3640	Li-ion	7.6	42/na	na	8.1
8	2017	Ford	Focus	EV	$29,120	3640	Li-ion	33.5	na/118	100	9.9
9	2017	Ford	Fusion S	HEV	$25,295	3668	Li-ion	1.4	43/na	na	9.1
10	2017	Honda	Accord	HEV	$29,605	3483	Li-ion	1.3	49/na	na	6.9
11	2017	Honda	Clarity	FCV	Lease	4148	Li-ion	1.7	69/69	366	8.1
12	2016	Honda	CR-Z	HEV	$20,295	2650	Li-ion	0.6	36/na	na	9.3
13	2014	Honda	FIT EV	EV	$36,625	3252	Li-ion	20	na/132	82	8.6
14	2017	Mitsubishi	i-MiEV	EV	$24,400	2552	Li-ion	16	na/121	66	13
15	2016	Nissan	LEAF S	EV	$30,680	3307	Li-ion	24	na/99	84	10.4
16	2016	Nissan	LEAF SV	EV	$34,200	3391	Li-ion	30	na/124	107	10.2
17	2012	Tesla	Model S	EV	$69,500	4150	Li-ion	40	na/88	160	5.6
18	2017	Tesla	Model S	EV	$74,500	4647	Li-ion	75	na/97	265	5.1
19	2016	Tesla	Model X 90D	EV	$95,500	5271	Li-ion	90	na/90	250	4.8
20	2017	Toyota	Avalon	HEV	$37,300	3461	NiMH	1.6	40/na	na	7.4
21	2017	Toyota	Camry	HEV	$27,625	3240	NiMH	1.6	42/na	na	7.2

(continued)

Table 5 (continued)

	Year	Brand	Model	Type	Approx. price in US$	Curb weight (lb)	Battery type	Battery (kWh)	City (MPG/MPGe)	Electric-only range (miles)	0–60 mph (s)
22	2017	Toyota	Highlander	HEV	$37,230	4134	NiMH	1.9	29/na	na	7.0
23	2017	Toyota	Mirai	FCV	$57,500	4078	NiMH	1.6	67/67	300	9
24	2017	Toyota	Prius II	HEV	$24,685	3075	NiMH	1.2	54/na	na	10.2
25	2017	Toyota	Prius c	HEV	$20,150	2530	NiMH	0.9	48/na	na	10.9
26	2017	Toyota	Prius v	HEV	$26,675	3340	NiMH	1.3	43/na	na	10.3
27	2017	Toyota	Prius Prime	PHEV	$27,965	3365	Li-ion	9	54/133	25	10.9

3 Quantification of Indicators

3.1 Emission and Energy Indicators

This section presents the assumptions for modeling emissions and energy indicators per life cycle stage. The indicators for each sustainability dimension shown in Table 3 were quantified based on the conditions described below.

Manufacturing. Manufacturing emissions and energy that are modeled in the Greenhouse Gases, Regulated Emissions, and Energy Use in Transportation (GREET) model include vehicle materials, batteries, fluids, and vehicle assembly. Specific input assumptions related to each vehicle and its components are extracted from the official specifications sheet of each vehicle (e.g., vehicle weight, battery weight, fluid weight, other material percentage by weight). The weight and battery properties of each vehicle are used as an input data in the model together with GREET's material percentage composition of each vehicle component (e.g., body, powertrain, chassis, transmission, generator). The percentages of material compositions for the EV components were calculated using the mass of each material for an EV, as these were estimated for electric compact vehicles by Argonne National Laboratory [17]. Additionally, two battery replacements are included for the ICEV, one for the HEV, and none for the EV, FCV, and PHEV [13]. Manufacturing emissions and energy inventory for transit buses are estimated by using the Economic Input–Output Life Cycle Assessment (EIO-LCA) method.

Fueling. GREET is used for the fuel life cycle ("well to wheel"). The model estimates the emissions and energy associated with primary energy production (feedstock recovery), transportation, and storage, and with fuel production (transportation, storage, and distribution). The fuel production option for conventional gasoline and low sulfur diesel is petroleum. Gaseous hydrogen is produced from natural gas via steam methane reforming at refueling stations. For electricity generation, the following mix is assumed: coal 50.4%, nuclear power 20.0%, natural gas 18.3, residual oil 1.1%, biomass 0.7%, other 9.5% (i.e., hydro, solar, wind, and geothermal).

Operation and idling. For the operation stage, MOBILE 6.2, GREET, and EIO-LCA were used to obtain estimates for all vehicle types. MOBILE 6.2 was used to estimate the emissions generated from gasoline vehicles. Urban average speeds of 28 miles per hour (mph) and 12 mph were used for passenger vehicles and buses, respectively [18, 19]. Energy consumption was estimated with GREET. Idling emissions were estimated based on the assumption that the 2.5 mph emission factors can be applied to the entire idling time [20]. The average passenger vehicle and transit bus were assumed to be idle for 7.5 and 35 min per day, respectively.

EIO-LCA estimates the materials and energy resources required for specific economic activities, and the environmental emissions resulting from these activities. Vehicle fixed costs have few supplier impacts, and therefore, the impact to environment is not going to be significant.

Maintenance. Vehicle maintenance includes the maintenance and disposal of vehicle parts. Disposal includes the dismantling process. The energy required for dismantling vehicles for disposal was estimated to be approximately 1.48 GJ per vehicle for a vehicle weighing 1360 kg (3000 lb) [13]. GREET examines the emissions and energy associated with vehicle disposal. EIO-LCA was used to estimate the emissions and energy inventory associated with automotive maintenance and the tire manufacturing services based on maintenance costs.

3.2 *Environmental Sustainability Indicators*

Emissions: The life cycle tools which are used to quantify the emission indicators were presented in Sect. 3.1.

Noise: At speeds greater than 30 mph, vehicles with advanced propulsion offer negligible noise benefits because at higher speeds, noise is generated mostly by the tire/road interaction [21] and vehicle aerodynamics.

3.3 *Technology Performance Sustainability Indicators*

Fueling frequency: Vehicle technologies with improved fuel consumption such as hybrids, fuel cell, or EVs may reduce fueling frequency and require different types of fuels and infrastructure. For comparing various technologies fairly, the fueling or charging time should be considered in the estimation of the total user time to use a vehicle over its lifetime. The time expenditure refers to the time a vehicle user spends to fuel or charge a vehicle over its lifetime. This loss of time reflects a loss of productivity. The lifetime fueling frequency per vehicle is calculated as follows:

$$\text{Fueling frequency} = \frac{\text{Lifetime miles traveled}}{\text{Fuel tank} \times \text{Vehicle fuel efficiency}} \left(\frac{\frac{\text{miles}}{\text{year}} \times \text{veh. lifetime in years}}{\text{gallons} \times \frac{\text{miles}}{\text{gallon}}} \right) \tag{1}$$

For the estimation of the time loss, it is assumed that it takes an average duration of six minutes for each user to complete the fueling procedure (i.e., to enter the fuel station, wait, fuel, pay, and leave the fuel station). In the EV case, the fuel tank is substituted by the battery. An analysis on the behavior of EV users showed that 82% of charging events are conducted at the EV users' homes using their residential EV supply equipment; the remainder of charging events occurred away from home. Nearly all locations away from home used for charging were shopping centers, health clubs and spas, bars and restaurants, office buildings and other homes;

therefore, EV users did not usually waste time to charge their vehicles [22]. Time losses for an EV user are estimated based on the assumption that 26 min are required to charge a depleted battery in order to compete the required trip to a charger at home or work [23], and this event will occur for 5% of the total charging cycles in urban driving conditions during a year. For the rest of the charging cycles, it is assumed that no time is wasted by users for charging batteries (i.e., charging occurs overnight or at destinations with charging stations).

Maintenance frequency: The maintenance frequency for parts replacement of a gasoline vehicle during its lifetime is estimated to be 22 times, and each owner is assumed to lose two hours per time for dropping off and picking up the vehicle [24]. Information on FCV, EV, and PHEV maintenance costs is limited because relatively few vehicles exist. The maintenance schedule of an HEV suggests frequency maintenance of 20 times. Reliable data are not yet available for PHEV maintenance costs; their internal combustion engine supplies only a fraction of the overall vehicle miles, maintenance intervals should be longer, and cost should be lower. PHEV maintenance frequency is assumed to be the same as an HEV. For an EV, the maintenance schedule (excluding severe operating conditions) proposes a maintenance frequency of 10 times for 144,000 miles or 10 years [23]. The FCV is the same as the EV with one extra servicing time added for the hydrogen equipment. Maintenance frequency for HEV, FCV, EV, and PHEV is 20, 11, 10, and 20 times, respectively. For transit buses, it was assumed that each one requires an average of 260 h per year for maintenance [25].

Engine power: This index is estimated as the ratio of torque to vehicle weight; using torque tends to favor EV motors and diesel engines, but also is more representative of the performance needed in an urban environment, which is the focus of this assessment.

3.4 Energy Sustainability Indicators

Quantification of life cycle energy indicators was presented in Sect. 3.1 under emission and energy indicators.

3.5 Economic Sustainability Indicators

All economic sustainability indicators are estimated for and converted to the year 2011.

Manufacturing cost: It represents the invoice price of a vehicle. The invoice price is the price a car dealer pays the manufacturer; it is constant for every dealer in the USA. For fuel cell technology, the Honda Clarity FCV is available in Southern

California, for leasing only, at a cost of $600 per month for 36 months including maintenance and insurance. Since a final retail cost is not available for a FCV, its market price, when it will be mass produced, is estimated by matching the lease cost of the FCV to a conventional vehicle with both leasing and purchasing options. The resultant MSRP is $39,000 in 2011$. For public transit buses, the invoice price was 90% of the manufacturer's suggested retail price to account for typical block orders by transit systems [10].

Operation cost: It includes operating (fueling/charging or using the vehicle); insurance, license, registration, and taxes. Fueling/charging costs are estimated by considering average US values for vehicle fuel efficiency and fuel cost. The ICEV is the base vehicle; the insurance, license, registration, and tax costs for the rest of the vehicles are estimated proportionally based on their weight [14]. Fixed costs for vehicle usage include insurance, license fees, and taxes. EIO-LCA estimates the required materials and energy resources, and the environmental emissions resulting from activities in the economy. Finance charges were assumed to have negligible impacts to the environment. Insurance cost for passenger vehicles is based on a full-coverage policy for a married 47-year-old male, commuting three to ten miles daily to work. The annual insurance cost for an ICEV is estimated to be $974 [14]. License, registration, and taxes costs include all governmental taxes and fees payable at time of purchase, as well as fees due each year to keep the vehicle licensed and registered. Vehicle annual registration, driving license, and taxes for an ICEV are estimated to be $591 [14]. Insurance, annual registration, driving license, and taxes costs for other vehicle types are extrapolated from vehicle weights.

Maintenance cost: It is the average cost for maintaining the vehicle over its lifetime. The maintenance costs of the passenger vehicles were estimated based on ICEV maintenance cost of 0.0432 $/mile (2011$). The ICEV requires a wide range of maintenance such as oil changes, filter replacements, less frequent replacements of brake parts, belts, hoses, tires and occasional component replacements such as pumps, exhaust mufflers. The main components of the EV and the PHEV that are different from an ICEV are the battery, motor, and power electronics, all of which are claimed to be maintenance free [26]. The EV's maintenance requirements are lower, and therefore, the maintenance costs are lower. Batteries used in current EVs are sealed and maintenance free. Based on the literature, it was assumed that the maintenance cost for EV is 50% less than for an ICEV [27]. The HEV embraces all the components of an ICEV, but due to its regenerative braking there is less brake wear. It is estimated that its maintenance cost is 0.0396 $/mile based on the maintenance schedule and costs for a HEV relative to an ICEV [28, 29]. The maintenance cost for a PHEV based on scheduled maintenance costs using average driving schedule and night charging is estimated to be $0.0371 per mile or 14% lower than the ICEV maintenance cost [30, 31]. FCV drive train is similar to EV with the fuel cell, hydrogen storage system, and DC converter added on. The maintenance costs of the FCVs are assumed to be equal to the EV maintenance costs when FCVs become mass produced. Due to uncertainties in expected life and costs associated with high-performance batteries, the cost to replace the battery pack has not been included. Currently, battery degradation warranties vary widely: BMW i3, Chevrolet Bolt, and Nissan Leaf with 30 KW battery come with 8-year, 100,000-mile warranty. The Nissan Leaf with the 24 KW battery comes with a 5-year, 60,000-mile warranty, and the Mitsubishi i-MiEV and all Teslas came with no warranty for battery degradation [32].

Tire cost for ICEV, HEV, FCV, EV, and PHEV is 0.0113 $/mile (2011$). The maintenance cost for transit buses was estimated to be 0.470 $/mile including tires [14]. Vehicle maintenance includes the maintenance and disposal of vehicle parts. EIO-LCA models the emissions inventory associated with automotive mechanical repair and maintenance and tire manufacturing services based on costs. Additional expenses for a replacement vehicle during the days that the vehicle remains under repair are not included.

Subsidies: For passenger vehicles, subsidy refers to the federal tax credits, and for public transit buses, it refers to taxpayer funds required to operate and maintain each vehicle [18]. Indirect costs to users in the form of property taxes to subsidize streets were excluded. We assumed that the same federal income tax credit will be applied to all alternative fuel vehicles including the FCV when they become available for purchase. Indicator values can be replaced with local data.

Parking cost: For ICEV and HEV, parking price is estimated based on the US national average [33]. For alternative fuel vehicle owners, it was assumed that free parking is offered in designated areas. Indirect costs such as city taxes used to subsidize the free stalls for alternative fuel vehicles were not included. Indicator values can be replaced with local data.

3.6 *Users Sustainability Indicators*

Global availability: It is estimated by dividing the total hours a vehicle is unavailable per year by the total number of hours in a year. The unavailable hours for vehicles were estimated by multiplying the time it is required to fuel/charge a vehicle by the fueling/charging frequency per year. We assumed that transit buses are not in operation for five hours per day (from midnight to 5 a.m.).

Reasonable availability: It is the time during which a vehicle is available to its potential users during the 19 h of a day (i.e., from midnight to 5 a.m.). It is assumed that an EV and a PHEV require 7 and 4 h, respectively, per charging cycle at 220 V home charging dock starting from a depleted battery [23, 34]. Public transit buses are assumed to be fueled upon start of service, and they do not require fueling until the end of their shift.

Comfort and convenience: It is expressed with three indicators: passenger space, cargo space, and legroom space available to each user of a vehicle. For transit buses, it was assumed that the space under the seat is the cargo space assigned to each passenger, and for passenger space, it was assumed that the internal height of buses scales from 96 in. to the front to 77 in. to the back and its width is 100 in. [35].

Fueling opportunities: It is expressed by the number of gas stations, hydrogen stations, or public electric stations [36]. This indicator is not applicable to public transit modes. Indicator values can be replaced with local data.

4 Aggregation of Indicators

The next major task is the summation of sustainability assessment from each indicator to a SI per dimension and an overall sustainability index (OSI) for each urban transportation vehicle. This is a sensitive task that must ensure that (1) the final result is understandable to decision makers and stakeholders, (2) all applicable sustainability indicators have been included, and (3) biases are minimized and explicitly reported. The sustainability assessment tool estimates and compares the sustainability performance of each vehicle. It can aggregate the results based on vehicle population per transportation project to provide project, area, corridor, or citywide estimates and support a decision making process.

Although the range of the indicator values is not known and it will vary as vehicle technologies and specifications evolve, their definitions provide clear guidance of the direction toward sustainability. The range of desirable values and/or thresholds for the proposed set of sustainability indicators may be determined by experts, laws, or standards [37] as sustainability assessments become more standardized and routine.

The ultimate sustainability framework requirement for urban transportation vehicles would be to produce minimal pollutants. However, the assessments must use reasonable prevailing averages, as follows. Based on national average driving conditions, the emission rates or "emission indicators" were estimated to be for CO_2 (368.4 g/mi), VOC (1.034 g/mi), CO (9.4 g/mi), NO_x (0.693 g/mi), and PM_{10} (0.0044 g/mi) [38]. The total CO_2 emission rates (i.e., including tailpipe emissions and the emissions associated with the production and distribution of gasoline) for the average gasoline car are estimated to be 480 g/mi [39]. The EPA GHG emission standards under the Clean Air Act are projected to result in an average passenger vehicle tailpipe CO_2 level of 212 and 143 g/mile in 2017 and 2025, respectively [40]. EVs have no tailpipe emissions; however, emissions are created during both the production and distribution of the electricity used to fuel the vehicle, which vary based on regional data. For example, while the national average total CO_2 emission rate (i.e., including production and distribution of electricity) for an EV with a 24 kWh battery pack is 190 g per mile, based on Hawaii and California data, the total CO_2 emission rates change to 260 and 120 g per mile, respectively [39].

4.1 Normalization and Vehicle Sustainability Indices

The proposed sustainability indicators have a positive (+) or a negative (−) impact. The larger the absolute value of the indicator, the more positive or negative impact it has on sustainability. Negative impact is assigned to economy indicators based on the objective of offering affordable transportation choices for all users. These impacts may change when the objective is to reduce vehicle utilization. For example, a positive impact may be assigned to the "parking price" indicator to represent that more expensive parking may reduce city pollution by incentivizing the use of mass transit.

Indicators are expressed in different units. Addition among indicators with different units is performed only after the different measurement units are normalized into a dimensionless scale. Each sustainability indicator is normalized by using Eqs. 2 and 3 [41].

$$N_{ij}^{+} = \frac{I_{ij}^{+} - I_{\min,j}^{+}}{I_{\max,j}^{+} - I_{\min,j}^{+}} \tag{2}$$

$$N_{ij}^{-} = \frac{I_{\min,j}^{-} - I_{ij}^{-}}{I_{\min,j}^{-} - I_{\max,j}^{-}} \tag{3}$$

where N_{ij}^{+} is the normalized indicator with positive impact achieved by the ith alternative with respect to the jth indicator of sustainability. I_{ij}^{+} is the indicator value achieved by the ith alternative when evaluated based on the jth indicator, $I_{\min,j}^{+}$ is the indicator with the worst value achieved by the jth indicator of sustainability, and $I_{\max,j}^{+}$ is the best value of jth indicator of sustainability obtained. The normalization process used herein estimates the sustainability performance of each vehicle relative to the best vehicle type considered. After a set has been developed, the consideration of a vehicle with better performance or of policies for the use of existing modes (e.g., pricing) will change the range of values. The normalization equations need to be applied to the modified full set, which, in turn, may change the final outcome.

The normalized values are dimensionless and range from 0 to 1; therefore, the greater the absolute value of the normalized indicator, the more sustainable it is. Hence, on a relative scale, the most sustainable vector for each vehicle type is $I_{\max} = (1, 1, 1, 1, 1)$ and the least sustainable vector is $I_{\min} = (0, 0, 0, 0, 0)$.

Problems that involve multiple criteria and alternatives, such as the ones associated with sustainable transportation, are defined as multi-criteria decision making (MCDM) problems. Several methodologies exist and have been adopted in the transportation sector for assessing projects and plans. MCDM methods assume that indicators are well defined, certain, and independent. The analytic hierarchy process (AHP) has been used in several sustainability and transportation studies. Paez and Currie [42] adopted the AHP method to develop an Integrated Transport Plan in Melbourne by considering six criteria and 45 values for each criterion, resulting in a 270 pairwise decisions. The AHP which is one of the most popular MCDM methods [43] is time- and cost-intensive.

Other methodologies, such as the Bayesian decision theory or fuzzy logic, can be used to account for nonlinear, interrelated, and stochastic aspects of transportation. However, the development of a fuzzy model becomes more complex as the number of indicators increases [44].

MCDM methods add a degree of subjectivity in the analysis since weights indicate relative preferences associated with specific indicators. Utilization of equal weights to minimize bias is less challenging and may be a preferred base assessment. Determination of weights is not the main objective of this study; thus, equal weights were assigned to each indicator and sustainability dimension. Zheng et al. [45] state that utilization of equal weights is a common practice for developing composite indices. Equal weights have been used in various past sustainability assessments; they indicate the same relative importance among sustainability indicators [8, 45, 46]. A Delphi forecast, which utilizes a panel of experts and their judgments, or a Pareto analysis in order to propose a wider set of final evaluations, may be more appropriate methodologies for the real-world application of the proposed sustainability assessment for specific regions or projects.

In this chapter, aggregation of normalized indicators into a SI per dimension and an OSI per vehicle type was performed by using the weighted sum method [47]. The weighted sum model (WSM) was used to aggregate sustainability indicators. The assumption that governs this model is the additive utility assumption (i.e., the total value of each alternative equals to the sum of products). Addition among criteria with different units is performed only after the different measurement units are normalized into a dimensionless scale, and the utility V^i for each alternative is estimated as follows:

$$V_i = \sum_{j=1}^{n} w_j N_{ij} \quad i = 1, \ldots, m \tag{4}$$

where w_j is the assigned weight for each indicator j for alternative I and N_{ij} is the normalized value of indicator j for alternative i.

5 Results and Discussion

Quantified indicator values for each vehicle type are shown in Table 6. The SI per dimension and the OSI per vehicle type summarize the sustainability performance for the seven vehicle types. The outcomes provide a comparison for estimating the total impact of any fleet mix scenario containing these seven vehicle types. The indicator values are weighted per passenger-mile traveled (PMT), and based on the OSI, the most sustainable transportation vehicle is found to be the HDEB. It achieved a sustainability level of 68%. The DB was ranked second. It achieved sustainability of 63 or 92% of HDEB's value. For passenger vehicles, the FCV, the HEV, the EV, the PHEV, and the ICEV achieved sustainability of 56, 56, 51, 43,

Table 6 Quantified sustainability indicators and relative sustainability indices

	Indicators	Code	Units	ICEV	HEV	FCV	EV	PHEV	DB	HDEB
Environment[a]	CO_2 (w/C in VOC and CO)[b]	–	g/PMT	515.8	275.7	241.8	325.4	447.7	326.0	258.7
	CH_4	–	g/PMT	0.713	0.430	0.742	0.523	0.515	0.323	0.286
	N_2O	–	g/PMT	0.018	0.015	0.003	0.006	0.016	0.003	0.002
	GHGs	–	g/PMT	538.7	290.7	261.1	340.0	465.1	338.3	270.6
	VOC	–	g/PMT	0.893	0.804	0.070	0.073	0.791	0.214	0.198
	CO	–	g/PMT	6.626	6.577	0.414	0.429	6.523	0.877	0.825
	NO_x	–	g/PMT	0.849	0.727	0.242	0.394	0.779	1.012	1.005
	PM_{10}	–	g/PMT	0.171	0.143	0.171	0.457	0.258	0.066	0.069
	SO_x	–	g/PMT	0.310	0.331	0.356	0.915	0.541	0.157	0.130
	Average noise level	–	dB	61.0	57.0	57.0	57.0	57.0	78.0	75.0
	Environment sustainability index			0.259	0.568	0.839	0.646	0.357	0.704	0.783
Technology performance	Fuel frequency	–	min/PMT	0.012	0.009	0.021	0.011	0.014	NA	NA
	Maintenance freq.	–	min/PMT	0.018	0.017	0.008	0.008	0.017	0.003	0.003
	Space occupied	–	Square meters (m^2)/passenger	7.27	6.46	7.42	6.53	6.42	3.01	3.08
	Engine power	+	lb.ft./lb	0.050	0.050	0.053	0.059	0.072	0.035	0.023
	Technology performance sustainability index			0.262	0.375	0.250	0.472	0.388	0.812	0.763
Energy[c]	Manufacturing energy	–	MJ/PMT	0.592	0.604	0.684	0.681	0.633	0.335	0.391
	Fueling energy	–	MJ/PMT	1.202	0.526	1.204	1.886	1.122	0.478	0.409
	Operation energy	–	MJ/PMT	4.680	2.379	1.751	1.372	3.313	3.599	2.579
	Maintenance energy	–	MJ/PMT	0.261	0.250	0.186	0.186	0.246	0.176	0.170

(continued)

Table 6 (continued)

	Indicators	Code	Units	ICEV	HEV	FCV	EV	PHEV	DB	HDEB
	Energy sustainability index			0.182	0.491	0.541	0.456	0.309	0.803	0.869
Economy[2]	Manufacturing cost	–	$/PMT	0.140	0.150	0.315	0.247	0.261	0.055	0.092
	Operate (user costs)	–	$/PMT	0.246	0.166	0.235	0.123	0.209	0.641	0.061
	Maintenance cost	–	$/PMT	0.045	0.042	0.027	0.027	0.043	0.045	0.043
	Any form of subsidy	–	$/PMT	0.000	0.000	0.052	0.052	0.052	0.023	0.023
	Parking cost	–	$/month/passenger	180.5	180.5	0.0	0.0	0.0	0.0	0.0
	Economy sustainability index			0.471	0.523	0.540	0.631	0.411	0.516	0.709
Users	Global availability	–	Hours of down time or not operable per year expressed as an annual % based on 24 h	0.03%	0.02%	0.05%	9.61%	1.45%	20.83%	20.83%
	Reasonable availability	–	Hours of down time or not operable per year expressed as an annual % based on 19 h	0.04%	0.03%	0.07%	3.47%	0.04%	0.00%	0.00%
	Passenger space	+	cu.ft/passenger	25.4	23.4	25.2	18.4	23.0	19.9	18.0
	Goods carrying (cargo) space	+	cu.ft/passenger	3.75	5.40	3.28	2.44	2.65	1.85	1.85
	Legroom front	+	Inches	41.7	42.5	41.9	42.1	42.0	27.0	27.0
	Fuel frequency	+	Number of stations in operation	121,446	121,446	58	626	121,446	121,446	NA
	Users sustainability index			0.783	0.820	0.620	0.367	0.686	0.322	0.286
	Overall sustainability index			39.1	55.5	55.8	51.4	43.0	63.1	68.2

Notes [a]Environment and energy indicators are not fixed but depend on project-specific or regional inputs of vehicle average lifetime, annual miles traveled, weight, and speed

[b]The carbon fraction in VOC and CO is considered in the total CO_2 emissions because carbon in VOC and CO will eventually be converted to CO_2 with further atmospheric chemical reactions (oxidation)

[c]All costs are converted in 2011$. All economy indicators are assumed to have negative impact to sustainability. Indicators are perceived from users' point of view; therefore, they reveal how vehicle monetary parameters may affect vehicle utilization and make sustainable or unsustainable transportation vehicle for a chosen network

and 39, respectively. Clearly, vehicles with modern propulsion systems that depend exclusively or partially on electric drive did better than the traditional internal combustion-powered vehicle.

Technology performance and economy have comparable results among all urban vehicles, which reveal that if a user or an agency is not particularly sensitive to issues related to environment and energy, then they would tend to choose a vehicle that satisfies mostly other needs. The results for users show that the vehicles that satisfy transportation user needs the most are the HEV and the ICEV largely due to their low purchase cost and convenient refueling compared to alternative fuel vehicles. Starting in 2011, the Corporate Average Fuel Efficiency (CAFE) standards will affect high fuel consumption vehicles the most. For example, the CAFE standard for the ICEV used in our analysis will increase from 27.3 MPG in 2011 to 54.5 MPG in 2025 [48]. This will lower operating costs and make ICEV and HEV competitive for several decades.

The FCV has the highest index for environment with a significant difference from all other passenger cars, and it surpasses DB and HDEB which are public vehicles. The FCV's results for environment show that utilization of alternative fuel or hybrid technologies combined with policies that increase vehicle occupancy has the potential to improve passenger vehicle sustainability performance faster. For example, if occupancy was increased to 2 for all passenger vehicle types, e.g., with extensive use of high-occupancy lanes, then the HEV would have the highest OSI of all vehicles, while DB would be sixth from the top.

Energy requirements for alternative fuel vehicle and hybrid operation are significantly lower due to improved fuel efficiencies relative to gasoline-powered vehicles. The FCV has the highest index for energy among passenger vehicles. The main reasons for this result are the low maintenance and operation energy requirements compared to vehicles that include an internal combustion engine.

The EV has the highest economic SI among passenger vehicles, and it exceeds the index for DB. The EV performs well for economy due to its low maintenance and fuel costs. In technology performance, the EV again has the highest index due to low maintenance frequency, engine performance, and long lifetime of batteries. Although EV performs relatively well, its OSI is reduced by the indicator "reasonable availability," which contributes significantly to its final ranking. Improvements in range performance and speed of charging will likely make EVs more reliable and thus more competitive.

Emission intensity and sources differ for each vehicle type. For example, while ICEV produces more CO_2 emissions during its operation, EV produces more CO_2 during the production of its fuel. Policy formulation for treating impacts related to emissions should be based on the number and intensity of emission sources for each region. Accounting for upstream emissions and energy requirements for alternative fuel vehicles, including the FCV, the PHEV, and the EV, is important. Without it, assessment of operation-only vehicle emissions and consumption create strong biases against ICEV, HEV, and DB. The upstream emissions and energy

Table 7 Vehicle rankings for different passenger vehicle occupancies

	ICEV	HEV	FCV	EV	PHEV	DB	HDEV
Ranking VMT	5	1	2	4	3	7	6
Ranking PMT (pax:average)	7	4	3	5	6	2	1
Ranking PMT (pax:2)	7	1	2	3	5	6	4
Ranking PMT (pax:3)	5	1	2	4	3	7	6

requirements for fuel production depend on the electricity mix used at each community. Therefore, they are likely to vary significantly for different geographical areas.

Results were weighted per PMT instead on a mile basis to relax differences between passenger and public transit. When results are weighted per vehicle mile traveled (VMT), the two types of buses in the assessment are ranked last (Table 7). Vehicle occupancy plays a critical role in estimating results per PMT. Table 7 shows how ranking between vehicles shifts when a different number of passengers (pax) are considered. It should be noted that for this assessment, HEV ranks first between passenger vehicles when an average vehicle occupancy ratio of 1.4 or higher is considered.

6 Conclusion

Transportation policies should be adjusted to promote a more sustainable transportation system in the short term and long term. A dynamic sustainability framework that can be updated with the latest data of a transportation system (i.e., fuel types, insurance, fees and taxes, vehicle weight, fuel efficiency, vehicle mileage, battery capacity, etc.) and adjusted based on local policies (i.e., fuel costs, parking pricing charges, etc.) is necessary to support transportation policy and planning.

The sustainability framework that was used in this study for the assessment of urban transportation vehicles provides a holistic approach as opposed to assessments based solely on the operational stage of vehicles. The indicators used herein were organized into five sustainability dimensions (environment, technology performance, energy, economy, and users) that do not cross sustainability dimension boundaries which occur in the traditional triple bottom-line approach (environment, society, economy).

All indicators are comparable as per PMT measures, which allow us to avoid the bias occurring from different vehicle sizes or classes. Buses are typically viewed as more sustainable compared to other passenger vehicles. This is based on their emission and energy life cycle outputs vis-a-vis their large passenger capacity which lowers their per-passenger numbers; thus, buses were used in this study as a basis for assessing the sustainability of seven light-duty vehicles. This assessment enables us to understand where passenger vehicles stand relative to public transit and vice versa, and explore their weaknesses and privileges. Sustainability assessment of one vehicle class should be performed exclusively when decision making is based on a set of requirements and/or restrictions. For example, determine which vehicle is the "best" for a fixed corridor, for a social excluded group, or for a specific route which has a narrow or a curved road.

In the short term, there are no barriers to be overcome to increase the penetration of HEV in the market, which was found to have the second best sustainability performance. Such barriers include infrastructure, maintenance and repair shops, and overall awareness and familiarity with alternative fuel systems. In the long term, electric drive and FCVs show that they have the potential to reduce environmental impact. For many reasons, automobile ownership per household in fully developed countries is likely to decrease due to aging and lower birth rates, the fading of the baby boomer effect in the USA, restrictions of certain types of automobiles (for pollution or congestion reasons) and transportation-as-a-service (TaaS) with driver-operated (e.g., Lyft, Uber) and autonomous taxis [49]. The question remains whether EVs will increasingly become the replacement of fossil fuel-powered light-duty vehicles.

EVs and HEVs are becoming more capable and are particularly suitable for polluted environments. They have the potential to become the dominant type of light-duty passenger service vehicle in large urban areas. The time that is required for this shift depends primarily on their purchase price, and regulations favoring them. As long as manufacturers of gasoline-powered vehicles offer substantial performance and technological and safety content for $25,000 or less (approx. 2017 US $), EVs with the same range and content priced above $35,000 (as is the case currently) will have difficulty in selling a large number of units. Meanwhile, light-duty vehicles costing above $15,000 are too expensive for most of the developing world. Therefore, EV and HEV are less likely to be popular in the developing world. Electric buses and light delivery vehicles (i.e., scooters and tricycles) are far more likely to be operated in large numbers in the large cities of developing countries (New Delhi, Dhaka, Ho Chi Minh City, Kathmandu, etc.) as currently observed in Beijing and Shanghai, China.

Combination of transportation policies and alternative fuel vehicles should support the creation of a sustainable transportation system. Policies should be focused on improving the fuel efficiency of vehicles and on lowering the purchase price of HEV while vehicle technologies advance.

Technology, fuel pricing and availability, and user habits are certain to change in the decades ahead. The sustainability assessment method is flexible, detailed, and manageable. It has the potential to enable comparisons by mode type (e.g., private vehicle with technology X vs. bus), by broad class (e.g., motorized vs. non-motorized vehicles), by system (e.g., BRT vs. light rail), by corridor (e.g., HOT lanes vs. mass transit), and by area (comparisons of sections in the same city, or comparisons among cities or metro areas). The results are both technology- and policy-sensitive and thus useful for both short- and long-term planning.

References

1. U.S. Department of Energy, The history of the electric car (2014), https://energy.gov/articles/history-electric-car. Accessed on 15 Aug 2017
2. Natural Gas Vehicle Knowledge Base, Current natural gas vehicle statistics (2017), http://www.iangv.org/current-ngv-stats/. Accessed 15 Aug 2017
3. Inside EVs (2017), http://insideevs.com/monthly-plug-in-sales-scorecard/. Accessed 15 Sept 2014
4. California Air Resources Board (2017), California's hydrogen transportation initiatives. https://www.arb.ca.gov/msprog/zevprog/hydrogen/hydrogen.htm. Accessed 15 Sept 2017
5. European Logistic Portal, Hydrogen fuel for cell electric cars on the rise in Germany (2017), http://www.eurologport.eu/hydrogen-fuel-for-cell-electric-cars-on-the-rise-in-germany/. Accessed 23 June 2017
6. Forbes, Japan's big carmakers gang up in support of hydrogen fuel cell vehicles, at least officially (2017), https://www.forbes.com/sites/bertelschmitt/2017/05/19/japans-big-carmakers-gang-up-in-support-of-hydrogen-at-least-officially/#731146c21a9d. Accessed 23 June 2017
7. IEEE Spectrum, Why the automotive future will be dominated by fuel cells (2016), https://spectrum.ieee.org/green-tech/fuel-cells/why-the-automotive-future-will-be-dominated-by-fuel-cells. Accessed 09 Mar 2017
8. L.K. Mitropoulos, P.D. Prevedouros, Trans. Res. Rec. **2344**, 88 (2013)
9. Bureau of Transportation Statistics, National household travel survey (2002), https://www.rita.dot.gov/bts/sites/rita.dot.gov.bts/files/subject_areas/national_household_travel_survey/index.html. Accessed 28 Nov 2012
10. Edmunds Official Website (2012), http://www.edmunds.com. Accessed 25 Dec 15
11. S.C. Davis, S.W. Diegel, R.G. Boundy, *Transportation Energy Data Book*, 30th edn. (U.S. Department of Energy, Oak Ridge National Laboratory, Oak Ridge, 2011)
12. Consumer Leasing Guide (2002), http://leaseguide.com/index2.htm. Accessed 05 Dec 2010
13. A. Burnham, M. Wang, Y. Wu, *Development and Applications of GREET 2.7* (Argonne National Laboratory, Argonne, 2006)
14. American Automobile Association, Your driving fixed costs (2012), http://www.aaa.com. Accessed 10 Mar 2012
15. U.S. Department of Energy, Alternative and advanced fuels—energy efficiency and renewable energy (2011), http://afdc.energy.gov/afdc/fuels/electricity_locations.html. Accessed 30 Dec 2012
16. J.M. Ogden, R.H. Williams, E.D. Larson, Energy Policy **32**, 7 (2004)
17. Center for Transportation Research, *Total Energy Cycle Assessment of Electric and Conventional Vehicles: An Energy and Environmental Analysis*, vol. 1 (Argonne National Laboratory, Argonne, 1998)
18. American Public Transportation Association, *Public Transportation Fact Book*, 60th edn. (2009)
19. Texas Transportation Institute, Urban Mobility Report (2009), http://tti.tamu.edu/. Accessed 15 June 2012
20. U.S. Environmental Protection Agency, User's guide to MOBILE6.1 and MOBILE6.2 mobile source emission factor model (2003)
21. G.G. Fleming, R.E. Armstrong, E. Stusnick, K. Polcak, W. Lindeman, Transportation-related noise in United States. Transp. New Millenn. Transp. Res. Board (2000)
22. J. Smart, S. Schey, SAE International **1**, 27 (2012)
23. Nissan (2014), http://www.nissanusa.com/electric-cars/leaf. Accessed Feb 2014
24. Toyota (2014), http://www.toyota.com. Accessed 10 Feb 2014
25. K. Chandler, N. Walkowicz, *King County Metro Transit Hybrid Articulated Buses: Final Evaluation Results* (National Renewable Energy Laboratory, Golden, 2006)
26. G. Duleep, H. Essen, B. Kampman, M. Grünig, Impacts of electric vehicles—Deliverable 2. Assessment of electric vehicle and battery technology. Delft, CE Delft (2011)

27. M.A. Delucchi, A. Burke, T. Lipman, M. Miller, Electric and gasoline vehicle lifecycle cost and energy-use model, report for the California Air Resources Board. USD-ITS-RR-99-04 (2000)
28. M. Duvall, *Comparing the Benefits and Impacts of Hybrid Vehicle Options for Compact Sedan and Sport Utility Vehicles* (EPRI, Palo Alto, 2002)
29. Toyota Service (2013), http://smg.toyotapartsandservice.com. Accessed 10 Sept 2013
30. M. Duvall, L. Browning, F. Kalhammer, W. Warf, D. Taylor, M. Wehrey, N. Pinsky, *Advanced Batteries for Electric-Drive Vehicles: A Technology and Cost-Effectiveness Assessment for Battery Electric Vehicles, Power Assist Hybrid Electric Vehicles, and Plug-In Hybrid Electric Vehicles* (Electric Power Research Institute (EPRI), Palo Alto, 2004)
31. B.M. Al-Alawi, T.H. Bradley, Appl. Energy **103**, 488 (2013)
32. Green Car Reports, Electric car battery warranties compared (2016), http://www.greencarreports.com/news/1107864_electric-car-battery-warranties-compared. Accessed 23 June 2017
33. Colliers International, *Parking Rate Survey* (North America, Central Business District, 2010), http://downtownhouston.org/site_media/uploads/attachments/2010-0716/CollierslInternational_ParkingRateSurvey2010.pdf. Accessed 15 Jan 2011
34. Chevrolet, Official website (2014), http://www.chevrolet.com/. Accessed 15 Sept 2014
35. S.L. Zimmerman, H. Levinson, J. Public Transp. **7**, 83 (2004)
36. U.S. Census Bureau, Industry statistics sampler—Gasoline stations (2002), http://census.gov/econ/census02/data/industry/E4471.htm. Accessed 30 Dec 2010
37. Y.A., Phillis, E. Grigoroudis, V.S. Kouikoglou, Ecol. Econ. **70**, 542 (2011)
38. Environmental Protection Agency, Average annual emissions and fuel consumption for gasoline-fueled passenger cars and light trucks (2008), http://www3.epa.gov/otaq/consumer/420f08024.pdf. Accessed 25 Dec 2015
39. U.S. Department of Energy, *Greenhouse Gas Emissions for Electric and Plug-in Hybrid Electric Vehicles* (Energy Efficiency and Renewably Energy 2015). https://www.fueleconomy.gov. Accessed 25 Dec 2015
40. Environmental Protection Agency, EPA and NHTSA set standards to reduce greenhouse gases and improve fuel economy for model years 2017–2025 cars and light trucks (2012), http://www3.epa.gov/otaq/climate/documents/420f12051.pdf. Accessed 25 Dec 2015
41. D. Krajnc, P. Glavic, Conserv. Recycl. **43**, 189 (2005)
42. D. Paez, G. Currie, Improving transport planning decision making—Adapting the analytical hierarchy approach to large number of options. In *Transportation Research Board Annual Conference*, Washington D.C., 2008
43. S.D., Pohekar, M. Ramachandran, Renew. Sustain. Energy Rev. **8**, 365 (2009)
44. L.K. Mitropoulos, P.D. Prevedouros, J. Urban Plan. Dev. **142**, (2016). https://doi.org/10.1061/(asce)up.1943-5444.0000336
45. J. Zheng, C. Atkinson-Palombo, C. McCahill, R. O'Hara, N.W. Garrick, Quantifying the economic domain of transportation sustainability. In *Transportation Research Board 90th Annual Meeting*, Washington, D.C., 2011
46. C.M. Jeon, A. Amekudzi, R. Guensler. Sustainability assessment at the transportation planning level: performances and measures and indexes. In *Transportation Research Board Annual Conference*, Washington, D.C., 2008
47. K.P. Yoon, C.L. Hwang, *Multiple Attribute Decision Making—An Introduction* (Sage University Paper, London, 1995)
48. National Highway Traffic Safety Administration, Model year light-duty vehicle GHG emissions and CAFE standards: supplemental notice of intent. Washington, D.C. (2010)
49. RethinkX, Rethinking transportation (2017), https://www.rethinkx.com/executive-summary/. Accessed 10 Sept 2017

Increasing the Fuel Economy of Connected and Autonomous Lithium-Ion Electrified Vehicles

Zachary D. Asher, David A. Trinko and Thomas H. Bradley

Abstract When the sensors and signals that enable connected and autonomous vehicle (CAV) technology are combined with vehicle electrification, new vehicle control strategies that improve fuel economy (FE) are possible through perception, planning, and a control request issued to the vehicle plant. In this chapter, each CAV technology that could contribute to planning is introduced and discussed. Next, the techniques for modeling and validating a vehicle plant and running controller are discussed. Then, three planning-based control strategies are developed: (1) an Optimal Energy Management Strategy (Optimal EMS), (2) Eco-Driving strategies, and (3) an Optimal EMS combined with Eco-Driving strategies. Each of these planning-based control strategies is evaluated using a validated model of a 2010 Toyota Prius in Autonomie so that engine power, battery state of charge, and FE results can be compared. The results indicate that a 40% + FE improvement is possible when an Optimal EMS is combined with Eco-Driving for city drive cycles. Overall, as more vehicles incorporate CAV technologies and electrification, these FE improvements will be easier to achieve and will have a greater impact on transportation sustainability.

1 Introduction

Modern vehicles are incorporating electrification to evolve from conventional vehicles (CVs) to hybrid electric vehicles (HEVs), plug-in hybrid electric vehicles (PHEVs), and fully electric vehicles (EVs) [1, 2]. At the same time, rapid advances

Z. D. Asher · D. A. Trinko · T. H. Bradley (✉)
Department of Mechanical Engineering, Colorado State University,
Fort Collins, CO 80524, USA
e-mail: Thomas.Bradley@ColoState.edu

Z. D. Asher
e-mail: asherzd@gmail.com

D. A. Trinko
e-mail: david.trinko@gmail.com

G. Pistoia and B. Liaw (eds.), *Behaviour of Lithium-Ion Batteries in Electric Vehicles*, Green Energy and Technology,
https://doi.org/10.1007/978-3-319-69950-9_6

in computational technology have provided vehicles with the ability to perceive their environment by use of sensor technologies and computer systems [3]. Combined, these two trends provide new possibilities for improving fuel economy (FE) from vehicle control.

Connected and Autonomous Vehicle Technology

Connected and autonomous vehicle (CAV) technology can be realized using sensors and signals currently available, but can be made more efficient and reliable using near-future sensors and signals. Currently available CAV technology includes camera systems (CS), radio detection and ranging (RaDAR), light detection and ranging (LiDAR), the global positioning system (GPS), and drive cycle databases. Near-future CAV technology includes an advanced global navigation satellite systems (GNSS), vehicle-to-vehicle communication (V2V), vehicle-to-infrastructure communication (V2I), and vehicle-to-everything communication (V2X). Each of these sensors and signals is discussed in Sect. 2 and each contributes to the new trend of vehicle sensing which can be leveraged in electrified vehicles to improve FE.

Lithium-Ion Vehicle Electrification

Vehicle electrification has facilitated new vehicle configurations, architectures, and control strategies. Additionally, lithium-ion vehicle batteries have provided improved battery capacity allowing increased vehicle powertrain operational freedom [4]. Hybrid vehicles such as HEVs and PHEVs provide the most freedom in vehicle powertrain control due to the two sources of propulsive power that can be used.

New Possibilities for Improving Fuel Economy

When CAV technology is combined with lithium-ion vehicle electrification, new FE improvement control strategies are possible. CAV technology enables vehicle environment perception, also known as a worldview, and consequently a prediction of future vehicle operation. This vehicle operation prediction can be leveraged by a planning algorithm to compute a vehicle control strategy to improve FE. This control strategy is issued as a request to the vehicle running controller, which implements the control request without violating operational constraints in the vehicle plant. The energy consumption from the vehicle can then be measured. A systems-level viewpoint of this process is shown in Fig. 1.

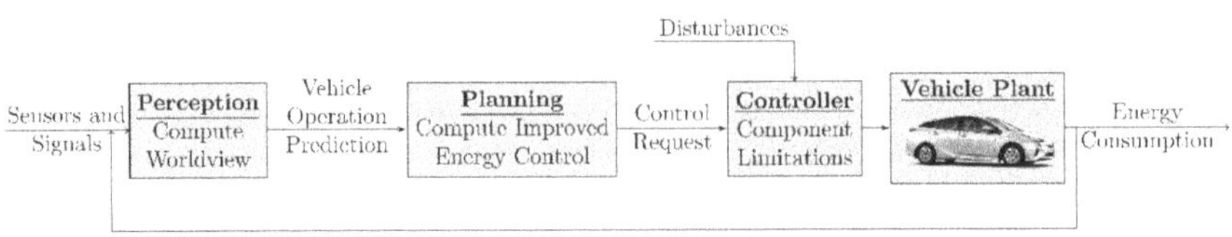

Fig. 1 A systems-level viewpoint presenting the subsystems required for advanced control strategies enabled by connected and autonomous technology

Improved vehicle powertrain operation, referred to as an Optimal Energy Management Strategy (Optimal EMS), is achieved by increasing the efficiency of the vehicle powertrain without modification of the drive cycle. Improved vehicle operation, referred to as Eco-Driving, is achieved by decreasing the energy output of the vehicle through modification of the drive cycle. Both of these strategies are enabled by CAV technologies, and their FE improvements are significant due to the powertrain flexibility of electrified vehicles.

2 Perception Enabled by Connected and Autonomous Vehicle Technology

CAV technologies such as CS, RaDAR, LiDAR, GPS, GNSS, drive cycle databases, V2V, V2I, and V2X enable perception of the vehicle environment, or worldview, which can be leveraged to predict vehicle routes, vehicle speeds, energy use, and driver behavior, thus enabling the FE improvement techniques of an Optimal EMS and Eco-Driving.

Currently available CAV technology that could be used to implement an Optimal EMS and Eco-Driving is shown in Fig. 2 and discussed in Table 1. Near-future CAV technology that could be used to implement an Optimal EMS and Eco-Driving is shown in Fig. 3 and discussed in Table 2. Utilization of only currently available CAV technology to improve FE through an Optimal EMS and/or Eco-Driving is a frequently debated topic in the literature, and ongoing efforts are an active subject of research [5]. But, the literature is consistent in the view that when near-future CAV technology is available, significant FE improvements through an Optimal EMS and Eco-Driving will be feasible. Note that the Optimal EMS and Eco-Driving FE impacts shown in Tables 1 and 2 are highly dependent on the vehicle type, architecture, and the drive cycle.

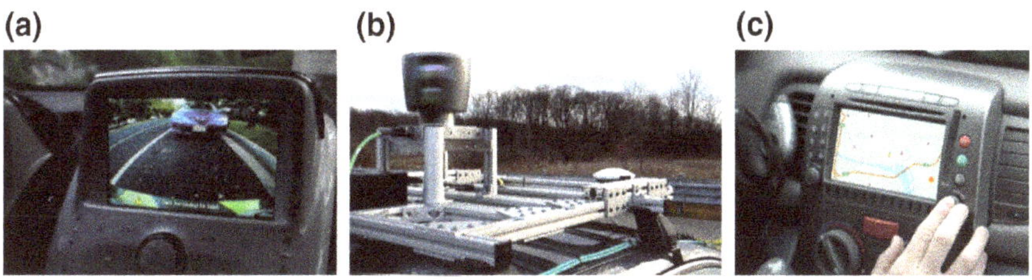

Fig. 2 Vehicle implementation of currently available CAV technologies: **a** CS, **b** LiDAR [6], and **c** GPS [7]

Table 1 Currently available CAV technologies and their potential usage with Eco-Driving and Optimal EMS FE improvement techniques

CAV technology	Eco-Driving impacts	Optimal EMS impacts	FE impact
Camera systems (CS)	Localized velocity modification helps enable adaptive cruise control	Localized prediction of future velocity through sign recognition	Small FE improvements from short predictions
Radio detection and ranging (RaDAR)	Localized velocity modification fully enables adaptive cruise control	Localized prediction of future velocity through object recognition	Small FE improvements on the highway
Light detection and ranging (LiDAR)	Higher accuracy localized velocity modification which could enable lane switching	Higher accuracy localized prediction of future velocity through object recognition	Better FE improvements on the highway
Global positioning system (GPS)	Velocity modification to coincide with speed limits along the route	Basic prediction of the full drive cycle using stop light and speed limit information	Prediction accuracy-dependent FE improvements along an entire route
Drive cycle database	Velocity modifications in historically costly sections of the drive cycle	Route length velocity predictions that improve with repeated trips	Prediction accuracy-dependent FE improvements along an entire route

Fig. 3 Conceptual examples of near-future CAV technologies: **a** V2V, **b** V2I, and **c** V2X

2.1 Camera Systems

CS were one of the first steps taken to increase vehicle environment awareness, enabling monitoring of other, less aware vehicles. CS can interpret immediate vehicle surroundings to provide a localized prediction of vehicle speed. They can recognize other vehicle locations [8], obstructions [9], traffic signs [10], and traffic signals [11], thus determining the driving vehicle's likely speeds in the next few seconds. This information can be used to increase FE through Eco-Driving [12, 13] and using an Optimal EMS. CS are currently in use by modern vehicles most commonly for backup assistance (shown in Fig. 2a), collision safety [14–17], and adaptive cruise control [18] which becomes more accurate and reliable when combined with RaDAR or LiDAR (as shown in Fig. 2b) sensors as discussed in Sect. 2.2.

Table 2 Near-future CAV technologies and their potential usage with Eco-Driving and Optimal EMS FE improvement techniques

CAV Technology	Eco-Driving impacts	Optimal EMS impacts	FE impact
Global navigation satellite systems (GNSS)	Velocity modification to coincide with speed limits along the route	Route length velocity predictions that improve with repeated trips	Prediction accuracy-dependent FE improvements along an entire route
Vehicle-to-vehicle comm. (V2V)	Opens numerous driving velocity modifications and enables cooperative adaptive cruise control	High accuracy of future velocity prediction along a busy road	Large FE improvements along busy roads
Vehicle-to-infrastructure comm. (V2I)	Enables velocity modifications along an entire route that coordinate with traffic signals	High accuracy of future velocity prediction near traffic lights	Large FE improvements near traffic lights
Vehicle-to-everything comm. (V2X)	Enables full velocity modification along an entire route while accounting for all drive cycle obstructions	High accuracy of future velocity prediction along the full route	Enables absolute optimal FE

2.2 *Radio/Light Detection and Ranging*

RaDAR is an inexpensive means of determining additional vehicle environment information and monitoring other, less aware vehicles. RaDAR and LiDAR provide similar information to CS about vehicle surroundings, but they interpret the vehicle surroundings differently. Comparing the transmitted and received radio waves (RaDAR) or light waves (LiDAR) provides advantages such as good performance in low visibility and disadvantages such as the inability to interpret street signs [19, 20]. The most robust localized prediction of vehicle driving speed can be obtained with the combination of RaDAR/LiDAR and CS [21–23], leading to improved FE increases from Eco-Driving [24, 25] and adaptive cruise control [26, 27].

2.3 *Global Navigation Satellite Systems*

In order to expand beyond immediate vehicle surroundings, information from GNSS can be used. GNSS technologies allow advanced knowledge of the vehicle route (and thus the speed limits along the route) and the current location of the vehicle along the route. This information can be leveraged to determine vehicle velocities that meet the speed limit but improve FE through Eco-Driving [28, 29]. Even though GNSS cannot predict traffic lights and sudden accidents, GNSS provides sufficient information for Optimal EMS FE improvements [30, 31].

The United States' GPS is an example of a GNSS and has already been successfully integrated into modern vehicles, as shown in Fig. 2c, for route calculations and traffic warnings. Current GPS technology requires improvements to identify vehicle orientation, velocity, and position in all environments [32–34], but upcoming GNSS technologies allow improved frequency and accuracy in determining these vehicle parameters [35–37].

2.4 Drive Cycle Databases

A database of previous drive cycles can improve autonomous navigation [38] and is particularly valuable when implementing an Optimal EMS. Polling a drive cycle database for the same drive cycle is unique because it provides a detailed velocity prediction for the entire drive cycle. But, for previously undriven drive cycles, there is no drive cycle database to poll and thus alternative perception methods are required. FE improvements can be realized by through Eco-Driving and an Optimal EMS [30, 31, 39, 40] when a drive cycle database is used in conjunction with the current vehicle state.

2.5 Vehicle-to-Vehicle Communications

Information gained from CS, RaDAR/LiDAR, drive cycle databases, and GNSS on one vehicle can be communicated to other vehicles wirelessly through V2V and is anticipated to occur over the DSRC 5.85–5.925 GHz band [41]. Direct knowledge of other vehicle information increases prediction accuracy when used with CS, RaDAR, and GNSS information for high FE gains from Eco-Driving [42, 43] and an Optimal EMS [44]. A special case of Eco-Driving, which is enabled by V2V communication, is known as "platooning" where multiple vehicles drive very close together to minimize air drag [45].

2.6 Vehicle-to-Infrastructure Communications

Advanced communication of traffic signal state can facilitate FE-improved vehicle drive cycles as well as powertrain control. Vehicles with advanced warnings of traffic lights can improve FE through Eco-Driving by maintaining more consistent vehicle speeds and planning efficient accelerations/decelerations [46–49]. Initial research demonstrates that significant FE improvements from an Optimal EMS are also possible once V2I has been implemented [39, 44, 50, 51].

2.7 *Vehicle-to-Everything Communications*

V2X involves communication with pedestrians (V2P), mobile devices (V2D), the cloud (V2C), and the grid (V2G). Despite being several decades away from real-world realization, near-perfect drive cycle prediction enabled by this technology would maximize FE gains from Eco-Driving [52–59] and an Optimal EMS.

3 Vehicle Plant and Controller Model

Real-world feasibility of perception and planning FE improvement control strategies must be investigated using high-fidelity models. These models can be validated by comparing the simulated vehicle parameters (such as engine power, battery state of charge (SOC), fuel consumption) with real-world vehicle parameters over the same drive cycle. When comparing the simulated FE to the real-world FE for a variety of fixed drive cycles, it is desirable to obtain a difference of no more than 3% across all drive cycles [60, 61].

There are several options for obtaining a high-fidelity model. The Autonomie software developed at Argonne National Labs is comprehensive and popular and can be interfaced with MATLAB through Simulink. Alternatively, the Future Automotive Systems Technology Simulator (FASTSim) developed by the National Renewable Energy Lab is less comprehensive but allows for faster simulations when considering vehicle fleets. Another option is to create a custom model using advanced modeling software, e.g., Modelica.

The Autonomie simulation tool has demonstrated close alignment with real-world Chevrolet Volt PHEV operation [60] and Toyota Prius HEV operation [61], thus demonstrating the model's effectiveness. The drawback of using Autonomie is that the specific vehicle model parameters used in those studies are not publically available and customers must manually override the preloaded vehicle models. To obtain a validated vehicle model, numerous parameters such as control logic, engine operation maps, and component efficiency maps must be modified.

An alternative method is to develop a custom high-fidelity vehicle model using the Modelica modeling language. Modelica is a free tool that uses a differential algebraic equation solver to simulate real-world stimulus responses and can be used in a wide variety of applications to develop models of desired fidelity [62–64]. This custom simulation tool is useful because modifications for FE improvement strategies are clear and transparent in comparison with Autonomie. The drawback of this simulation technique is that Modelica does not receive as much support, resulting in program crashes and arduous troubleshooting.

The first step in the analysis of improved FE control strategies is to establish a Baseline Energy Management Strategy (Baseline EMS) that mimics existing vehicle operation. This Baseline EMS should be validated against physical vehicle operation characteristics such as engine power, battery SOC, and FE over a variety of drive cycles (city, highway, aggressive, etc.). Typically, the city-focused Urban Dynamometer Driving Schedule (UDDS), the highway-focused Highway Fuel Economy Test (HWFET), and the aggressive US06 drive cycles are used, while the New York City Cycle (NYCC) can also be added. These drive cycles are shown in Fig. 4. They are used to validate the model and investigate alternate planning methods in Sects. 4–6. A 2012 Toyota Prius PHEV Autonomie model validation is shown in Table 3. Because the Baseline EMS is in close agreement with physically measured values, the model is considered validated

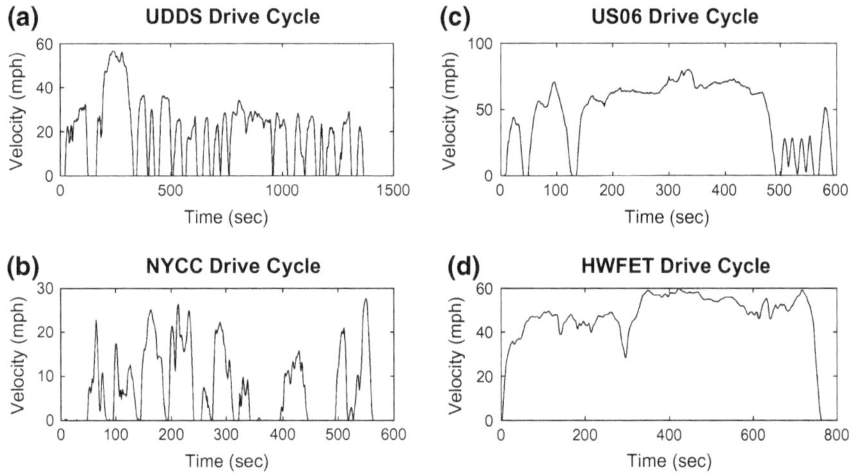

Fig. 4 Four US drive cycles frequently used by the Environmental Protection Agency (EPA)

Table 3 Simulated and measured FE and battery energy usage comparison for a 2012 Toyota Prius PHEV model developed using Autonomie

EPA drive cycle	Simulated fuel economy (mpg)	Measured fuel economy [65] (mpg)	Percent diff. (%)	Simulated battery net energy (Wh)	Measured battery net energy [65] (Wh)	Percent diff. (%)
UDDS	79.3	81.5	2.8	366.2	237.6	35.1
HWFET	86.5	88.8	2.7	582.2	549.0	5.7
US06	53.4	54.3	1.7	450.2	472.7	5.0

4 Planning Method 1: An Optimal Energy Management Strategy

An Optimal EMS enables FE improvements by increasing powertrain efficiency along a fixed drive cycle. Methods used to determine the global Optimal EMS include dynamic programming (DP) [66, 67] and Pontryagin's Minimization Principle (PMP), which is based on calculus of variations [68, 69]. Note that alternate Optimal EMS exists that makes optimality trade-offs to improve robustness such as stochastic dynamic programming (SDP) [70] and adaptive equivalent consumption minimization strategy (a-ECMS) [71] as well as Optimal EMS that makes optimality trade-offs for computation time such as optimized rule-based control [72], equivalent consumption minimization strategy (ECMS) [73], and model predictive control (MPC) [74]. Despite the numerous Optimal EMS derivation strategies, DP remains the overwhelming favorite due to its ease of use, robustness, and lack of dependence on derivatives or analytic expressions [75].

4.1 Deriving the Optimal EMS

DP finds the optimal solution using backward recursion, which avoids solutions that are not optimal as defined by the Bellman principle of optimality [67]. For every feasible state variable value, the optimal solution is stored. The globally optimal control is derived using the standard DP formulation of

$$\text{Dynamic Equation}: S(k+1) = S(k) + f(S, u, w, k)\,\Delta t \tag{1}$$

$$\text{Cost Function}: J = \sum_{k=0}^{N-1} f(S, u, w, k, \Delta t) \tag{2}$$

$$\text{State Feasibility Constraints}: S_{\min} \le S(k) \le S_{\max} \quad (k = 0, \ldots N) \tag{3}$$

$$\text{Control Feasibility Constraints}: u_{\min} \le u(k) \le u_{\max} \quad (k = 0, \ldots N-1) \tag{4}$$

where S is the state, u is the control, w is the exogenous input, k is the timestep number, Δt is the timestep length, J is the cost, and N is the final timestep number.

Depending on the problem discretization, hundreds of thousands of calculations may be required to determine the globally optimal control through DP and typically a low-fidelity, low computational cost vehicle model is needed. Numerous low computational cost vehicle models have been employed in research, and thorough descriptions of model development are available in the literature [76, 77].

To implement an Optimal EMS in a 2012 Toyota Prius PHEV, an approximate model of a power-split PHEV is required (shown in Fig. 5). The model consists of a force balance in the longitudinal direction to capture vehicle dynamics, a propulsion

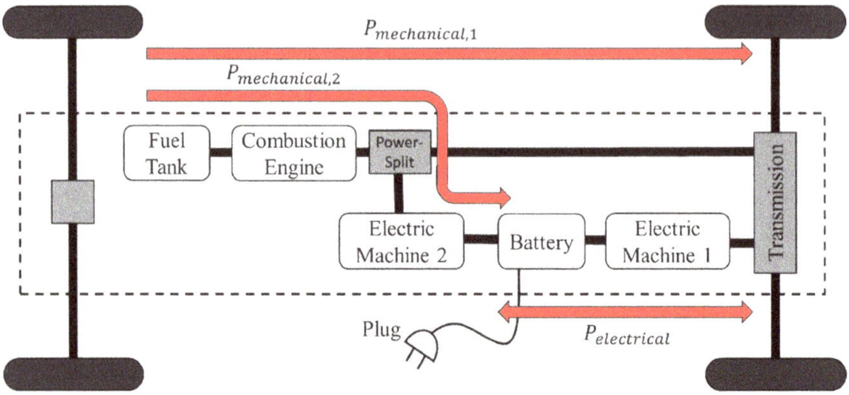

Fig. 5 A power-split PHEV schematic showing operation in parallel mode (propulsion power from mechanical, electric, or both) and series mode (propulsion power from the engine stored in the battery)

equation accounting for energy conversion efficiencies, a lithium-ion battery model, a brake-specific fuel consumption engine map (typically approximated with a response surface), and drivetrain torque and speed constraint equations all using appropriate vehicle parameters. A comprehensive derivation of the low computational cost vehicle model is available in a previous publication from our research group [5].

For a PHEV Optimal EMS DP derivation, the state is chosen to be the SOC, the control is chosen to be the engine power (P_{ICE}), the exogenous input is the vehicle (v), and the cost is chosen to be the fuel mass required (m_{fuel}). This formulation with the added feasibility constraints for a PHEV yields the following modified equations:

$$\mathrm{SOC}(k+1) = \mathrm{SOC}(k) + f(\mathrm{SOC}, P_{\mathrm{ICE}}, v, k)\Delta t \tag{5}$$

$$\mathrm{Cost} = \sum_{k=0}^{N-1} f(\mathrm{SOC}, P_{\mathrm{ICE}}, v, k, \Delta t) \tag{6}$$

$$\mathrm{SOC}_{\min} \leq \mathrm{SOC}(k) \leq \mathrm{SOC}_{\max} \quad (k = 0, \ldots N) \tag{7}$$

$$P_{\mathrm{ICE,min}} \leq P_{\mathrm{ICE}}(k) \leq P_{\mathrm{ICE,max}} \quad (k = 0, \ldots N-1) \tag{8}$$

A timestep of $\Delta t = 1$ s and a discretization of $\Delta\mathrm{SOC} = 0.02\%$ $\Delta P_{\mathrm{ICE}} = 0.1$ kW were chosen, and the resulting control map solution is incorporated into the high-fidelity Autonomie simulation using a 2-D lookup table Simulink block.

Note: The initial SOC value is chosen to be 23% with a charge sustaining value of 20%. This ensures there will be a surplus of battery power to be used while also ensuring that engine power must be used for all drive cycles investigated. FE will be improved by using all excess battery power (ending the drive cycle at 20% SOC) while using engine power only when necessary.

4.2 *Optimal EMS Results*

When comparing the desired engine power of the Baseline EMS to the desired engine power of the Optimal EMS (shown in Fig. 6), it is apparent that for each of the drive cycles, an engine power around 20 kW is often optimally efficient. However, an application of 20 kW of engine power is only efficient at certain speeds, which are known to the Optimal EMS from drive cycle prediction.

When comparing the battery SOC from the Baseline EMS and the Optimal EMS (Fig. 7), it is apparent that the Optimal EMS ends the drive cycle at the minimum allowable value. Since the Baseline EMS does not use drive cycle end information, it is at a significant disadvantage.

FE improvements are shown in Fig. 8 and are calculated as

$$\text{FE Improvement} = \frac{\text{New FE} - \text{Baseline FE}}{\text{Baseline FE}} \tag{9}$$

An Optimal EMS provided the largest FE increase in city drive cycles (UDDS and NYCC) which have frequent accelerations and decelerations. For aggressive

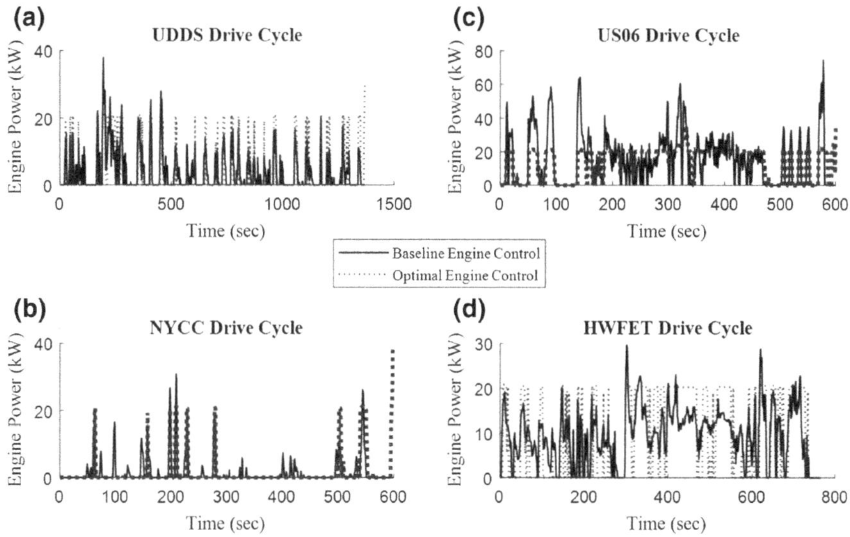

Fig. 6 A comparison of the engine power used by the Baseline EMS and the engine power used by the Optimal EMS

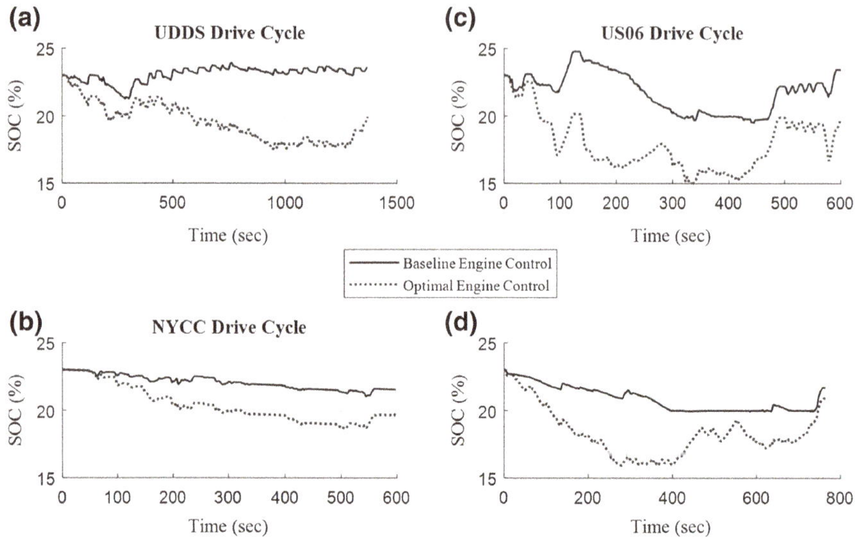

Fig. 7 A comparison of the battery SOC results from the Baseline EMS and the battery SOC results from the Optimal EMS

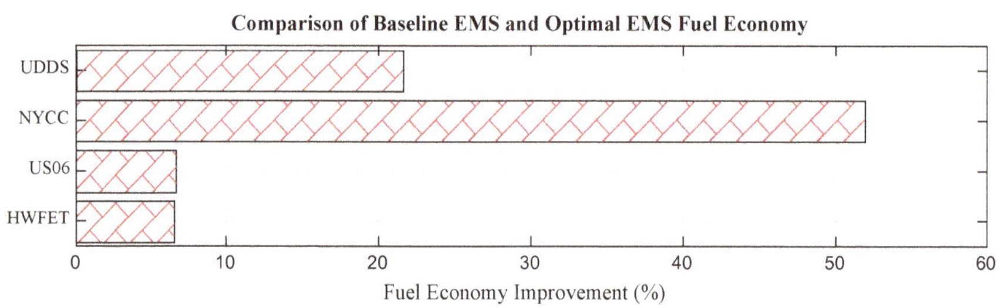

Fig. 8 A comparison of the Baseline EMS fuel economy results and the Optimal EMS fuel economy results

and highway drive cycles, there is less freedom in powertrain operation due to high power demand, resulting in less dramatic FE improvements.

5 Planning Method 2: Eco-Driving

Eco-Driving enables FE improvements by modifying vehicle speed along a fixed drive cycle. Eco-Driving can be communicated and encouraged to the driver in a variety of ways including driver training, vehicle dashboards, smartphone applications, and pedal feedback [78] (examples shown in Fig. 9).

Fig. 9 Eco-Driving achieves a FE improvement through lower energy driving which is encouraged by driver feedback. An existing implementation of this feedback from a 2012 Chevrolet Volt dashboard is shown (**a**) as well as a proposed general advice image from the literature [79] (**b**)

Eco-Driving can be formulated using an optimal control approach as

$$\min: \quad P_{\text{prop}}(k) \tag{10}$$

$$\text{subject to}: \quad \text{time constraints (e.g.} \quad t_{\text{total}} \leq t_{\max}) \tag{11}$$

$$\text{safety constraints (e.g.} \quad v \leq v_{\text{speed limit}}) \tag{12}$$

$$\text{operational constraints (e.g.} \quad a \leq a_{\text{limit}}) \tag{13}$$

However, in this formulation, it can be difficult to incorporate real-world constraints such as traffic lights, other vehicles, and pedestrians. Because of this difficulty, rules have been extracted from studying the results of optimal control problems. The rules are typically generalized as eliminating full stops, maintaining a constant speed, limiting acceleration, and smoothing the velocity profile [78–82]. Each of these rules will be examined independently by eliminating stops in the UDDS drive cycle, maintaining a more constant speed in the NYCC drive cycle, limiting acceleration and deceleration in the US06 drive cycle, and smoothing the velocity in the HWFET drive cycle. The FE results will not be as drastic as when all methods are combined, but instead will shed light on the relative importance of each rule.

5.1 *Deriving Eco-Driving Strategies*

To study the effect of removed stops from drive cycles, the UDDS drive cycle was modified by removing velocities below 15 mph while preserving the overall drive cycle distance. This modified drive cycle is compared to the original as shown in Fig. 10a. Note that in Fig. 10b, the acceleration magnitudes remain approximately the same except for a few reductions. The engine power also remains relatively unaffected as shown in Fig. 10c, and the battery SOC has a similar profile but ends at a higher value of SOC.

To study the effect of a more constant velocity, the NYCC drive cycle was modified by increasing the speed in some sections and reducing speeds in other sections while preserving the overall drive cycle distance. This modified drive cycle is compared to the original in Fig. 11a. These modifications significantly affect the acceleration, as shown in Fig. 11b, but the reductions in acceleration contribute to the improved performance. Figure 11c shows a drastically different engine power used by the Baseline EMS for the modified NYCC drive cycle. Engine power is increased in some areas and decreased in other areas. Figure 11d shows that the SOC ends at a significantly higher value, which is not ideal for a PHEV seeking to end the drive cycle at the lowest possible value of battery SOC.

To lower the propulsive power required in aggressive driving, the acceleration and deceleration magnitudes can be reduced. To study this effect, the US06 aggressive drive cycle was modified by limiting the acceleration magnitude to below 1.5 m/s^2 and above the deceleration rate of −1.5 m/s^2 while preserving the overall drive cycle distance, which is shown in Fig. 12b. Reducing acceleration magnitudes has the favorable effect of reducing peak engine power (Fig. 12c). Figure 12d shows a smoother SOC profile resulting from the reduced acceleration magnitudes.

To study the effect of velocity smoothing, the HWFET drive cycle can be modified by aggressively smoothing the velocity profile while preserving the overall drive cycle distance. This modified drive cycle is compared to the original as shown in Fig. 13a. This drive cycle modification has a drastic effect on the drive

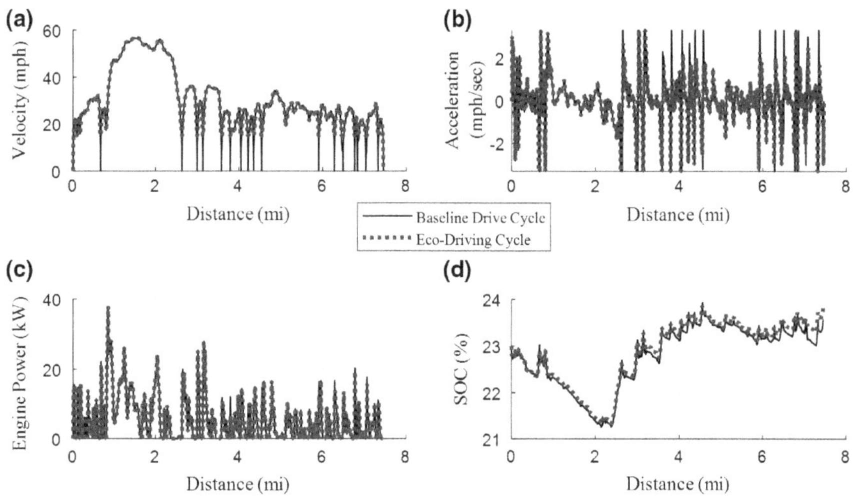

Fig. 10 A comparison of the changes that occur along the drive cycle and in vehicle operation when removing stop and go driving from the UDDS drive cycle

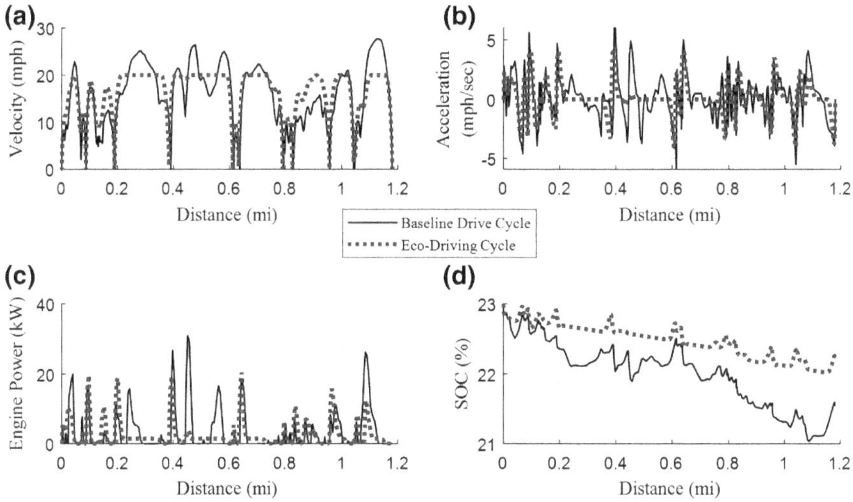

Fig. 11 A comparison of the changes that occur along the drive cycle and in vehicle operation when making vehicle speed more consistent along the NYCC drive cycle

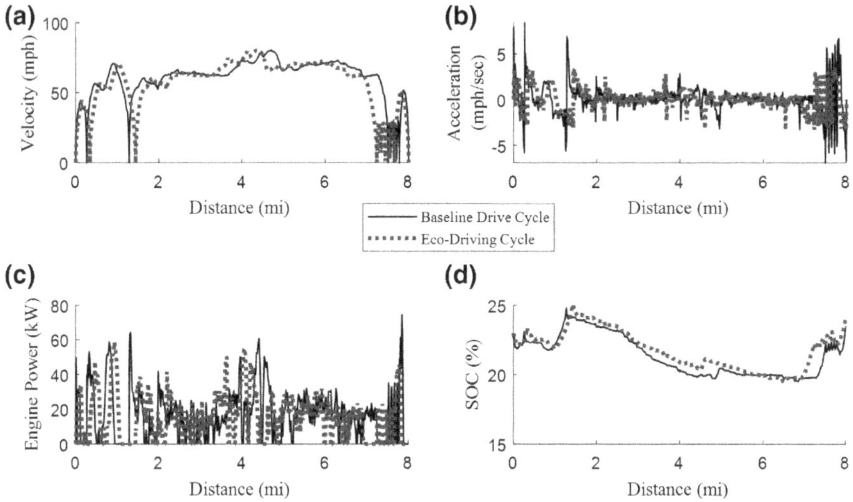

Fig. 12 A comparison of the changes that occur along the drive cycle and in vehicle operation when limiting acceleration and deceleration rates along the US06 drive cycle

cycle acceleration as shown in Fig. 13b. For much of the drive cycle, the acceleration magnitude is almost completely eliminated. When the drive cycle is smoothed, the engine operation is also smoothed as shown in Fig. 13c, although the overall shape of engine power remains consistent. Also, by smoothing the drive cycle, the battery SOC remains consistent as shown in Fig. 13d.

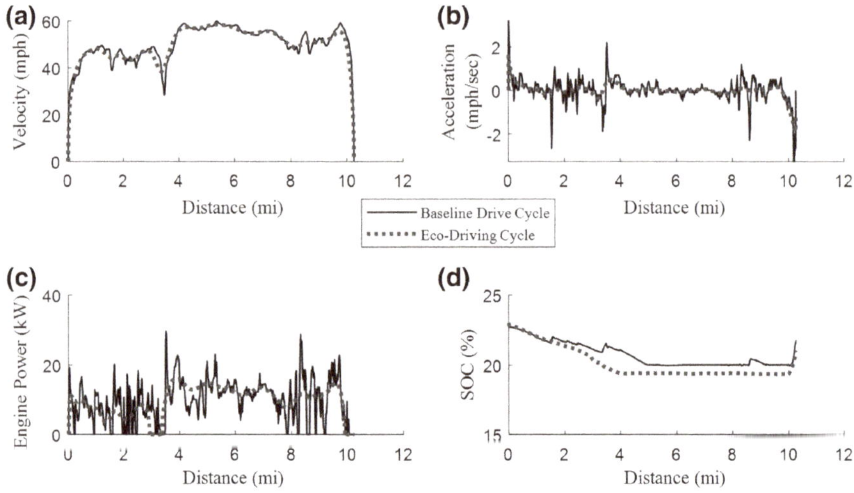

Fig. 13 A comparison of the changes that occur along the drive cycle and in vehicle operation when smoothing the drive cycle velocity along the HWFET drive cycle

5.2 *Eco-Driving Results*

Figure 14 demonstrates that significant FE improvements can be realized by eliminating stops and by reducing acceleration/deceleration magnitudes, both of which reduce the total energy expended during the drive cycle. Driving at a more constant speed, as demonstrated in the NYCC drive cycle, can provide moderate FE improvements, while smoothing the velocity profile provides a small FE improvement.

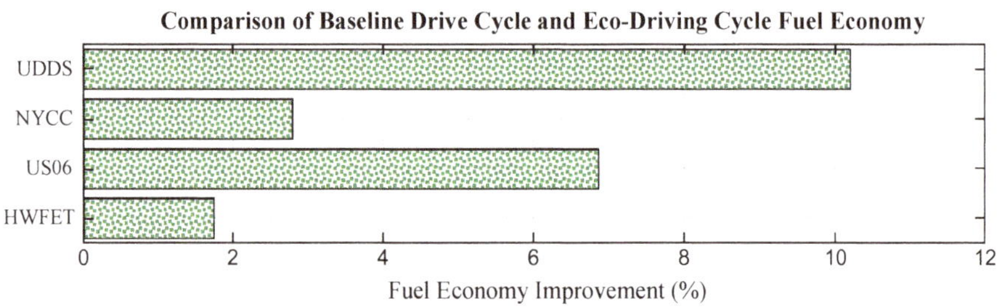

Fig. 14 Fuel economy results for removing stops along the UDDS drive cycle, creating a more constant speed along the NYCC drive cycle, limiting acceleration and deceleration rates along the US06 drive cycle, and smoothing the velocity profile for the HWFET drive cycle

6 Planning Method 3: Eco-Driving and an Optimal Energy Management Strategy

Combining Eco-Driving with an Optimal EMS has received little attention in the literature, but could result in FE increases beyond what is possible from either strategy alone. Removing drive cycle power restrictions through Eco-Driving provides an Optimal EMS with even greater freedom for FE improvements.

6.1 *Eco-Driving and an Optimal EMS Results*

For the Eco-Driving drive cycles, as with the baseline drive cycles, the Optimal EMS most often seeks an engine power of approximately 20 kW. In the UDDS and US06 drive cycles, significant engine power reductions are achieved, while in the NYCC and HWFET drive cycles, significant engine power increases are achieved as shown in Fig. 15.

Knowledge of the entire drive cycle in advance allows a final battery SOC of 20% to be achieved for all drive cycles. Since there is no adjusted FE penalty that must be calculated for PHEVs since they are designed to end at the minimum SOC, this provides a significant FE benefit. Note that in the highway drive cycle, there is significantly more SOC fluctuation from only running the engine at 20 kW.

The FE improvements from Eco-Driving combined with an Optimal EMS are significant. For the NYCC drive cycle, the regions of constant vehicle speed

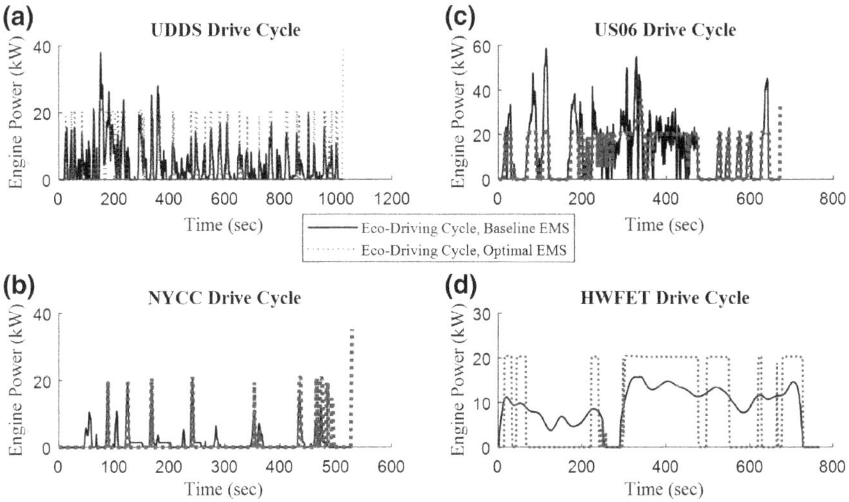

Fig. 15 A comparison of the engine power used by the Baseline EMS along the Eco-Driving drive cycle and the engine power used by the Optimal EMS along the Eco-Driving drive cycle

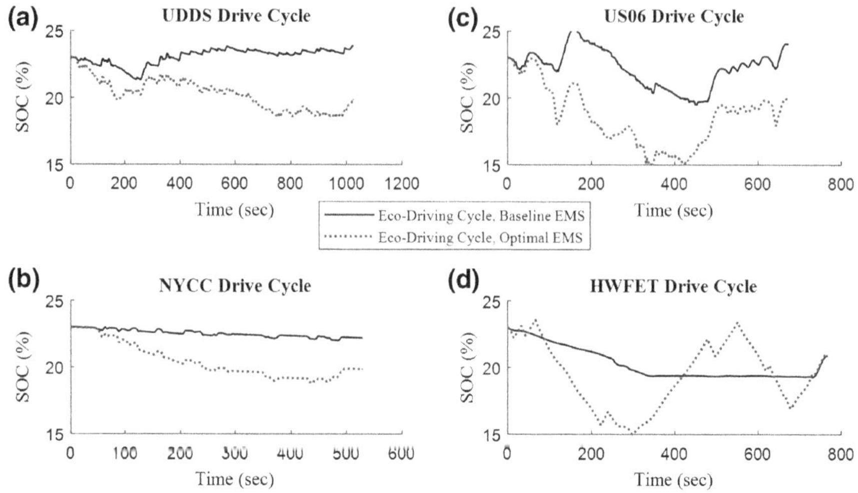

Fig. 16 Battery SOC for **a** UDDS, **b** NYCC, **c** US06, and **d** HWFET

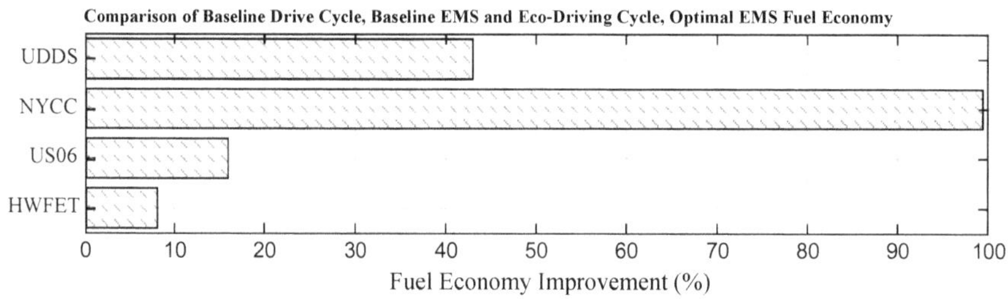

Fig. 17 Fuel economy results for various aspects of Eco-Driving combined with an Optimal EMS along the UDDS, NYCC, US06, and HWFET drive cycles

provided from Eco-Driving allow the Optimal EMS to achieve high powertrain efficiency and realize significant FE benefits. Additionally, knowledge of the drive cycle allows perfect utilization of battery energy, which is lost from Eco-Driving alone (Fig. 16b). A similar phenomenon is observed for the UDDS and US06 drive cycles. Conversely, the HWFET drive cycle realized a FE benefit but at the expense of wild battery SOC fluctuations which may impact battery life (Fig. 17).

6.2 *Comparing Planning Strategies*

When comparing all three vehicle control strategies enabled by connected and autonomous lithium-ion electric vehicles, it is apparent (from Fig. 18) that the largest FE improvements are possible when combining Eco-Driving with an

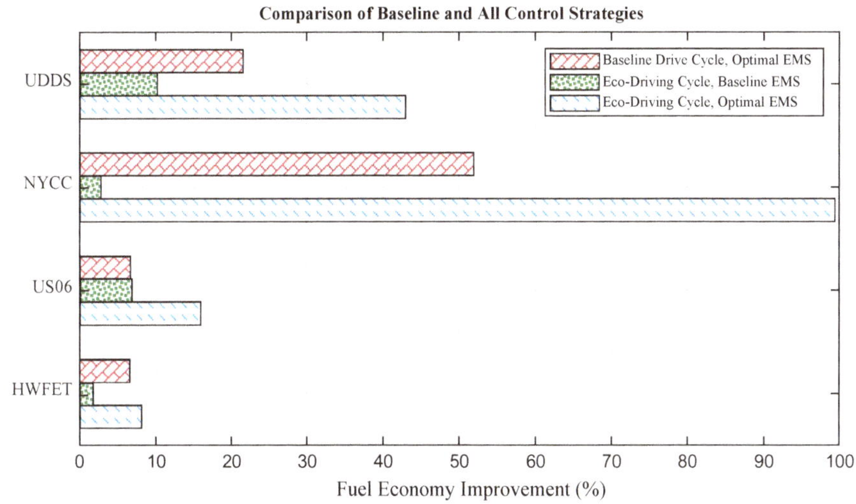

Fig. 18 FE results comparing the FE improvements from a baseline drive cycle with a Baseline EMS to: (top bar) a baseline drive cycle with an Optimal EMS, (middle bar) an Eco-Driving drive cycle, with a Baseline EMS, and (bottom bar) an Eco-Driving drive cycle with an Optimal EMS

Optimal EMS while the next largest FE improvements are achieved using the Optimal EMS on the baseline drive cycle. Note that the Eco-Driving FE improvements are from individual investigation of Eco-Driving rules, and if all rules were to be combined, the Eco-Driving FE improvement would be more significant. Additionally, city drive cycles with velocities modified to be consistent speeds or with stop elimination provide large FE improvements from an Optimal EMS. Limiting acceleration and deceleration rates in aggressive drive cycles provides double the FE improvement from an Optimal EMS. Smoothing vehicle velocity during highway drive cycles provides a marginal FE improvement increase from an Optimal EMS.

7 Conclusions

When CAV technologies are combined with lithium-ion electric vehicle technology, new vehicle control strategies that improve FE are possible. In this chapter, a review of CAV technologies, lithium-ion electric vehicle modeling techniques, and three control strategies for improved FE were presented. The results demonstrate that significant FE gains can be achieved through the realization of an Optimal EMS, Eco-Driving, and from the combination of an Optimal EMS and Eco-Driving.

Although there are several CAV technologies that are not currently available, it may be possible to achieve significant portions of these FE benefits today. As more vehicles incorporate electrification and CAV technologies, these FE improvements will be easier to achieve, thus helping to achieve transportation sustainability.

References

1. *Annual Energy Outlook 2017*. Energy Information Administration (5 Jan 2017) AEO2017
2. Z.D. Asher, V. Wifvat, A. Navarro, S. Samuelsen, T.H. Bradley, SAE Technical Paper 2017-26-0106 (2017), https://doi.org/10.4271/2017-26-0106
3. D.J. Fagnant, K. Kockelman, Transp. Res. A Policy **77**, 167–181 (2015)
4. M. Broussely, in *Electric and Hybrid Vehicles: Power Sources, Models, Sustainability, Infrastructure and the Market*, ed. by G. Pistoia (Elsevier, Amsterdam, 2010), pp. 305–347
5. Z.D. Asher, D.A. Baker, T.H. Bradley, IEEE Trans. Control Syst. Tech. PP **99**, 1–14 (2017). https://doi.org/10.1109/tcst.2017.2747502
6. A. Iliaifar, LIDAR, lasers, and logic: anatomy of an autonomous vehicle (2013), www.digitaltrends.com
7. M. Trimarchi, *Do Car GPS Devices Cause Accidents*? (Howstuffworks.com., Atlanta, 2017)
8. M. Bertozzi, A. Broggi, A. Fascioli, Rob. Auton. Syst. **32**, 1–16 (2000)
9. Z. Sun, G. Bebis, R. Miller, IEEE Trans. Pattern Anal. Mach. Intell. **28**, 694–711 (2006)
10. A. Ruta, F. Porikli, S. Watanabe, Y. Li, Mach. Vis. Appl. **22**, 359–375 (2011)
11. F. Lindner, U. Kressel, S. Kaelberer, Robust recognition of traffic signals. IEEE Intell. Veh. Symp. 49–53 (2004)
12. M.A.S. Kamal, M. Mukai, J. Murata, T. Kawabe, IEEE Trans. Intell. Transp. Syst. **4**, 244–251 (2010)
13. M. Munoz-Organero, V.C. Magana, IEEE Trans. Intell. Transp. Syst. **14**, 1023–1028 (2013)
14. E. Ohn-Bar, A. Tawari, S. Martin, M.M. Trivedi, Comput. Vis. Image Underst. **134**, 130–140 (2015)
15. E. Raphael, R. Kiefer, P. Reisman, G. Hayon, SAE Int. J. Passeng. Cars Mech. Syst. **4**, 467–478 (2011)
16. M.M. Trivedi, T. Gandhi, J. McCall, IEEE Trans. Intell. Transp. Syst. **8**, 108–120 (2007)
17. E. Dagan, O. Mano, G.P. Stein, A. Shashua, Forward collision warning with a single camera. IEEE Intell. Veh. Symp. Proc. 37–42 (2004)
18. G.P. Stein, O. Mano, A. Shashua, Vision-based ACC with a single camera: bounds on range and range rate accuracy. IEEE Intell. Veh. Symp. Proc. 120–125 (2003)
19. S. Clark, H. Durrant-Whyte, Autonomous land vehicle navigation using millimeter wave radar, in *Proceedings 1998 IEEE International Conference on Robotics and Automation (Cat. No. 98CH36146)*, vol. 4, May 1998, pp. 3697–3702
20. D.J. Belgiovane, C.C. Chen, S.Y.P. Chien, R. Sherony, IEEE Trans. Intell. Transp. Syst. **99**, 1–10 (2017). http://ieeexplore.ieee.org
21. Z. Ji, M. Luciw, J. Weng, S. Zeng, IEEE Trans. Intell. Transp. Syst. **12**, 402–411 (2011)
22. B. Steux, C. Laurgeau, L. Salesse, D. Wautier, Fade: a vehicle detection and tracking system featuring monocular color vision and radar data fusion. IEEE Intell. Veh. Symp. Proc. **2**, 632–639 (2002)
23. A. Gern, U. Franke, P. Levi, Robust vehicle tracking fusing radar and vision, in IEEE *International Conference on Multisensor Fusion and Integration* for *Intelligent Systems (MFI'01)* (2001) pp. 323–328
24. P. Themann, T. Uhrner, D. Kuck, M. Müller, M. Klubal, Ecodriving support based on cooperative prediction models, in *19th ITS World Congress* (Vienna, Austria, 2012), pp. 22–26

25. F. Mensing, E. Bideaux, R. Trigui, H. Tattegrain, Transp. Res. Part D Transp. Environ. **18**, 55–61 (2013)
26. S. Tokoro, K. Kuroda, A. Kawakubo, K. Fujita, H. Fujinami, Electronically scanned millimeter-wave radar for pre-crash safety and adaptive cruise control system. IEEE Intell. Veh. Symp. Proc. 304–309 (2003)
27. G.R. Widmann et al., SAE Technical Paper Series, No. 724 (2000)
28. F. Rahman, C. O'Brien, K. Manning, Let EcoDrive be your guide: development of a mobile tool to reduce carbon footprint and promote green transport, in *Proceedings of the 27th Annual ACM Symposium on Applied Computing* (2012), pp. 519–524
29. K. Jakobsen, S.C.H. Mouritsen, K. Torp, Evaluating eco-driving advice using gps/canbus data, in *21st ACM SIGSPATIAL International Conference on Advances in Geographic Information Systems SIGSPATIAL'13* (2013), pp. 44–53
30. D.A. Baker, Z.D. Asher, T.H. Bradley, SAE Technical Paper (2017), http://papers.sae.org/2017-01-1262/, https://doi.org/10.4271/2017-01-1262
31. F.A. Bender, M. Kaszynski, O. Sawodny, IEEE Trans. Veh. Tech. **62**(8), 3581–3592 (2013). http://ieeexplore.ieee.org
32. A.G. Adinolfi, H.K. Kuga, M.L. de Oliveira e Souza, SAE Technical Paper 2013-36-0650 (2013)
33. Z. Popovic, A. Soloviev, Y. Mochizuki, SAE Technical Paper 2011-01-1035 (2011)
34. Z. Popovic Y. Mochizuki, SAE Technical Paper 2011-01-1036 (2011)
35. H. Kitano, H. Kazuo, H. Tanaka, An estimation method of vehicle position for automated driving with GNSS, in *SAE World Congress and Exhibition* (2016)
36. E. Stenborg, L. Hammarstrand, Using a single band GNSS receiver to improve relative positioning in autonomous cars, in *Intelligent Vehicles Symposium* (2016)
37. F. Meng, S. Tan, S. Wang, B. Zhu, New techniques to enhance the performance of stand-alone GNSS positioning. in *International Conference on Communication and Networks* (2015), pp. 548–552
38. M. Campbell, M. Egerstedt, J.P. How, R.M. Murray, Phil. Trans. Series A Math. Phys. Eng. Sci. 368 (1928), 4649–4672 (2010). http://rsta.royalsocietypublishing.org
39. C. Sun, S.J. Moura, X. Hu, J.K. Hedrick, F. Sun, IEEE Trans. Control Syst. Tech. 23(3), 1075–1086 (2015). http://ieeexplore.ieee.org
40. C. Sun, F. Sun, H. He, App. Energy **185**, 1644–1653 (2017)
41. C. Smith, *The Car Hacker's Handbook: A Guide for the Penetration Tester* (No Starch Press, Inc., San Fransico, 2016)
42. M. Kerper, C. Wewetzer, H. Trompeter, W. Kiess, M. Mauve, Driving more efficiently—The use of inter-vehicle communication to predict a future velocity profile, in *IEEE Vehicular Technology Conference* (2011), pp. 1–5
43. D. Lang, R. Schmied, L. Del Re, SAE Int **7**, 14 (2014)
44. M.A. Mohd Zulkefli, J. Zheng, Z. Sun, H.X. Liu, Transp. Res. Part C Emerg. Technol. **45**, 41–63 (2014)
45. M. Brackstone, M. McDonald, Transp. Res. Part F Traffic Psychol. Behav. **2**, 181–196 (1999)
46. B. Asadi, A. Vahidi, in *IFAC Proceedings Volumes* **42**(15), 484–489 (2009)
47. B. Asadi, A. Vahidi, IEEE Trans. Control Syst. Technol. **19**, 707–714 (2011)
48. S. Mandava, K. Boriboonsomsin, M. Barth, Arterial velocity planning based on traffic signal information under light traffic conditions, in *Proceedings of IEEE Conference* on *Intelligent Transportation Systems* (ITSC) (2009), pp. 1–6
49. E. Koenders, J. Vreeswijk, Cooperative infrastructure, in *IEEE Intelligent Vehicles Symposium, Proceedings* (2008), pp. 721–726
50. Q. Gong, Y. Li, Z.R. Peng, IEEE Trans. Veh. Tech. **57**, 3393–3401 (2008)
51. Q. Gong, Y. Li, Z. Peng, Power Management of plug-in hybrid electric vehicles using neural network based trip modeling, in *American Control Conference* (2009), pp. 4601–4606. http://ieeexplore.ieee.org

52. J. Gonder, A. Brooker, E. Burton, L. Wang, E. Wood, Mobility patterns informing V2X research projects: eco-routing and electrified roadway project examples, in *Task 28 Workshop, Home Grids and V2X Technologies* (2016)
53. S.E. Li, S. Xu, X. Huang, B. Cheng, H. Peng, IEEE Trans. Veh. Technol. **64**, 5439–5449 (2015)
54. T. Edwards, IEEE Intell. Transp. Syst. Mag. **7**, 4 (2015)
55. D. Krajzewicz, M. Heinrich, M. Milano, COLOMBO: investigating the potential of V2X for traffic management purposes assuming low penetration rates, in *9th ITS European Congress*, Dublin (IE), 4–7 June 2013
56. Y. Chen, D. Zhang, K. Li, Enhanced eco-driving system based on V2X communication, in *Proceedings of IEEE Conference on Intelligent Transportation Systems (ITSC)* (2012), pp. 200–205
57. S. Diewald, A. Möller, L. Roalter, M. Kranz, in *Proceedings of Mensch und Computer*, Konstanz, Germany, 1–8 Sept 2012
58. C. Weiß, Comput. Netw. **55**, 3103–3119 (2011)
59. B. Schünemann, J.W. Wedel, I. Radusch, Tamkang. J. Sci. Eng. **13**, 637–641 (2010)
60. N. Kim, M. Duoba, A. Rousseau, SAE Int. **6**, 985–992 (2013). https://doi.org/10.4271/2013-01-1458
61. N. Kim, A. Rousseau, E. Rask, SAE Int **2012**, 1–14 (2012). https://doi.org/10.4271/2012-01-1040
62. P. Fritzson, P. Aronsson, H. Lundvall, K. Nyström, A. Pop, L. Saldamli, D. Broman, The OpenModelica modeling, simulation, and development environment, in *Conference on Simulation and Modelling of the Scandinavian Simulation Society (SIMS2005)* (Trondheim, Norway, 2005)
63. H. Tummescheit. Ph.D. dissertation, Lund University (2002)
64. B.M. Geller, T.H. Bradley, SAE Technical Paper (2011). https://doi.org/10.4271/2011-01-0943
65. Argonne National Lab, Downloadable dynamometer database (D3)—Test 37; summary sheet. PHEV 2012 Toyota Prius, 72°F. https://www.anl.gov/energy-systems/group/downloadable-dynamometer-database/plug-hybrid-electric-vehicles/2012-toyota
66. C. Lin, J. Kang, J.W. Grizzle, H. Peng, Energy management strategy for a parallel hybrid electric truck, in *American Control Conference (Cat. No. 01CH37148)*, vol. 4 (2001), pp. 2878–83. http://ieeexplore.ieee.org
67. R. Bellman, Dynamic programming and lagrange multipliers. Proc. Natl. Acad. Sci. USA, **42**(10), Nat. Acad Sci. **76** (1956)
68. N. Kim, S. Cha, H. Peng, IEEE Trans. Control Syst. Technol. **19**, 1279–1287 (2011)
69. V.G. Boltyanskii, R.V. Gamkrelidze, L.S. Pontryagin, *The Theory of Optimal Processes. I. The Maximum Principle* (TRW Space Technology Laboratories, Los Angeles, California, 1960). http://www.dtic.mil/docs/citations/AD0264147
70. C. Lin, H. Peng, J.W. Grizzle, A stochastic control strategy for hybrid electric vehicles, in *American Control Conference*, vol. 5 (2004), pp. 4710–4715. http://ieeexplore.ieee.org
71. S. Onori, L. Serrao, G. Rizzoni, Adaptive equivalent consumption minimization strategy for hybrid electric vehicles, in *2010 ASME Dynamic Systems and Control Conference*, vol. 499 (American Society of Mechanical Engineers, 2010)
72. D. Bianchi, L. Rolando, L. Serrao, S. Onori, G. Rizzoni, N. Al-Khayat, T. Hsieh, P. Kang, A rule-based strategy for a series/parallel hybrid electric vehicle: an approach based on dynamic programming, in *ASME 2010 Dynamic Systems* and *Control Conference,* vol. 507 (American Society of Mechanical Engineers, 2010)
73. G. Paganelli, S. Delprat, T.M. Guerra, J. Rimaux, J.J. Santin, Equivalent consumption minimization strategy for parallel hybrid powertrains, in *IEEE Vehicular Technology Conference VTC Spring 2002 (Cat. No. 02CH37367)*, vol. 4 (2002), pp. 2076–2081. http://ieeexplore.ieee.org

74. H.A. Borhan, A. Vahidi, A.M. Phillips, M. L. Kuang, I.V. Kolmanovsky, Predictive energy management of a power-split hybrid electric vehicle, in *American Control Conference* (2009), pp. 3970–3976. http://ieeexplore.ieee.org
75. P. Zhang, F. Yan, C. Du, 2015/8. Renew. Sust. Energy Rev. **48**, 88–104 (2015)
76. I. Husain, *Electric and Hybrid Vehicles: Design Fundamentals*, 2nd edn. (CRC Press, Boca Raton, 2010)
77. R. Rajamani, *Vehicle Dynamics and Control*, 2nd edn. (Springer, New York, 2011)
78. J. Gonder, M. Earleywine, W. Sparks SAE Int. J. Passeng. Cars Electron Electr. Syst. **5**, 450–461 (2012-01-0494) (2012). http://papers.sae.org
79. P. Michel, D. Karbowski, A. Rousseau, SAE Technical Paper (2016). http://papers.sae.org/2016-01-0152/
80. Y. Saboohi, H. Farzaneh, App. Energy **86**, 1925–1932 (2009)
81. S. Mandava, K. Boriboonsomsin, M. Barth, arterial velocity planning based on traffic signal information under light traffic conditions. IEEE Intell. Trans. Syst. 1–6 (2009). http://ieeexplore.ieee.org
82. D. Wissam, A. Chasse, A. Sciarretta, P. Moulin, Optimal energy management compliant with online requirements for an electric vehicle in eco-driving, in *Workshop on Engine and Powertrain Control, Simulation and Modeling*, Rueil-Malmaison, France, Oct 2012

Electric Commercial Vehicles in Mid-Haul Logistics Networks

Maximilian Schiffer, Sebastian Stütz and Grit Walther

Abstract Recent research on electric commercial vehicles (ECVs) has mostly been limited to short-haul applications and single planning perspectives. Especially in mid-haul logistics networks where recharging on routes is necessary, integrated planning approaches become inevitable due to interdependent decisions on network design and vehicle operations. This chapter provides an overview of planning approaches for ECVs that have been presented so far and presents a generic modeling approach for integrated TCO analysis, taking strategic network design and operational vehicle routing and recharging decisions into consideration. This approach is then applied to a real-world case study of a large German retail company. We discuss results with respect to the competitiveness of ECVs compared to ICEVs. Herein, we study economic and ecological benefits. Furthermore, we analyze battery degradation effects from a technical point of view. Results show that ECVs are on the verge of breaking even in mid-haul logistics for certain application cases.

1 Introduction

Transport activities cause environmental harm by greenhouse gas (GHG) emissions at global level and other noxious emissions as well as noise at local level. In detail, transportation causes 25% of CO_2 emissions, 30% of particulate matter, and 60% of other noxious emissions in the European Union (EU) [1]. Furthermore,

M. Schiffer (✉) · G. Walther
Chair of Operations Management, School of Business and Economics,
RWTH Aachen University, Aachen, Germany
e-mail: maximilian.schiffer@om.rwth-aachen.de

G. Walther
e-mail: walther@om.rwth-aachen.de

S. Stütz
Fraunhofer Institute for Material Flow and Logistics, Dortmund, Germany
e-mail: Sebastian.Stuetz@iml.fraunhofer.de

G. Pistoia and B. Liaw (eds.), *Behaviour of Lithium-Ion Batteries in Electric Vehicles*, Green Energy and Technology,
https://doi.org/10.1007/978-3-319-69950-9_7

transportation remains the only sector in which the energy consumption and emissions are still rising [2, 3], due to increasing transport volumes in the course of globalization and thriving e-commerce [4]. To mitigate the environmental harm of transportation, the EU aims at a reduction of GHG emissions by 20% until 2020 and by 40% until 2030 compared to the emissions produced in 1990 [5]. Herein, electric vehicles (EVs) are seen as one of the cleanest means of transportation, and thus, it is assumed that they can contribute significantly to achieve these goals. However, the adoption rate of EVs is still low and challenges in different fields have to be solved, e.g., by technological enhancements, installation of infrastructure, and optimal operation of vehicles, to accelerate this market uptake.

In this course, EVs with their respective challenges have gained the attention of different research areas. For instance, engineering has solved and discussed technical challenges and enhancements, operations research has been focusing on decision support for optimal allocation of charging infrastructure and energy efficient operation of vehicles, and business administration has been focusing on business models of EVs for private as well as commercial use. While all of these fields tackle important aspects that have to be solved to succeed in establishing EVs, publications that draw a comprehensive picture by combining insights and limitations from all of these fields are still sparse. While engineers analyze technical issues in high detail, operations researchers often simplify technical aspects for operational and strategic decision support. Also, economists state total cost of ownership (TCO) calculations based on (very) aggregated values neglecting (essential) details. However, a profound analysis is needed that at the same time accounts for technical details and limitations, but also regards economic aspects at an adequate level, as well as estimates the integration and future operation of EVs within transportation networks. This holds especially in a commercial context as is given by the usage of electric commercial vehicles (ECVs) in logistics fleets. Here the challenge is to persuade logistics fleet operators of adopting ECVs and changing the technological split in their car pools in order to accelerate the market uptake of ECVs.

In the last years, large logistics companies carried out first pilot projects on ECVs. In 2011, the Deutsche Post DHL group (DPDHL) launched a first electric fleet of 12 Renault Kangoo ZE [6] as well as a large pilot project on zero-emission deliveries in an urban area with 310,000 inhabitants in 2013. DPDHL succeeded in this ambitious project by using custom-build ECVs [7–9] and finally bought the company that manufactured these ECVs [10]. Further pilot projects have been launched by the United Postal Service (UPS) [11]. However, all of these pilot projects are focusing on short-haul transportation. To increase the contribution of ECVs on emission reduction, a usage of ECVs in mid-haul transportation is necessary. So far, studies and projects in this field are very scarce as large logistics fleet operators still question the operational applicability and economic viability of ECVs in mid-haul logistics networks due to limited driving ranges (given the higher distances) as well as time inefficient recharging processes. However, a first research project called ELMO—'Elektromobile urbane Wirtschaftsverkehre' ('electrified commercial transport in urban areas') (cf. [12]), that was partly founded by the

German government, delivered promising results that called for a more detailed analysis.

Against this background, we present a case study in order to show that ECVs can be used in mid-haul logistics with nearly no limitations compared to internal combustion engine vehicles (ICEVs). To change the perception of practitioners and fleet operators in this context, we analyze all potential retentions mentioned above. We present an integrated planning approach that combines TCO calculations with strategic network design and operational design components. Furthermore, we analyze technical characteristics regarding the ECV's batteries with respect to charging cycles and discuss the battery stress caused by a utilization of the battery in mid-haul logistics. Based on these analyses, we gain insights on viable planning horizons for (integrated) TCO analysis, since the charging cycles of a battery are limited. Our analyses are based on real-world data and insights from the pilot project ELMO and the mathematical and algorithmic fundamentals as well as TCO findings presented in [13].

The remainder of this chapter is structured as follows: First, we give a brief overview about the state of the art in related research streams. Then we provide some information about the ELMO project and introduce our case study before we present a mixed integer program that is the formal basis for our analysis. Next, we describe our design of experiments and discuss the corresponding results. Finally, we conclude this chapter with a summary of the main findings and remaining challenges for both practitioners and future research.

2 State of the Art

In the following, we present a brief overview on related literature of the different fields. Herein, we do not aim at giving detailed explanations, but rather at providing a comprehensive overview on all streams for the interested reader. First, we focus on related literature from business administration and operations research. Herein, we provide an overview on existing approaches for planning of charging station infrastructure, routing of EVs, and aggregated (TCO) feasibility studies. Second, we give a very brief overview on results of battery degradation studies from the field of electrical engineering.

2.1 Strategic Planning on Charging Station Infrastructure

Different mathematical optimization approaches have been developed to determine optimal configurations of charging infrastructure. These approaches can basically be differentiated into vertex-based, flow-based, and location-routing models. While the first two approaches are mainly used to design public charging infrastructure for private cars, the latter are used to design charging station infrastructure for logistics fleets.

In vertex-based planning approaches, charging demand is predefined and assigned to certain vertices, i.e., to specific points on a map. This is for instance given if charging demand is known in advance or if parking space is used for charging locations. An overview on these approaches is given in [14]. Regarding applications, charging infrastructure is planned for different fleets, e.g., determining a charging infrastructure for taxi fleets [15], electric scooters [16], or passenger vehicles [17–20].

Within flow-based planning approaches, charging demand is not assumed to be predefined for single points, but is estimated for traffic flows. These approaches have for instance been used for planning of classical gasoline networks [21] and have also been extended for alternative fuel vehicles [22–25]. State-of-the-art solution methodologies for these approaches can be found in [26, 27].

In general, the planning of charging infrastructure depends on the existing or expected charging demand, which in turn depends on the vehicles' routes. These routes are exogenously given for public charging infrastructure. However, logistics fleet operators decide on both, the charging infrastructure of the logistics networks as well as the routes of the operated vehicles. Therefore, integrated planning approaches have been developed that simultaneously determine route plans and a corresponding optimal charging infrastructure. First approaches were provided by [28] for battery swapping and by [29] for conventional charging of ECVs. Extensions focused on application cases [30], uncertain customer patterns [31], and generic solution methods [32, 33] that can be used for further analyses on a wide range of application cases. However, these approaches usually aim at minimizing the number of vehicles or charging stations. A comprehensive TCO calculation has not yet been considered in this context.

2.2 *Electric Vehicle Routing*

Vehicle routing problems (VRPs) are commonly used to design route plans for logistics fleets. Herein, the objective is either to route the vehicles in a cost-optimal way or with minimum total traveled distance. In recent years, several publications extended state-of-the-art VRP approaches to approaches for ECVs, considering intermediate stops for recharging on routes. Herein, [34] introduced the green VRP that focused on refueling stops for alternative fuel vehicles, while [35] introduced the electric VRP with time windows (EVRPTW) that focused explicitly on recharging of ECVs. From then on, several extensions of the EVRPTW have been published, focusing on charging technologies [36], energy consumption representations [37, 38], battery swapping [39], exact algorithms [40], and partial recharging [41, 42], as well as synchronization potential with hours of service regulations [43]. For a profound overview of these approaches, we refer the interested reader to [44]. Although these approaches take into account some of the characteristics of ECVs, this is mostly done at a rather aggregated level. Technical details, e.g., with regard to battery degradation, are usually not regarded.

2.3 Competitiveness Analysis

There are also publications that conduct competitiveness analyses of ECVs against ICEVs. These analyses are mainly based on aggregated estimations of the total driven distance and charging cycles, as well as on assumptions for further operational parameters. Estimated parameters are then discussed within a sensitivity analysis to analyze break even. A TCO analysis of medium-duty ECVs was presented by [45]. However, driving distances were limited between 48 and 96 km. A comparison between ECVs and ICEVs with a sensitivity analysis on fleet and battery replacement scenarios can be found in [46]. Davis and Figliozzi [47] consider at least approximated routing constraints. Another TCO analysis that focuses on a sensitivity analysis between vehicle utilization and battery degradation was presented in [48]. However, even if these analyses may provide a rough estimation of the applicability and benefit of ECVs in logistics fleets, they are not sufficient to eliminate the doubts of practitioners as not all practical and technical constraints are taken into consideration at an adequate level of detail (cf. [13]).

2.4 Battery Degradation

To keep the discussion concise, we assume in the following that the reader has a basic knowledge of battery technologies, especially of lithium-ion batteries, from previous or following chapters.

Within lithium-ion batteries, (dis-)charging processes are only partially reversible so that the technical properties of batteries decrease due to battery degradation effects. These degradation effects comprise two different kinds of degradation: (i) *calendric* and (ii) *cyclic* aging.

Calendrical battery degradation results from chemical side reactions during storage, mainly independent from the battery's utilization profile. These side reactions increase the storages inner resistance and, thus, worsen the specific energy and specific power characteristics [45, 46]. Calendrical degradation effects are the smaller:

- the less the time is in which a battery remains at high SOCs during its lifetime,
- the less the time is in which a battery is exposed to high temperatures.

Cyclic battery degradation results from the degeneration of the crystal structure of the storage medium due to mechanical stress caused by charging and discharging processes. Herein, the mechanical stress is proportional to the depth of discharge and the overall time span of charging and discharging cycles. Thus, deep discharge states should be avoided and charging and discharging cycles should be limited to a minimum to keep cyclic degradation as small as possible. Furthermore, charging processes at low temperatures lead to increasing degradation effects. Concluding, cyclic degradation effects can be reduced by avoiding:

- large charging and discharging cycles,
- deep discharging, and
- charging at low temperatures.

Within an ECV, more than 1000 lithium-ion batteries are connected to a battery pack to provide a sufficient energy storage. Herein, a battery management system (BMS) ensures a distortion-free mode of operation. The BMS automatically performs operations to avoid some of the mentioned battery degradation effects, such as overcharging, deep discharging, and charging at low temperatures. However, the time a battery is remaining at high SOC, the time a battery is exposed to high temperatures, and large charging and discharging cycles are not covered by the BMS. From these factors, temperature can be assumed as an exogenous factor that cannot directly be influenced by decisions on vehicle operations. Thus, (i) *the time a battery remains at high SOC* and (ii) *large charging and discharging cycles* remain as factors of interest for our analyses as these can directly be influenced by routing and charging decisions.

3 Real-World Application

In the following, we detail our real-world application case that provides the basis of our studies. First, we focus on the underlying field project ELMO from which we gained real-world data on charging processes and consumption profiles. Second, we illustrate our case study, which is based on the planning tasks of the logistics network of a large German retail company.

3.1 The ELMO Project

In 2011, the German government initiated the field project ELMO 'Elektromobile urbane Wirtschaftsverkehre' ('electrified commercial transport in urban areas'). At this time, data and experiences on ECVs within logistics networks (especially in mid-haul logistics) were still sparse, since most manufacturers were focusing on electric vehicles for the private sector or at small commercial vehicles that were based on the same chassis as private cars. Additionally, field tests were only focusing on short-haul logistics with vehicle payloads smaller than 3.5 tons.

Therefore, the intention of the ELMO project was to test ECVs in a mid-haul logistics setting and to gain empirical insights on overall applicability and technical or operational limitations of ECVs. Also, an overall assessment of ECVs and a comparison with ICEVs were aimed at as were recommendations for best practice and deductions on worthwhile application fields.

In ELMO, several companies collaborated with the Fraunhofer IML to integrate and track medium-duty ECVs in their daily logistics tasks. More than 3000 transportation records were retained, covering a total mileage of 158,209 km and yielding an energy consumption of 108,543 kWh. The field test provided first promising results, suggesting that medium-duty ECVs may get economically and technically competitive against ICEVs in certain mid-haul logistics settings.

However, only single vehicles were replaced and tracked. Also, routing and handling processes were changed as less as possible in order to allow for a comparison between (single) ECVs and ICEVs and to ensure acceptance for the operation of this new technology. Also, aspects like range and recharging were not regarded in routing decisions, and potentials and limitations of a further electrification of the logistics fleets were not considered. Against this background, the target of our study is to assess the advantage of an electrification of a complete mid-haul logistics network, i.e., not electrifying only single ECVs but as much vehicles as possible. We conduct our study for the specific case of the retail logistics company TEDi.

3.2 The TEDi Case

One company that took part in (and partly initiated) the ELMO project was the German non-food retailer TEDi, who operates more than 1400 stores all over Europe. Using ECVs instead of ICEVs in mid-haul distribution is of special interest for retail companies such as TEDi, because stores are often located in inner cities so that fleet operators have to cope with emission restrictions and fear a future ban of ICEVs from these areas.

In our case study, we focus on TEDi's logistics network within the federal state of North Rhine-Westphalia, the state in which the ELMO project was conducted. In North Rhine-Westphalia, 302 stores are supplied from a central warehouse. The stores are (nearly) equally distributed with a lensing number of stores in the outer regions and a higher number of stores in the central city regions. The most distant store is located 190 km away from the warehouse. The considered network can be assumed to be representative for a majority of mid-haul logistics networks in retail logistics.

The planning task within this network results as follows: Each store has to be supplied with new goods once in a week. The goods are transported on pallets or roller containers, and empty pallets or containers are picked up while delivering new goods and are then backhauled to the depot. For these tasks, a fleet of 12-tonne medium-duty trucks is used that start and end their tours at the central warehouse and are operated in a two-shift mode.

Within the ELMO project, two of these trucks were replaced by electric 12-tonne trucks of the model CM1216 of the manufacturer EMOSS. These electric trucks were used within a vicinity of 70 km around the central warehouse. However, the trucks were neither operated on specifically tailored routes, nor were larger vicinities considered, nor was recharging on routes addressed. Despite neglecting all these aspects, the results of the field tests within the ELMO project looked promising as the electrified EMOSS trucks had slight advantages compared to ICEVs.

In the following, we investigate if and how the whole logistics network of TEDi can be operated by ECVs. Other than existing approaches, we do not use aggregated analysis based on predefined route plans, but we use an integrated planning approach based on optimized strategic and operational decisions on charging station infrastructure, vehicle fleet size, and route plans using a mixed integer location-routing model. The results of this model allow for detailed TCO calculations, but also for an analysis of the batteries' SOC during the resulting tours regarding the depth of discharge and the number of charging cycles. Our analysis is backed by real-world data regarding the demand of the stores as well as the travel times, energy consumption and charging times that were gathered during the ELMO field test.

4 Mixed Integer Program

The planning task of TEDi can be described as a pickup and delivery, multi-period, multi-shift location-routing problem with intra-route facilities. To provide a formal basis for our analysis, we present a mixed integer program (MIP) for this problem, considering a TCO-based objective function as described in [13].

Given a set of vertices $\mathsf{V}_{0,n+1}$ and a set of arcs A, this MIP is defined on a directed, complete graph $G = (\mathsf{V}_{0,n+1}, \mathsf{A})$. Herein, $\mathsf{V}_{0,n+1}$ consists of a subset of customer vertices C, a set of vertices at which recharging facilities can be sited F, and a set of dummy vertices D to allow for multiple visits to charging stations. D is indexed by κ if it refers to a subset of D belonging to a single vertex κ. The depot is represented by vertices 0 for outgoing and $n+1$ for ingoing arcs. To model multiple shifts and periods, we use sets S and P respectively. To keep the model as concise as possible, we use cutsets (1)–(3) within the formal notation. Herein, (1) includes all arcs of an arbitrary subset B, while (2) refers to its outgoing and (3) to its ingoing arcs:

$$\delta(\mathsf{B}) = \{(i,j) \in \mathsf{A} : i \in \mathsf{B}, j \in \mathsf{B}\}, \tag{1}$$

$$\delta^{+}(\mathsf{B}) = \{(i,j) \in \mathsf{A} : i \in \mathsf{B}, j \notin \mathsf{B}\}, \tag{2}$$

$$\delta^{-}(\mathsf{B}) = \{(i,j) \in \mathsf{A} : i \notin \mathsf{B}, j \in \mathsf{B}\}. \tag{3}$$

Analogously, $\gamma(i)$ denotes all succeeding shifts of i.

Table 1 provides an overview of the variables and parameters that are necessary to state our MIP: To model routing decisions, binary x_{ij}^{s} states whether arc (i,j) is traversed or not in shift s. Accordingly, binary y_i depicts if a charging station is located at vertex i or not. The arrival time τ_i^s, the battery load q_i^s, the delivery freight load f_i^s, and the pickup freight load u_i^s at vertex i in work shift s, as well as the amount of energy w_i recharged at vertex i are used to trace arrival times, the battery's SOC and the freight load of a vehicle. Furthermore, z denotes the maximum number of vehicles used in one shift, i.e., the fleet size. Stores are characterized by an earliest e_i^s and latest l_i^s arrival time in-between which goods can be delivered and picked up. Delivering and picking up goods lasts s_i time units. Herein, the service time s_i depends on the demand of delivery freight p_i and pickup freight v_i at store i. Any arc (i, j) is characterized by a travel time t_{ij}, a distance d_{ij}, and an amount of consumed energy h_{ij}. Furthermore, a vehicle is characterized by its battery capacity Q and its freight capacity F. Assuming fast recharging, recharging of 1 kWh lasts r time units. To account for a TCO objective, we consider discounted cost terms for vehicles with respect to vehicle's investment costs $\overline{c}^{\text{inv,v}}$ and fixed costs $\overline{c}^{\text{fix,v}}$, investment costs for charging stations $\overline{c}^{\text{inv,s}}$, driver costs $\overline{c}^{\text{oper,dr}}$ and operational costs $\overline{c}^{\text{oper,di}}$.

Table 1 Variables and parameters of the MIP formulation

x_{ij}^s	Routing binary: arc (i, j) is traveled
y_i	Siting binary: recharging station is sited at vertex i
τ_i^s	Vehicle arrival time at vertex i
w_i	Recharged energy at vertex i
q_i^s	Vehicle's battery load at vertex i
f_i^s	Amount of delivery freight at vertex i
u_i^s	Amount of picked up freight at vertex i
z	Fleet size
e_i^s	Earliest time of service at store i
l_i^s	Latest time of service at store i
s_i	Service time at store i
p_i	Delivery freight demand at store i
v_i	Pickup freight demand at store i
t_{ij}	Driving time on arc (i, j)
d_{ij}	Distance of arc (i, j)
h_{ij}	Energy consumed by traversing arc (i, j)
r	Recharging rate
Q	Battery capacity
F	Freight capacity

Using this notation, our TCO objective function (4) results based on the determined fleet size and the calculated number of sited charging stations each multiplied with its cost coefficient, as well as the operational costs for traversing arcs and the number of driver shifts times the shift costs. The number of driver shifts is determined based on the traversed arcs that leave the depot. The resulting fleet size is given by constraints (5) and (6).

The remaining constraints focus on demand fulfilment at stores, herein complying with time restrictions and keeping vehicles operational with respect to battery limitations: Constraints (7)–(9) ensure that each store is served (7) and vertices are traversed at most once (7)–(8), while each vehicle finally returns to the depot (9). Freight capacity violations are prohibited by (10)–(12). While (10) contains the delivery freight balance, (11) accounts for the pickup freight balance, and (12) secures that the total freight on a vehicle does not exceed the vehicle's freight capacity on a tour. The time balance is given in constraints (13) and (14), accounting for service times at stores (13) and recharging times at charging stations (14). Note that charging is allowed to take place in parallel while a vehicle serves a store (cf. Sect. 2). Time window violations for all stores are prohibited by constraint (15). Constraints (16)–(19) obtain battery restrictions. While (16) and (17) obtain the energy balance without (16) and with (17) recharging, (18) and (19) secure that the battery's SOC always remains greater or equal to zero and below the battery's capacity limit. Constraints (20)–(22) comprise charging constraints and charging station location decisions. While (20) secures that charging is only possible if a charging station is available at vertex i, (21) secures that a charging station is only sited if a charging demand must be fulfilled. Constraint (22) mirrors the siting decisions to dummy vertices. Finally, (23) provides the ranges for the decision variables.

$$\begin{aligned}\text{minimize}\quad & \\ \mathrm{TCO}^{\mathrm{ECV}} = {} & \left(\bar{c}^{\mathrm{inv,v}} + \bar{c}^{\mathrm{fix,v}}\right) z + \bar{c}^{\mathrm{inv,s}} \sum_{i \in \mathsf{C} \cup \mathsf{F}} y_i + \bar{c}^{\mathrm{oper,dr}} \sum_{(i,j) \in \delta^+(0), s \in \mathsf{S}} x_{ij}^s \\ & + \bar{c}^{\mathrm{oper,di}} \sum_{(i,j) \in \delta\left(\mathsf{V}_{0,n+1}\right), s \in \mathsf{S}} d_{ij} x_{ij}^s \end{aligned} \tag{4}$$

s. t.

$$z \geq \sum_{j \in \mathsf{V}_{n+1}} x_{0j}^s \quad \forall s \in \mathsf{P} \tag{5}$$

$$z \geq \sum_{j \in \mathsf{V}_{n+1}} x_{0j}^s \quad \forall s \in \gamma(\mathsf{P}) \tag{6}$$

$$\sum_{(i,j) \in \delta^+(i), s \in \mathsf{S}} x_{ij}^s = 1 \quad \forall i \in \mathsf{C} \tag{7}$$

$$\sum_{(i,j)\in\delta^{+}(i)} x_{ij}^{s} \leq 1 \quad \forall i \in \{\mathsf{V}\backslash\mathsf{C}\}, s \in \mathsf{S} \tag{8}$$

$$\sum_{(j,i)\in\delta^{-}(i)} x_{ji}^{s} - \sum_{(i,j)\in\delta^{+}(i)} x_{ij}^{s} = 0 \quad \forall i \in \mathsf{V}, s \in \mathsf{S} \tag{9}$$

$$0 \leq f_j^s \leq f_i^s - p_i x_{ij}^s + F\left(1 - x_{ij}^s\right) \quad \forall (i,j) \in \delta\left(\mathsf{V}_{0,n+1}\right), s \in \mathsf{S} \tag{10}$$

$$0 \leq u_j^s \leq u_i^s + v_i x_{ij}^s + F\left(1 - x_{ij}^s\right) \quad \forall (i,j) \in \delta\left(\mathsf{V}_{0,n+1}\right), s \in \mathsf{S} \tag{11}$$

$$0 \leq f_i^s + u_i^s \leq F \quad \forall i \in \mathsf{V}_{0,n+1}, s \in \mathsf{S} \tag{12}$$

$$\tau_j^s \geq \tau_i^s + \left(t_{ij} + s_i\right)x_{ij}^s - l_0^s\left(1 - x_{ij}^s\right) \quad \forall (i,j) \in \delta^+(\mathsf{C}_0), s \in \mathsf{S} \tag{13}$$

$$\tau_j^s \geq \tau_i^s + t_{ij}x_{ij}^s + r\, w_i - \left(l_0^s + rQ\right)\left(1 - x_{ij}^s\right) \quad \forall (i,j) \in \delta^+(\mathsf{V}), s \in \mathsf{S} \tag{14}$$

$$e_i^s \leq \tau_i^s \leq l_i^s \quad \forall i \in \mathsf{V}_{0,n+1}, s \in \mathsf{S} \tag{15}$$

$$q_j^s \leq q_0^s - h_{ij}x_{ij}^s + Q\left(1 - x_{ij}^s\right) \quad \forall (i,j) \in \delta^+(0), s \in \mathsf{S} \tag{16}$$

$$0 \leq q_j^s \leq q_i^s + w_i - h_{ij}x_{ij}^s + Q\left(1 - x_{ij}^s\right) \quad \forall (i,j) \in \delta(\mathsf{V}_{n+1}), s \in \mathsf{S} \tag{17}$$

$$0 \leq q_0^s \leq Q \quad \forall s \in \mathsf{S} \tag{18}$$

$$0 \leq q_i^s + w_i \leq Q \quad \forall i \in \mathsf{V}, s \in \mathsf{S} \tag{19}$$

$$w_i \leq Q y_i \quad \forall i \in \mathsf{V} \tag{20}$$

$$y_i \leq w_i \quad \forall i \in \mathsf{D} \tag{21}$$

$$y_i \geq y_j \quad \forall i \in \{\mathsf{V}\backslash\mathsf{D}\},\ j \in \mathsf{D}_i \tag{22}$$

$$x_{ij}^s \in \{0;1\} \quad \forall (i,j) \in \delta\left(\mathsf{V}_{0,n+1}\right) \qquad y_i \in \{0;1\} \quad \forall i \in \mathsf{V} \qquad z \in \mathsf{N} \tag{23}$$

This MIP (4)–(23) is tailored to ECVs, but it can also be used for ICEVs by neglecting charging station costs in the objective and skipping constraints (14) and (16)–(22).

The MIP presented in this section provides a formal basis for our following studies. However, enhanced methods are necessary to solve large-sized instances and case studies. To keep this paper concise and focus on managerial and technical insights, we refer to [13] for an explanation of the matheuristic that has been used for our studies.

5 Design of Experiments

In the following, we describe our design of experiments. Herein, we stick to the basic experimental design presented in [13]. In [13], insights were derived into (i) *the competitiveness of ECVs* in mid-haul logistics compared to *ICEVs* and into (ii) *the ecological benefit* of integrating ECVs into mid-haul logistics. However, (iii) *technical limitations, especially battery degradation effects* for ECVs in mid-haul logistics are still missing. To keep this chapter concise, we will briefly discuss the insights from [13] on (i) and (ii), but focus our analysis on (iii).

In the ELMO project, TEDi has been operating ECVs only within a vicinity of 70 km around a central warehouse, without using tailored routes that consider the special characteristics of ECVs. Our experimental design focuses on the stepwise electrification of the whole fleet, i.e., serving customers by ECVs in increasing vicinities around the warehouse. Since the potential to operate ECVs within a vicinity of 70 km has already been proven in the ELMO project (without even tailoring routes), we start our experiments at a vicinity of 80 km and end at a vicinity of 190 km with which the whole distribution area could be covered. As Fig. 1 shows, the number of customers for the mentioned vicinities increases from 144 to 302. As can be seen, the number of stores grows less than proportional to increasing vicinities from the depot. This shows the typical structure of a retail logistics network in which a central warehouse is used and only single customers are located near the boundary of the service area.

We use real-world data regarding the stores' weekly demand, energy consumption profiles, recharging profiles, and travel speed. In our TCO analysis, we use the data of a MAN TGL 12.250 for a 12-tonne ICEV truck and the data of an EMOSS CM1216 as a representative 12-tonne ECV truck. The considered cost values hold as stated in Table 2, wherein we use a discount rate of 5% and a planning horizon of 5 years for the TCO calculations. We refer to [13] for a detailed discussion of the used data.

Fig. 1 Number of stores dependent on the service area vicinity

Table 2 Cost data for the representative trucks

	MAN TGL	EMOSS CM1216
Purchase price	75,000 €	160,000 €
Yearly taxes	534 €/a	0
Driver wage	29,000 €/a	29,000 €/a
Battery capacity	–	160 kWh
Consumption	0.19 l	0.73 kWh/km
Energy price	1.18 €/l	0.07 €/kWh
CO_2 equivalent	2624.89 $gCO_{2,eq}$/l	507.97 $gCO_{2,eq}$/kWh

6 Results

In the following, we discuss results for the TEDi case. As stated above, we present results on (i) *economic* and (ii) *ecological* viability of ECVs in mid-haul logistics regarding increasing customer vicinities. Afterward, we analyze the number of recharging stops on routes with resulting (dis-)charging cycles to highlight (iii) *battery degradation aspects* in ECV mid-haul applications.

6.1 Economic and Ecological Benefits

The TCO and CO_2 assessments are based on the network operation structure that results after solving the corresponding optimization problem for ECVs and ICEVs. Thus, we briefly present this structure in Fig. 2 showing the number of stores C, the number of vehicles Z, the number of necessary driver work shifts W, the weekly distance driven D (km 10^2), the average distance per tour $\overline{D}$ (km), and the number of charging stations Y built for each service area vicinity.

As can be seen, all quantities increase monotonously with an increasing service area vicinity. While the total distance driven increases proportionally to the number of stores, all other quantities increase with a lower gradient. As can be seen, the installation and usage of charging stations is not necessary up to a vicinity of 100 km. Moreover, (nearly) no operational disadvantages exist in the TEDi network by using ECVs instead of ICEVs, since the number of necessary vehicles and the number of driver shifts remain equal for both technologies, and the total driven distance is equal for all but the 180 km instance.

Based on the quantities of Fig. 2, Table 3 and Fig. 3 detail, the total cost of ownership for the respective service area vicinities served by ICEVs (TCO ICEV) or served by ECVs (TCO ECV). For ECVs, we conduct a sensitivity analysis with respect to the investment cost for vehicles and charging stations. In the l-scenario, our calculations are based on low investment costs for charging stations of 4500 € per station as was the case in the TEDi project. However, for different fields of

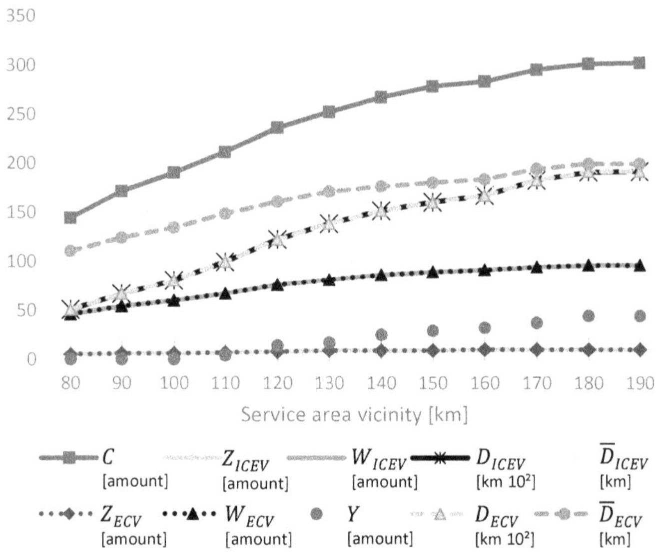

Fig. 2 Fleet structure for ECVs and ICEVs depending on the service area vicinity

Table 3 Total cost of ownership for ICEVs and different ECV scenarios

Service area vicinity (km)	80	90	100	110	120	130	140	150	160	170	180	190
TCO ICEV (m€)	1.72	2.06	2.28	2.61	3.01	3.28	3.47	3.59	3.74	3.89	3.98	3.98
Scenario l: low charging station costs												
TCO ECV (100%) (m€)	1.69	2.01	2.18	2.49	2.88	3.13	3.31	3.41	3.58	3.70	3.79	3.79
TCO ECV (90%) (m€)	1.65	1.95	2.12	2.42	2.80	3.05	3.22	3.33	3.48	3.60	3.69	3.69
TCO ECV (80%) (m€)	1.60	1.89	2.06	2.35	2.72	2.96	3.13	3.24	3.39	3.50	3.59	3.59
Scenario h: high charging station costs												
TCO ECV (100%) (m€)	1.69	2.01	2.18	2.59	3.24	3.57	3.95	4.15	4.40	4.64	4.91	4.91
TCO ECV (90%) (m€)	1.65	1.95	2.12	2.52	3.16	3.48	3.86	4.06	4.30	4.54	4.81	4.81
TCO ECV (80%) (m€)	1.60	1.89	2.06	2.45	3.08	3.39	3.77	3.98	4.20	4.45	4.72	4.72

application or logistics networks investment costs for charging stations might be significantly higher. To account for this varying cost structure, the h-scenario considers high investment costs for charging stations of about 30,000 € per station.

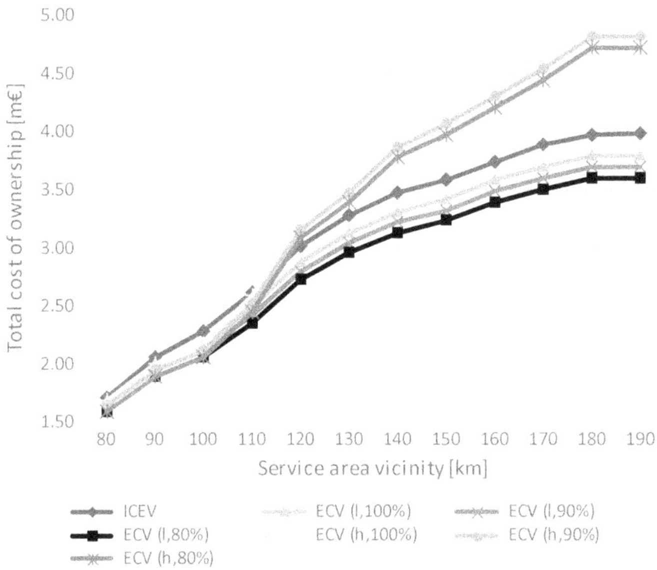

Fig. 3 Total cost of ownership for ECVs and ICEVs depending on the service area vicinity

As can be seen, for the basic scenario from the TEDi application case [ECV (l, 100%)], total costs are lower for ECVs than for ICEVs for all service area vicinities. Herein, the usage of ECVs instead of ICEVs yields cost savings between 1.33 and 4.90%. These cost savings increase to up to 10%, if the investment costs for vehicles decrease by 20% and investments for charging stations remain constant. However, if investment costs for charging stations increase to 30,000 € as in the h-scenario, the cost savings of ECVs vanish for service area vicinities above 110 km or 100 km depending on the vehicle investment costs. In the worst case, the usage of ECVs yields 23.32% higher costs than the operation of ICEVs within the network.

Concluding this sensitivity analysis shows that ECVs become competitive compared to ICEVs for special application cases and for particular cost structures. However, the potential benefit of operating ECVs depends heavily on the network structure and cost parameters at hand. If high investment costs for charging stations are required, ECVs are not competitive within large service are vicinities where charging on routes is indispensable. This even holds if the investment costs for ECVs decrease significantly. In this case, the number of necessary charging stations increases due to limited battery capacities and the cost savings in vehicle investment costs are not sufficient to compensate the additional investment costs for charging stations.

Figure 4 shows results on the total CO_2 emissions for both the operation of the network with a fleet of ICEVs and with a fleet of ECVs. ECVs emit around 25% less CO_2 emissions compared to ICEVs for a well-to-wheel system boundary

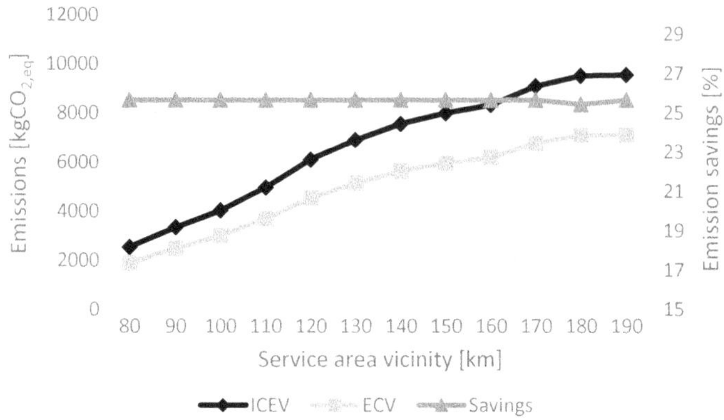

Fig. 4 Emissions and emission savings depending on the service area vicinity

(i.e., regarding all fuel/electricity emissions from the generation to the combustion respectively transmission). Since the operational quantities, especially the total driven distance, remain equal between ICEVs and ECVs for all but the service area vicinity of 180 km (cf. Fig. 3), the emission savings reflect the higher efficiency of ECVs due to the CO_2 conversion coefficients from Table 2.

6.2 *Battery Degradation Effects*

In the following, we analyze the impact of increasing service area vicinities on the factors that influence battery degradation that are subject to routing decisions and are not controlled by the BMS, i.e., times in which batteries remain at a high SOC, large charging and discharging cycles, and the number of discharging cycles. Herein, we analyze the number of recharges and the deepest discharge for each vicinity and tour. Doing so, we are capable of deriving insights into the number of discharging cycles (that correlate with the number of recharges on routes). Furthermore, implications of large (dis-)charging cycles and high SOCs are derived based on information on the deepest discharge.

Figure 5 shows the distribution of the deepest discharges on routes for each service area vicinity. For each group of bars, the percentages vary from discharges between 0–10% and 90–100%. As can be seen, the share of routes with a discharge of more than 90% of the battery's capacity increases significantly with increasing service area vicinities, starting at 2.17% (80 km) and ending with up to 53.13% (190 km). One could assume that similar trends exist for the other ranges of deepest discharges. However, this is not the case as can be seen for discharges between 20 and 90% of the battery's capacity. To explain this effect, we have to take into account the relationship between the number of stores and their distribution in the service area (cf. Figs. 1 and 2). As can be seen, the number of stores that is added to the service area decreases with increasing vicinities, while the added customers are

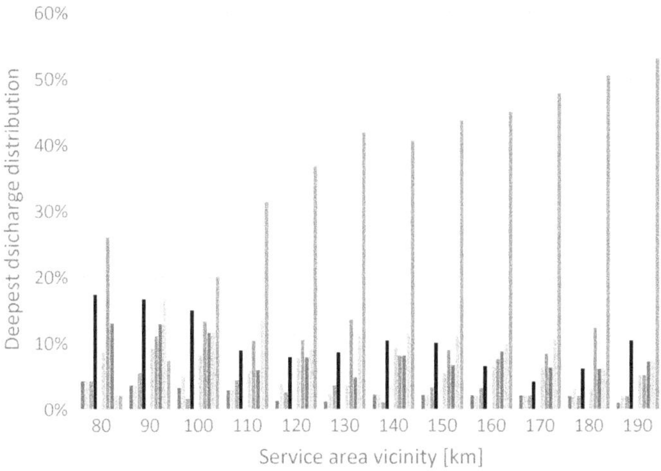

Fig. 5 Distribution of deepest discharges on routes with respect to the service area vicinity

still distributed over the whole plane. As the objective of our MIP is to minimize total costs [cf. (4)] (which is the most likely objective used in practice), the added stores are integrated into existing tours, i.e., tours that were designed for a smaller vicinity are split up to integrate new stores. This means that tours get longer and, thus, the proportion of tours where the maximum battery capacity is used increases. Thus, a shift to deeper discharges and a trend for deepest discharges above 90% can be observed, but other levels of discharge vary irregularly. Shares remain below 5% for deepest discharges between 0–10% or 10–20% independent of the service area vicinity. Note that results would change if another objective function would be used, i.e., minimizing the number of tours with deep discharges.

To analyze the shares of deepest discharges that accelerate the battery degradation in more detail, Fig. 6 shows the share of routes for which the battery's SOC remains at a high level (0–20%) or for which a large discharge occurs (80–100%). As can be seen, the share of routes with degradation accelerating SOCs or discharges increases significantly with increasing service area vicinities from 13.04% (80 km) to 63.54% (190 km). Herein, a very steep increase up to 50.57% can already be observed within the vicinity extensions up to 110 km. For all but the 80 km vicinity, routes that show a deep discharge constitute the majority of routes with accelerated battery degradation.

Figure 7 shows the share of routes with zero, one, or two recharging stops for each service area vicinity. Coherently to the findings of Fig. 2, no recharging stops occur on routes for service area vicinities up to 100 km. For larger vicinities, the number of routes with recharging stops increases proportionally to the increase of charging stations in Fig. 3. Routes with more than one recharging stop are rather scarce and occur only for vicinities of 180 km or 190 km.

Concluding, the findings of this section can be summarized as follows: All of the discussed impacts on accelerated battery degradation arise as soon as ECVs are

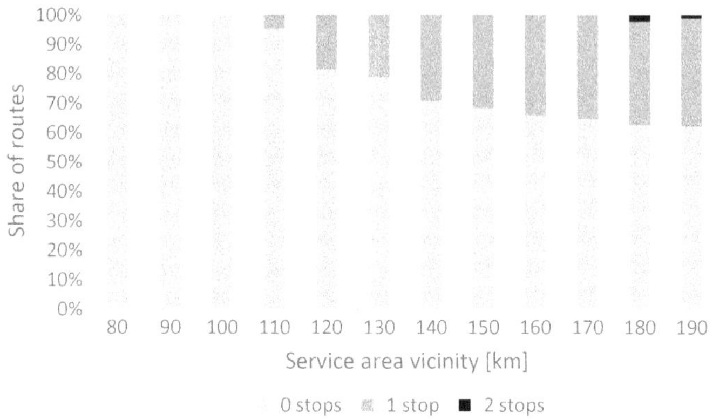

Fig. 6 Shares of routes with recharging stops with respect to the service area vicinity

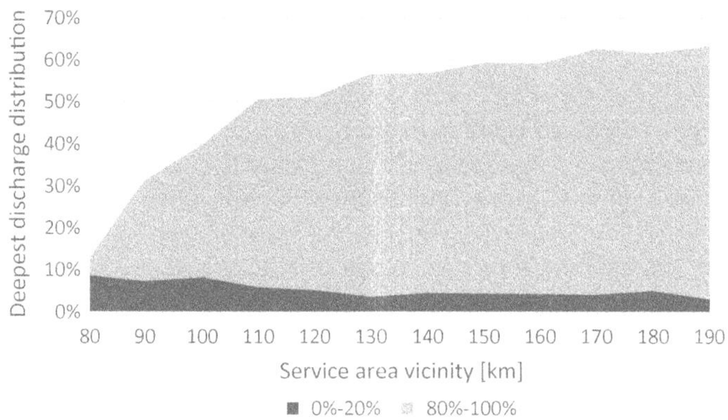

Fig. 7 Battery degradation accelerating shares of deepest discharge

used in mid-haul transportation, i.e., if distances of more than 100 km have to be covered. However, an increase in the number of (dis-)charging cycles occurs for less than 40% of the routes. This is significantly lower than assumed by aggregated TCO analysis that suspects additional recharges on (nearly) all routes. Also, accelerated battery degradation due to batteries remaining at high SOCs affects only 3–9% of the routes, dependent on the service area vicinity. The main accelerator of battery degradation is deep discharges due to increasing tour lengths. For large service area vicinities of more than 130 km, these discharges occur for more than 50% of the routes. Although an increased number of discharges occurs, the number of discharges per vehicle can easily be kept below 3000 cycles for the assumed planning period of 5 years by shifting vehicles between routes over time. Thus, the planning horizon of our experiments is appropriate to the assumption to exclude battery replacement costs that may arise if more than 3000 cycles are needed (cf. [47]).

7 Conclusion

In this chapter, we studied the usage of ECVs in mid-haul transportation networks. Herein, we introduced a case study that considers the logistics network of a large German retail company for a federal state of Germany. This case study is based on real-world data from the company itself and ECV characteristics and consumption profiles collected during a large field project. We introduced a MIP as a formal basis that allows for a detailed analysis of ECVs in mid-haul logistics networks, considering strategic aspects, i.e., the design of the charging station network and the vehicle fleet size, as well as operational limitations, i.e., routing and recharging decisions.

We conducted several experiments to overcome the limited perspective of studies that are restricted to a single aspect, either scientifically (e.g., a pure economic or technical perspective) or from the planning perspective (e.g., a pure strategic network design or a pure operational routing decision). We studied the competitiveness of ECVs against ICEVs with respect to economic and ecological benefits. Furthermore, we studied battery degradation accelerating factors for the resulting networks and operation modes.

Results show that ECVs are on the verge of breaking even for special application cases, yielding savings in total costs of up to 5%. However, this cost savings potential depends on the characteristics of the underlying network and the cost structure of the vehicles and charging stations. Especially, results may be heavily affected by the setup cost for charging stations. Emission savings result due to a higher efficiency and lower CO_2 emission coefficients for ECVs. Herein, a win–win situation occurs for the presented case study. Focusing on battery degradation effects, deep discharges reveal to be the most crucial limitation for large service area vicinities above 130 km.

These results highlight the potential of ECVs to contribute to sustainable transportation networks even in mid-haul logistics with large driving ranges. However, the presented results are case specific and have to be recalculated for different networks and planning scenarios. The methodology presented in this chapter provides a generic basis for these further studies.

References

1. European Environment Agency, EU. Energy, transport and GHG emissions trends to 2050 (2013). Available from: https://ec.europa.eu/transport/sites/transport/files/media/publications/doc/trends-to-2050-update-2013.pdf. Accessed 07.06.17
2. European Commission, EU. Transport in figures, Statistical pocketbook (2014). Available from: https://ec.europa.eu/transport/facts-fundings/statistics/pocketbook-2014_en. Accessed 07.06.17
3. European Union, Urban access regulation in Europe (2016). Available from: http://urbanaccessregulations.eu/userhome/how-many-access-regulations. Accessed 07.06.17

4. Statista, Annual retail e-commerce sales growth worldwide from 2014 to 2020 (2016). Available from: https://www.statista.com/statistics/288487/forecast-of-global-b2c-e-commerce-growt. Accessed 07.06.17
5. European Commission, Report from the commission to the European Parliament and the council—progress towards achieving the Kyoto and EU 2020 objectives (2014). Available from: http://ec.europa.eu/transparency/regdoc/rep/1/2014/EN/1-2014-689-EN-F1-1.Pdf. Accessed 07.06.17
6. DPDHL, Deutsche Post testet 12 neue Renault-Elektrofahrzeuge (2011). Available from: http://www.dpdhl.com/de/presse/pressemitteilungen/2011/deutsche_post_testet_12_neue_renault_elektrofahrzeuge.html. Accessed 07.06.17
7. M. Pieringer, Deutsche Post wechselt in Bonn und Umland komplett auf Elektrofahrzeuge (2013). Available from: http://www.logistik-heute.de/Logistik-News-Logistik-Nachrichten/Markt-News/10480/Deutsche-Post-DHL-wechselt-in-Bonn-und-Umland-komplett-auf-Elektro fahrzeuge. Accessed 07.06.17
8. DPDHL, Deutsche Post DHL Group macht Bonn zur Musterstadt für CO2-freie Zustellfahrzeuge (2013). Available from: http://www.dpdhl.com/de/presse/veranstaltungen/co2freie_zustellfahrzeuge.html. Accessed 07.06.17
9. DPDHL, Electric vehicles in inner city distribution traffic (2014). Available from: http://www.haw-hamburg.de/fileadmin/user_upload/FakLS/07Forschung/FTZ-ALS/Veranstaltungen/Fuelling_the_Climate/Lohmeyer_FTC2014_VOE.pdf. Accessed 07.06.17
10. DPDHL, Deutsche Post DHL übernimmt StreetScooter GmbH (2014). Available from: http://www.dpdhl.com/de/presse/pressemitteilungen/2014/deutsche_post_dhl_uebernimmt_street scooter_gmbh.html. Accessed 07.06.17
11. UPS: UPS to Rollout Fleet of Electric Vehicles in California (2013). Available from: http://www.pressroom.ups.com/Press+Releases/Archive/2013/Q1/UPS+to+Rollout+Fleet+of+Electric+Vehicles+in+California. Accessed 07.06.17
12. S. Stütz, A. Bernsmann, T. Baltzer, N. Hentschel, K. Pommerenke, B. Rogmann, P. Wunderlin, Elmo-Elektromobile Urbane Wirtschaftsverkehre (2016). Available from: http://www.iml.fraunhofer.de/content/dam/iml/de/documents/OE%20320/Infoseiten%20Abteilung%20und%20Gruppen/ELMO-Abschlussbericht_(%C3%96ffentliche_Fassung).pdf. Accessed 07.06.17
13. M. Schiffer, S. Stütz, G. Walther, Are ECVs breaking even? Competitiveness of electric commercial vehicles in retail logistics (2017). Available from: https://www.gerad.ca/en/papers/G-2017-47
14. F. Baouche, R. Billot, N.-E. El Faouzi, R. Trigui, Electric vehicle charging stations allocation model. ROADEF-15ème congrès annuel de la Société française de recherche opérationnelle et d'aide à la decision (2014)
15. H. Cai, X. Jia, A.S. Chiu, X. Hu, M. Xu, Transp. Res. D Transp. Environ. **33**, 39 (2014)
16. Y.-W. Wang, Transp. Res. D Transp. Environ. **13**, 193 (2008)
17. T.D. Chen, K. Kockelman, Transp. Res. Rec. **1254**, 28 (2013)
18. Y.-W. Wang, C.-C. Lin, Transp. Res. E Logist. Transp. Rev. **45**, 16 (2009)
19. Y.-W. Wang, C.-R. Wang, Transp. Res. E Logist. Transp. Rev. **46**, 791 (2010)
20. J. Cavadas, G.H. de Almeida Correia, J. Gouveia, Transp. Res. E Logist. Transp. Rev. **75**, 188 (2015)
21. M.J. Hodgson, Geogr. Anal. **22**, 270 (1990)
22. M. Kuby, S. Lim, Socio-Econ. Plan. Sci. **39**, 125 (2005)
23. M. Kuby, S. Lim, Netw. Spat. Econ. **7**, 129 (2007)
24. S. Lim, M. Kuby, Eur. J. Oper. Res. **204**, 51 (2010)
25. C. Upchurch, M. Kuby, S. Lim, Geogr. Anal. **41**, 85 (2009)
26. I. Capar, M. Kuby, V.J. Leon, Y.J. Tsai, Eur. J. Oper. Res. **227**, 142 (2013)
27. S.A. MirHassani, R. Ebrazi, Transp. Sci. **47**, 617 (2013)
28. J. Yang, H. Sun, Comput. Oper. Res. **55**, 217 (2015)
29. M. Schiffer, G. Walther, Eur. J. Oper. Res. **260**, 995 (2017)

30. M. Schiffer, S. Stütz, G. Walther, Are ECVs breaking even? Competitiveness of electric commercial vehicles in medium-duty logistics networks (2016). Available from: https://www.om.rwth-aachen.de/data/uploads/om-022016.pdf
31. M. Schiffer, G. Walther, Strategic planning of electric logistic networks: a robust location routing approach. Omega (2017)
32. M. Schiffer, G. Walther, An adaptive large neighborhood search for the location-routing problem with intra-route facilities. Transp. Sci. (2017)
33. M. Schiffer, M. Schneider, G. Laporte, Eur. J. Oper. Res. **265**, 517 (2017)
34. S. Erdogan, E. Miller-Hooks, Transp. Res. E Logist. Transp. Rev. **48**, 100 (2012)
35. M. Schneider, A. Stenger, D. Goeke, Transp. Sci. **48**, 500 (2014)
36. A. Felipe, M.T. Ortuno, G. Righini, G. Tirado, Transp. Res. E Logist. Transp. Rev. **71**, 111 (2014)
37. D. Goeke, M. Schneider, Eur. J. Oper. Res. **245**, 81 (2015)
38. J. Lin, W. Zhou, O. Wolfson, Transp. Res. Proc. **12**, 508 (2016)
39. A. Verma, K. Lamsal, S. Keough, 2015 Electric vehicle routing problem with time windows, recharging stations and battery swapping stations (2015). Available from: https://www.researchgate.net/publication/283225325_Electric_Vehicle_Routing_Problem_with_Time_Windows_Recharging_Stations_and_Battery_Swapping_Stations
40. G. Desaulniers, F. Errico, S. Irnich, M. Schneider, Oper. Res. **64**, 1388 (2016)
41. M. Keskin, B. Catay, Transp. Res. C Emerg. Technol. **65**, 111 (2016)
42. A. Montoya, C. Guret, J.E. Mendoza, J.G. Villegas, Transport. Res. B Methodol. **103**, 87 (2017)
43. M. Schiffer, G. Laporte, M. Schneider, G. Walther, The impact of synchronizing driver breaks and recharging operations for electric vehicles (2017). Available from: https://www.gerad.ca/en/papers/G-2017-46
44. M. Schiffer, M. Schneider, G. Walther, G. Laporte, Vehicle routing and location-routing with intermediate stops: a review (2017). Available from: https://www.gerad.ca/en/papers/G-2017-45
45. D. Aurbach, K. Gamolsky, B. Markovski, G. Salitra, Y. Gofer, U. Heider, R. Oesten, M. Schmidt, J. Electrochem. Soc. **147**, 1322 (2000)
46. A. Wezdanz, W. Jossen, Moderne Akkumulatoren richtig einsetzen (2006)
47. B.A. Davis, M.A. Figliozzi, Transp. Res. E Logist. Transp. Rev. **49**, 8 (2013)

Mechanical Design and Packaging of Battery Packs for Electric Vehicles

Shashank Arora and Ajay Kapoor

Abstract Safety and reliability are the two key challenges for large-scale electrification of road transport sector. Current Li-ion battery packs are prone to failure due to reasons such as continuous transmission of mechanical vibrations, exposure to high impact forces and, thermal runaway. Robust mechanical design and battery packaging can provide greater degree of protection against all of these. This chapter discusses design elements like thermal barrier and gas exhaust mechanism that can be integrated into battery packaging to mitigate the high safety risks associated with failure of an electric vehicle (EV) battery pack. Several patented mechanical design solutions, developed with an aim to increase crashworthiness and vibration isolation in EV battery pack, are discussed. Lastly, mechanical design of the battery pack of the first fully electric bus designed and developed in Australia is presented. This case study showcases the benefits of adopting modularity in the design of EVs. In addition, it highlights the importance of packaging space for EVs, particularly in low-floor electric buses, as weight distribution becomes a challenge in these applications.

Keywords Modular design · Thermal runaway · Vehicle impact and crash protection · Vibration isolation · Gas exhaust/venting mechanism · Electric bus

S. Arora (✉)
School of Engineering, Aalto University, 02150 Espoo, Finland
e-mail: shashank.arora@aalto.fi

A. Kapoor
Faculty of Science, Engineering and Technology, Swinburne University of Technology, Hawthorn, VIC 3122, Australia
e-mail: akapoor@swin.edu.au

G. Pistoia and B. Liaw (eds.), *Behaviour of Lithium-Ion Batteries in Electric Vehicles*, Green Energy and Technology,
https://doi.org/10.1007/978-3-319-69950-9_8

1 Introduction

Paris agreement has united 195 countries of the world who share a common goal of limiting the greenhouse gas emissions and gradually building a carbon-free society. Significant efforts are thus being focussed on increasing the share of renewable energy in the total energy generated in these regions [1]. However, the problem of intermittency affects all renewable energy resources. Use of battery packs to add an energy buffer and increase flexibility of the electric grids is considered a reliable as well as a sustainable solution for the problem of intermittency associated with renewable energy sources [2–4]. Also, battery-powered vehicles have the potential to substantially cut the greenhouse gas emissions from the transport sector. Electrification of transportation sector is thus integral to the long-term climate control policies of all nations.

Among the commercially available battery chemistries, Li-ion batteries offer features such as high efficiency, high gravimetric and volumetric densities, longer lifespan and low maintenance requirements that are all essential for setting up an efficient energy storage system [5–7]. Currently, the cost of manufacturing an EV battery pack is about $500 per kWh. However, with efforts to modify the microstructure of electrode materials for Li-ion batteries [8–12], the cost is expected to decrease to $200 per kWh by 2020 and $160 by 2025 [13]. Lastly, Li-ion batteries containing non-toxic metals such as iron, nickel, manganese, cobalt, have been classified as "non-hazardous waste and safe for disposal in the normal municipal waste stream" by the US government [14]. For these reasons, they are the preferred choice for the majority of high energy or high power applications in present times.

Li-ion batteries have long been used in a single-cell format for small portable electronic devices. Due to the fairly limited energy content of such cells, it was believed that failure of a single cell, which has been thoroughly investigated and relatively well understood [15–19], would have little impact beyond its surroundings. However, as these cells are now being scaled up and configured to find applicability in energy storage system for electric grids and vehicle electrification programmes, single-cell failures affecting neighbouring cells and damaging the entire battery pack are regularly reported. A gap lies in our understanding of the behaviour of large battery packs under abusive conditions [20, 21]; therefore, careful consideration must be given to design a Li-ion battery-based energy storage system for the targeted application.

2 Design Considerations

A simplified representation of an electric bus is presented in Fig. 1. It shows in a block format, various electromechanical systems such as electric motors, electric HVAC unit, electric air compressor and various types of controls that demand energy or act as load for the installed battery packs.

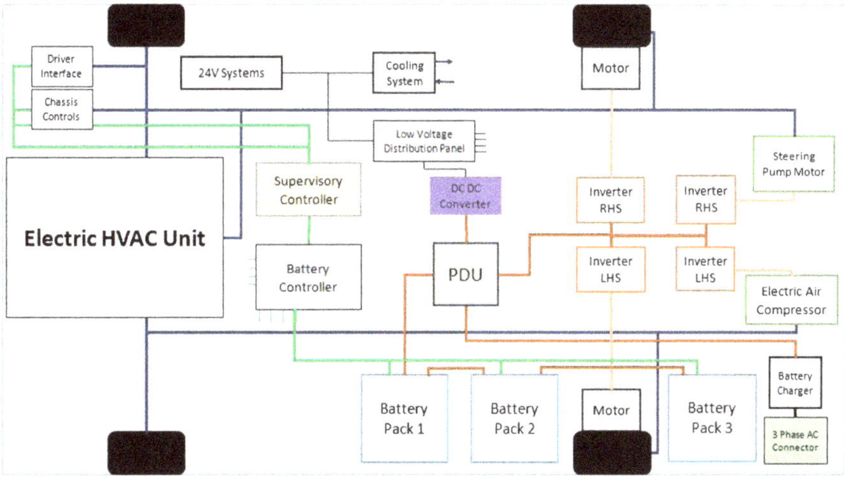

Fig. 1 Simplified representation of an electric vehicle. HVAC heating, ventilation, air conditioning; PDU power distribution unit; L/RHS left-/right-hand side

Table 1 Definition of various system interfaces in an EV battery pack [22]

Interface	Definition	Formed by
Mechanical	Mechanical design features included for safety reasons	Cell spacers, damping pads, gaskets, valves
Structural	Members that provide required protection and isolation	Case, cover, end plates, tie rods, cross-members
Thermal	Regulates battery cell temperature	Coolant, fans, pumps, heat exchangers
Electrical	Transmits power from, and to, the battery pack	Bus bars, cables, contactors, fuse, relays
Control	Monitor and regulate the state of battery pack	Battery management system, various sensors
Support	Vehicle body parts providing additional crash worthiness	Axles, chassis, seats, vehicle floor

It can be easily inferred from this basic schematic that an EV battery pack communicates with different sub-systems and multiple parameters simultaneously through various interfaces. Different interfaces, as visualised in the case of an EV battery pack, are briefly described in Table 1 [22].

Communication through each of these interfaces can influence reliability and safety of the battery pack and needs regulation. For example, it has been suggested that the battery temperature must be maintained below 50 °C for safe operation [23, 24]. The vibration frequencies of the battery pack should also be suppressed to avoid resonance at typical natural frequencies of the vehicle suspension system and sprung mass from 0 to 7 Hz, the vehicle power train, i.e. driveline and gearbox, from 7 to 20 Hz, and the vehicle chassis system from 20 to 40 Hz [25]. Marginal deviations from the designed boundary can compromise the cycle life of the battery pack. It can also set in motion an uncontrolled chain of exothermic reactions resulting in the release of smoke or toxic gas and the development of high-pressure events leading to premature failure, fire and explosions. These marginal deviations can be caused by excessive heat build-up or physical abuse of battery packs that includes puncturing or crushing the packs [26].

A reliable battery packaging design should address issues relating to thermal stability, vibration isolation and impact resistance at micro- as well as macro-level. Further, it should minimise thermal and mechanical interactions between different units of the battery pack at each level, i.e. at cell and module level, thus reducing the probability of failure of the battery pack itself. Design elements that can be optimised readily to achieve the required level of protection without much impact on available resources are called control factors [22]. Some of the most critical control factors of an EV battery pack are: battery cells and cell spacer type, number and location of gas exhaust nozzles, battery cooling system and insulation coating thickness. A rule of thumb for identification of control factors is: any factor that lies outside the system boundary is not regarded as a control factor.

As aforesaid, battery cell type has a significant influence on design of the battery packs. For example, it has been found that packing density of a battery pack with *18,650* type cells is 114 times more than that of a pack comprising large prismatic cells. Moreover, the packing density of a pouch cell is approximately two times lesser than that of a prismatic cell of similar nominal capacity mainly because of its smaller thickness and large surface area. It is therefore relatively easier to improve volumetric efficiency of the battery pack by packaging large quantities of smaller cylindrical cells in the available space than to use large prismatic or pouch cells [27].

Compactness of packaging design also has an appreciable impact on thermal performance of the battery pack. Research shows that increasing the cell-to-cell spacing for a battery pack from 1 to 10 mm can lead to a loss of approximately 1 ° C in the steady-state cell core temperature, for all the three physical formats [28]. According to NASA-Battery Safety Requirements Document (JSC 20,793 Rev C), cell spacing is more critical for pack designs employing battery cells of gravimetric energy density greater than 80 Wh/kg [29]. It has further been ascertained that to alleviate cell-to-cell heat propagation in the instance of a single-cell failure or a thermal runaway event, a minimum spacing of 2 mm is required for cylindrical cell formats. In addition, a physical barrier between neighbouring cells is required for the same reasons in battery packs that employ cell formats with side vents [30]. Other important design requirements are specified by various international standards; SAE standards applicable to mechanical design and testing of automotive battery packs are listed in Table 2 [26].

Table 2 SAE standards governing mechanical design of automotive battery packs [26]

Standard	Title	Scope
SAE J240	Life test for automotive storage batteries	Life test simulates automotive service when the battery operates in a voltage-regulated charging system
SAE J1766	Recommended practice for EV and hybrid vehicle battery systems crash integrity testing	Specifies test methods and performance criteria which evaluate battery spillage, retention and electrical isolation during specified crash tests
SAE J1797	Packaging of electric vehicle battery modules	Provides for common battery designs through the description of dimensions, termination, retention, venting system and other features required in an EV application
SAE J1798	Recommended practice for performance rating of electric vehicle battery modules	Common test and verification methods to determine EV battery module performance. Document describes performance standards and specifications
SAE J2185	Life test for heavy-duty storage batteries	Simulates heavy-duty applications by subjecting the battery to deeper discharge and charge cycles than those encountered in starting a vehicle
SAE J2289	Electric-drive battery pack system: functional guidelines	Describes practices for design of battery systems for vehicles that utilise a rechargeable battery to provide or recover traction energy
SAE J2344	Technical guidelines for electric vehicle safety	Defines safety guideline information that should be considered when designing electric vehicles for use on public roadways
SAE J2380	Vibration testing of electric vehicle batteries	Describes the vibration durability testing of an EV battery module or battery pack
SAE J2464	Electric vehicle battery abuse testing	Describes a body of tests for abuse testing of EV batteries
SAE J2929	Electric and hybrid vehicle propulsion battery system safety standard	Safety performance criteria for a battery system considered for use in a vehicle propulsion application as an energy storage system galvanically connected to a high-voltage power train

2.1 Thermal Runaway Protection

Manufacturing defects or events such as physical abuse and internal short circuit can push a battery cell into a state of thermal runaway. Thermal runaway is categorised as an exothermic chain reaction in which self-heating rate of a battery cell is more than 0.2 °C/min [31]. It can cause the battery to vent large quantities of flammable gases, emit jet of effluent materials and even combust spontaneously [32].

High-temperature gases and effluents emanating from the damaged cell pose safety risk to material/property in close proximity of it and also to vehicle passengers and first responders. High-pressure build-up inside the pack enclosure, due to the flammable gases, can also cause explosive failure of battery packaging if the gases cannot readily escape from the enclosure. It is thus recommended to include at least one pressure release valve, designed to set off at a pre-specified pressure, to minimise the safety risks posed to the EV and its passengers by an unknown point of failure. The damage to property and safety risks to the vehicle passengers and the first responders can be minimised by strategically controlling the direction of release of hot fumes and gases from the packaging. The damage can also be controlled by restricting thermal interactions between different entities of the battery pack. The point of egress of hot gases is controlled by incorporating one or more gas exhaust nozzles, shown in Fig. 2, that are designed to open during a battery thermal runaway event, while the spread of thermal runaway to larger area of the battery pack can be prevented by placing appropriate thermal barriers in the packaging.

2.1.1 Point of Egress

A battery cell does not necessarily need to be in a state of thermal runaway to emit hot gases and effluents. An exhaust gas nozzle can minimise the vehicle damage and safety risks by directing the hot material in a direction where no one would get affected by the hot gases leaving the battery pack. During EV's normal operation, a seal keeps the nozzle assembly closed and restricts entry of road debris and moisture into the battery pack. A pressure equalisation valve with cracking pressure in the range of 0.5–1.0 psi, i.e. much less than the pressure encountered during a thermal runaway event, is integrated into the exhaust nozzle to provide a means for handling pressure differentials due to non-thermal events (e.g. altitude variations). Hollow structural elements are included in the battery pack configuration to guide the flow of hot gas and material from the cell experiencing the thermal event to the exhaust nozzle.

Nozzle seal that keeps the gas exhaust nozzle closed during normal operations is held in its place by a nut. During thermal runaway, both the pressure and the temperature within the battery pack enclosure increase. Eventually, the nut melts and/or sufficiently deforms to allow the pressure within the pack enclosure to force the nozzle seal out of the nozzle. However, as the nozzle and its mounting assembly are fabricated from high-temperature materials such as steel or ceramic, they do not get affected by the increasing temperature and continue to guide the hot gases in a direction that minimises any personal loss or property damage.

Efficiency of the thermal design can be further increased by using perforated battery compartments. Effluents generated by the battery cells enter the hollow guideways formed within the battery pack through these perforations. The guideways direct the effluents to a gas exhaust nozzle, which releases it out of the battery pack [33].

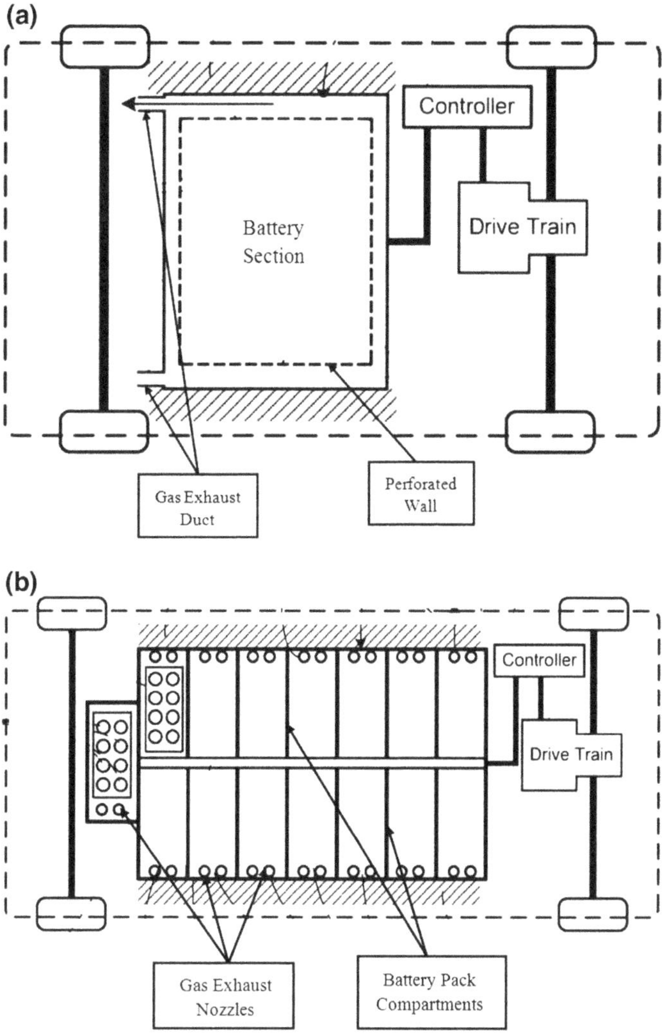

Fig. 2 Battery pack system with **a** hollow guideways or exhaust gas ducts and **b** multiple exhaust nozzle assemblies fitted in each battery compartment to exhaust hot gases and effluents generated during pack operation and/or thermal runaway events

2.1.2 Thermal Barrier

The increased temperatures associated with thermal runaway may cause the mounting brackets in close proximity of the battery region undergoing thermal runaway to melt or vaporise. As a result, the battery may no longer be held rigidly in its original position. As the affected battery cell/module moves, the spacing between battery components may be diminished, leading to decreased resistance to thermal runaway propagation. Battery cell/module movement may also compromise the battery pack cooling system, thus further increasing the thermal runaway propagation rate. Lastly, it should be noted that if the affected cell/module moves sufficiently, it may come to rest against an adjacent cell/module. If it does, the heat transfer process between the two regions would switch from radiation and convection to a combination of radiation, convection and thermally more efficient process of conduction. Further, in applications where a stacked-type battery configuration is used, i.e. a layer of battery cells arranged vertically over another layer, gravitational forces may expedite the movement of the top layer once the bracket(s) begins to melt and/or vaporise. It is therefore important to restrict the movement of the battery cell or module undergoing thermal runaway to minimise the risk of thermal runaway propagation.

Firstly, cross-members can be used to divide the battery pack into multiple compartments. The packaging design presented by US Patent No. 8663824 also demonstrated how a central battery pack member can be employed to further separate the right and the left compartments in addition to providing a channel for connecting power and data lines. In the design, module mounting flange of the battery module is captured by the upper and the lower cross-members of the packaging frame. The arrangement allows easy positioning and holding of the battery modules at their place in the compartment. It also creates an air gap between the top and bottom surfaces of the packaging and the battery modules [33]. The air gap reduces the probability of occurrence of conductive heat transfer between the neighbouring battery modules.

Secondly, battery cells can be held in their pre-specified location by using rigid spacers that are friction fit or bonded between neighbouring cells of the battery module. In general, the spacer assembly selected for integration within the cell-mounting bracket depends on the type and shape of the cells employed within the battery pack. Since the primary function of cell spacers is to keep the cells fixed in place during thermal runaway, to save mass and attain a higher specific energy rating for the battery pack a pair of much smaller spacers with an upper spacer and a lower spacer is preferred over one long spacer running from top to bottom of the cells [34]. Although just one spacer can be used, such as one located near the top or bottom, or near the centre of the cells, use of one spacer is not preferred as it still permits some movement. Height of the spacers used is usually in the range 1–5% of the overall battery height. Cell spacers perform a dual role in the case of prismatic cells and pouch cells. Besides their primary function, i.e. providing cell-holding functionality, they provide the binding pressure necessary to counteract the internal spring forces and to prevent the cell windings from expanding as a result of it.

Battery cell spacers create sufficient binding on the cell sides without covering so much of the cell surface area that cooling becomes ineffective.

2.2 Structural Stability

In the absence of adequate compressive force needed to maintain a uniform contact, delamination of electrode layers occurs in pouch cells and prismatic cells, which affects their performance and reliability. Delamination of the electrode layers can be avoided through usage of external structures that may include either hard plates stacked on each side of the battery cell or clamps made of thread rods. Although the stacking plate method provides significant advantage during manual assembly of battery packs, it is more expensive on a mass production basis. Also, holding clamps may make the pouch cells more vulnerable to mishandling during assembly process and to localised stress development due to unbalanced clamping force [27].

The solid structure created through metallic or rigid plastic casings typically used for the prismatic and the cylindrical battery cells prevents foreign objects such as nails from penetrating the electrochemical system. The metallic casings provide a greater degree of tolerance to pressures generated inside the battery cell because of gas generation and venting, a safety feature absent in pouch cells owing to their soft packaging.

Main structural issue with the prismatic cells is that their corners can be left vacant due to elliptical windings. It results in uneven pressure distribution in electrodes but the problem can be alleviated by filling vacant corners with solid material. Table 3 compares different battery cell formats according to structural characteristics considered important from safety perspective.

Table 3 Comparison of structural characteristics of different types of battery cells [35]

Criteria	Small cylindrical	Large cylindrical	Prismatic	Pouch
Casing	Metal	Metal	Semi-hard plastic or metal	Aluminium soft bag
Connections	Welded nickel or copper strips or plates	Threaded stud for bolt or threaded hole for bolt	Threaded hole for bolt	Tabs that are clamped, welded or soldered
Retention against expansion	Inherent in cylindrical shape	Inherent in cylindrical shape	Requires retaining plates at ends of battery	Requires retaining plates at ends of battery
Appropriateness for production runs	Good: welded connections are reliable	Good	Excellent	Excellent
Field replacement	Not possible	Possible	Possible	Not possible
Delamination	Not possible	Not possible	Possible	Highly possible
Compressive force holding	Excellent	Excellent	Poor	Extremely poor
Local stress	No	No	No	Yes
Safety	Good, integrated with PTC	Good, integrated with PTC	Good, integrated with PTC	Poor, no safety features included
Heat shrink wrapping	Yes	Yes	Depend on casing material	No

2.2.1 Crash Protection

Maintaining structural integrity of the battery pack during crash conditions is another challenge for EV designers. For this purpose, two packaging architectures—the "T-shaped" architecture and the "floor" configuration—are primarily utilised for EV battery packs.

The "T-shaped" architecture shown in Fig. 3a is used in GM Chevrolet Volt. It enables the battery modules to be arranged inside the primary safe zone of the vehicle, i.e. the area beneath the rear passenger seats and extending along the tunnel between the two rows of seats. It prevents the battery pack from direct frontal impact and side impact loads through usage of vehicle structure as a crash barrier at the expense of interior cabin space and sometimes passenger comfort as well.

In contrast, the "floor" configuration used in Tesla Model S and Audi e-tron Sportback concept, shown in Fig. 3b, involves arranging the battery cells in a slab format under the vehicle floor. It maximises the available cabin space to be used either by the vehicle occupants themselves or for storing their luggage. In addition, such configuration increases the vehicle stability during various driving manoeuvres by lowering its centre of gravity. However, it also reduces the ground clearance of the vehicle, thus exposing the battery pack to dangers of ground or bottom impact.

Battery cells are traditionally protected against the bottom impact via metal or plastic shell casing enclosures in conjunction with module and battery pack housings and vehicle body structure including transverse cross-members, doors and floor. Furthermore, as floor panel can only resist impact from small stones on a gravel road, armour made of 1–6 mm thick metallic sheet, with a monolithic or a sandwich structure or even their combination, is used as a protection against bottom impact. Polymeric coating is applied to it for rust protection.

Research has shown that severity of the damage to the protective armour plate is affected by the tip radius of the impacting body, the distance of the indentation point to the nearest boundary of the battery module and the exponent of the power law hardening curve. It has further been reported that other thinner protective members of the battery pack rupture soon after the armour is breached, thereby exposing the battery cells to damage from road debris and other sharp objects [38].

To restrict this damage to a minimum, a multifunctional granular battery assembly (GBA) pack, in which the battery cells are organised in a bimodal packing arrangement along with collapsible and sacrificial metal tubes, has been proposed. GBA can function as an energy storage system and a stress control plus energy dissipation unit simultaneously. Simulation studies rate it as 2.6 times more effective than a metal foam structure of equivalent density in reducing the probability of battery pack failure during crash conditions. A reduction of over 5% in the head injury criterion of EVs due to the use of GBAs has also been observed as opposed to likelihood of head injury arising from an impact to an EV occupant employing traditional battery packs. However, disadvantages of employing GBA in place of a conventional battery pack include a decline of 35 and 13% in the volumetric capacity and gravimetric capacity of the vehicle, respectively.

Fig. 3 Battery packaging architectures: **a** T-shaped architecture used in GM Chevrolet Volt [36] and **b** floor architecture used in Audi e-tron Sportback concept [37]

More importantly, the metal tubes of a GBA add approximately 3% to the gross vehicle weight that could influence not only EV driving range but also its rolling resistance [39].

In addition, side impact resistance of battery packs can be increased by including a pair of collapsible side sill assemblies and multiple cross-members in the battery pack design. In general, hollow cross-members should be used to gain benefits from high strength to weight ratios of hollow structures. However, both the material and the configuration used for the cross-member can vary with its location in the battery pack. For example, cross-members located in the centre of the battery pack are thicker than other cross-members to provide additional strength at the areas that are used for seat mounting assembly [40].

However, in EVs with air-cooled batteries, due to the large cross-sectional area of the air ducts, minimal packaging space is available to provide cross-braces for the battery assembly. US Patent 8276696 demonstrates a packaging design in which the inlet/outlet ducts for an air-cooled battery are modified and utilised as structural members to increase the impact resistance of the battery pack. As per the design, the forced air system includes an inlet duct for providing air to the battery and an outlet duct for directing exhaust air from the battery and a fan. At least one of those ducts is configured as a structural member to provide structural support and protect the battery assembly. The duct arrangement that extends between opposite sides of the vehicle is attached to the shock tower on each side providing support and protection to the battery assembly. The ducts, which are traditionally made of plastic, can be made of steel, aluminium, carbon fibre or any other suitable material in EV applications. Due to fewer parts being used, it also provides a more efficient and compact packaging solution [41]. In other words, benefits in terms of energy density (Wh/L) can be realised with this design, but the compromise in terms of specific power (W/kg) and cost of the system needs to be made.

2.2.2 Vibration Isolation

Vertical low-frequency vibrations are transmitted continuously to an EV structure as it is driven on a highway. Also, travelling over uneven surfaces, such as holes, grade crossings or bridge abutments, produces shocks that cause vertical vibrations. As a result, dynamic mechanical loads develop on the electrical sub-systems such as terminal connectors and bus bars in a battery pack, which can result in loss of electrical continuity and fatigue failure of the casing [42].

In order to prevent this, a compressive force is usually applied to the top surface of the battery packs through tensioning bolts and retainer frame. US Patent 7507499 illustrates one such design for stabilising a battery pack in EVs by using a cover-pad-tray retention arrangement. The design comprises of four beams, coupled at right angle to one another through four connectors to form a rectangular frame structure. Each beam engages one of the four sides of the battery pack. Positive connection between frame and the battery pack is maintained through tensioning bolts. The arrangement uses two types of damping pads, flat and L-shaped, to absorb vibration and prevent movement of the modules with respect to one another along the Z-axis. The L-shaped damping pads are placed adjacent to each of the corner connectors. They bear against the frame structure to provide relatively small pressure areas at the corners and push the separate battery modules of the battery pack laterally towards one another; on the other hand, the flat damping pads are positioned at the lower and upper corners of facing sides of the adjacent battery modules. A tray that could be bolted to a part of the vehicle structure provides the support to the battery pack.

Tensioning bolts are fastened after assembling the frame so that the beams are drawn against the corner pads in the longitudinal and lateral directions to peripherally squeeze the battery modules of the battery pack towards one another. Fastening the bolts also compresses the damping pads placed between the individual battery modules making them stationary with respect to one another. Figure 4 presents a perspective view of the design [43].

Weight distribution of the vehicle can also influence the degree of vibration isolation and ride quality. A battery mounting frame structure for achieving uniform vehicle weight distribution and to maintain a low centre of gravity was presented by US Patent 8561743. As seen in Fig. 5, the rectangular mounting frame is divided into two sections, front and rear by a girder that has been welded to the frame [44]. Furthermore, a beam member divides the front section into two equal rectangular areas. In the two front rectangular areas, the batteries are arranged in a vertical direction such that the long side is oriented in transverse direction and the short side is oriented in longitudinal direction of the vehicle, whereas the batteries in the rear rectangular section are arranged such that the shortest side is oriented in the vehicle transverse direction.

As a result of this arrangement, the weight of group of batteries mounted in the rear section is substantially equal to the total weight of group of batteries mounted in the two front columns. Subsequently, centre of gravity of the battery assembly is located around an intersecting point of median of the group of batteries in vehicle

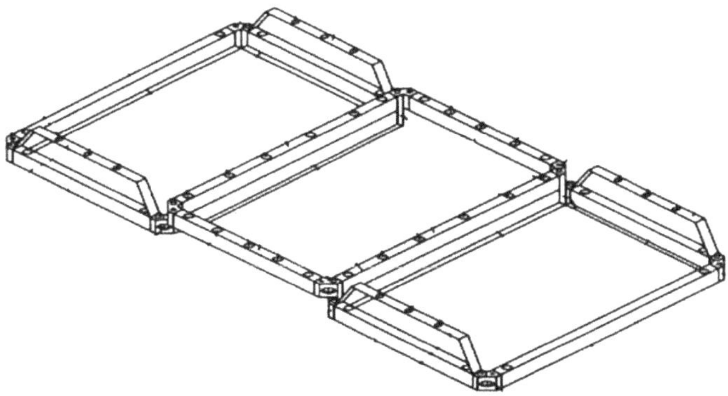

Fig. 4 Perspective view of a framing arrangement employed with a compact battery pack design [43]

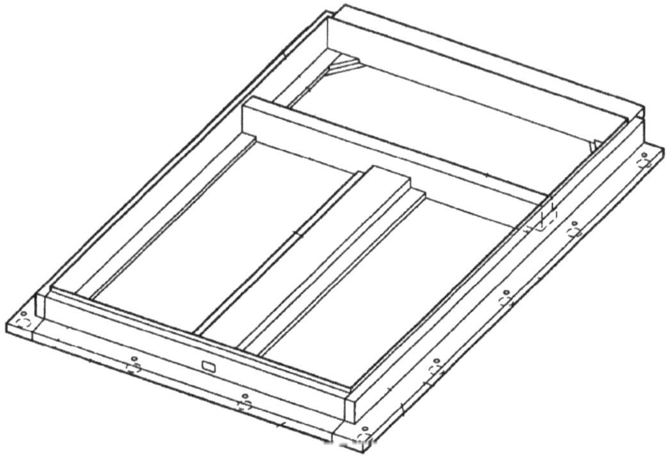

Fig. 5 Perspective view of battery mounting frame [44]

transverse direction and median of the same group in longitudinal direction. This point is located to the rear of the graphical centre of the vehicle and is preferable in terms of weight balance of the vehicle in the front–aft direction, considering that the electric motor, the battery charger and the inverter are housed in the front compartment [44]. A clear advantage of this configuration is that the mounting frame can be used for various types of vehicle. This means that in case seating layout of the EV is changed, then optimum weight distribution and vibration isolation can be achieved simply by modifying the number of battery stacks in the group and without making any major modifications to dimensions of the battery mounting frame disclosed here.

3 Case Study—Swinburne eBus Battery Pack

In 2014, a collaboration was formed between Swinburne University of Technology, Bustech, Malaysian Automotive Institute (MAI) and Excellerate Australia to build a prototype demonstrator electric bus (eBus). The latter two stakeholders were the relevant government bodies for Malaysia and Australia, respectively, and provided significant funding for the project. Bustech is a bus manufacturer based in Queensland, Australia, and has been designing and building buses since 1995, with a current production rate of 250 buses per year. The demonstrator eBus had ultra-low-floor architecture and was built using off-the-shelf hardware, where available, so that new and innovative technology opportunities could be identified that address the integration for a modular systems architecture approach. It was intended to be the first step in a bus development programme and enabled the development of sub-systems and integration of the driveline system and functional features.

The battery pack could not be sourced "off the shelf" and was therefore designed by the Swinburne engineering team from cell level. The design was based on available packaging space in the demonstrator vehicle platform and availability of suitable off-the-shelf components. This section is about mechanical design of the first fully electric city bus built in Australia.

3.1 Battery Cell Selection

Table 4 represents prominent battery manufacturers for automotive applications. K2 energy batteries are small cells. They have more thermal durability and increased life cycle as claimed by the manufacturer. However, these cells are smaller in size resulting in 27% volumetric losses when arranged within the packs. The A123 provides a modular concept where seven cells within module are arranged in series, while three are in parallel. A123 module is found easy to package and due to its excellent characteristics, such as high energy-to-weight ratio, can operate at high voltages and show low self-discharge rates. The CALB CAM72 battery cells are similar to Thunder Sky batteries, and their ready availability and meeting the capacity requirement make them suited to this research.

CALB CAM72 cells were selected for the demonstrator eBus requirement, with specification as shown in Table 5. These cells are aluminium alloy shelled, rechargeable lithium–iron–phosphate energy cell. They are widely used in high-speed energy storage systems for frequency control, high power renewable energy integration and other high power applications.

3.2 Battery Pack Design

Several iterations of battery cell layout inside the pack were considered to meet packaging space, operation and safety requirements of the eBus. A list of various design options is presented in Table 6, and related weight calculations for the pack are summarised in Table 7.

3.2.1 Final Design: Base Plate

Due to the constraint of loading from underneath the eBus, a "tooth" mounting system was proposed as the best strategy to increase package space and provide secure attachment to the vehicle. This mounting system is similar to bus fuel tank mounting, which is designed to hold a similar mass with full fuel tank. Also, it requires short development time. Bolt sizing was based on restraining the pack under the required loading with a safety factor of 1.5. High safety factor accounts

Table 4 Analysis of some commercial Li-ion battery cells considered

		A123 3P7S module	SINOPOLY LFP300 AH	CALB CAM72	K2 K226650E cell
Cell data	Cell capacity (Ah)	60	300	72	3.2
	Cell voltage (V)	23.1	3.2	3.2	3.2
	Cell capacity (kWh)	1.386	0.960	0.230	0.010
	Cell max voltage (V)	25.2	3.6	3.6	3.65
	Cell min voltage (V)	17.5	2.5	2.5	2.5
	Peak C rating (discharge) (A/Ah)	5	1	3	4.0625
	Cont. C rating (discharge) (A/Ah)	3	0.33	2	1
	Cell mass (kg)	12.3	9.5	1.9	0.082
	Cell volume (L)	8.8	5.896	0.849	0.036
	Cell internal resistance (mΩ)	8.17	0.8	1	19
Pack data	No. in series	27	200	200	200
	No. in parallel	7	1	6	134
	Total no. of cells	189	200	1200	26,800
	Pack capacity (kWh)	262	192	276	274
	Continuous current per cell (A)	60.7	401.8	67.0	3.0
	Peak current per cell (A)	82.8	548.0	91.3	4.1
	Peak C value (A/Ah)	1.4	1.8	1.3	1.3
	Continuous C value (A/Ah)	1.0	1.3	0.9	0.9
	Nominal pack voltage (V)	623.7	640.0	640.0	640.0
	Max pack voltage (V)	680	720	720	730
	Min pack voltage (V)	473	500	500	500
	Peak pack current (A)	580	548	548	548
	Continuous pack current (A)	425	402	402	402
	Pack internal heat generation (kW)	6	26	5	5
	Pack cell mass (kg)	2325	1900	2280	2198
	Pack cell volume (L)	1668	1179	1019	964

for reduced validation via physical testing and limited information about internal cell architecture (Fig. 6).

The "tooth" mounting system requires both base plate and frame to be laser/water cut from same sheet to ensure minimal tolerances; however, bus frame limitations require individual mounting on each side, which must be controlled for.

Table 5 CALB CAM 72 battery specifications

Capacity	72 Ah
Nominal voltage	3.2 V
Cycle life	$\geq$ 2000 cycles
Internal resistance	$\leq$ 1 mΩ
Max. charge rate	1 C
Charging cut-off voltage	3.65 V
Max. discharge rate	2 C
Discharging cut-off voltage	2.5 V
Charge time	4 h nom, 1 h fast
Weight	1.9 ± 0.1 kg
Dimensions, L × W × H (in mm)	135 × 29 × 222
Charging temperature	0–45 °C
Discharging temperature	–20 to 50 °C
Ambient humidity	<70%
Shell material	Aluminium alloy

Base plate has cut-outs to reduce weight and a thin sheet welded on top to seal from external elements.

The "tooth" mounting system allows the battery pack to vertically pass the bus mounting area and then, with lateral translation, line up the mounting holes. It requires a lifting tool/device to raise the pack into the vehicle and allow small adjustments to assist with lining up position for both the vertical movement and the lateral movement. The positioning of the top packs requires a 0.5 m vertical raise inside the bus frame, which means the dimensions for the tool will be restricted to within the pack dimensions (700 × 850—allowing additional clearances).

3.2.2 Final Design: Casing

The outer casing is not a load-bearing part, it is in place to assist with sealing and mounting of external attachments, and this requires some strength component; however, it cannot be so rigid as to cause loss of sealing or support too much weight. Composite materials can provide a lot of flexibility in material choice, with mixtures of chopped strands and more; however, lead time was found to be a severe problem in this case due to the requirement of a mould. A metal such as stainless steel was therefore considered as an option through a much simpler design.

Another important issue is that tight packaging constraints in an EV restrict the space available for mounting casing to the base plate while providing a good seal to the pack. Good sealing requires a uniform compression load around the outer edge of the pack to ensure the sealant (compression material) is always engaged and functioning. This, in turn, requires analysis of mount spacing to ensure this requirement is met. Manual silicon sealing can always be used around the battery box to ensure a proper seal is achieved.

Table 6 Various design options considered for battery pack construction

Design options			
Design option 1	Parts	Weight (kg)	Details
	Al base plate, 10 mm	22.03356	Initial Aluminium 6061, 10 mm base plate, considers CALB individual plastic cell protection; initial design with 4 corner supports and 2 fixes for top and bottom and not sufficient for restraining the movement (longitudinal, lateral and vertical) of the cells inside the pack and also does not provide the ease of manufacturing and mounting solution 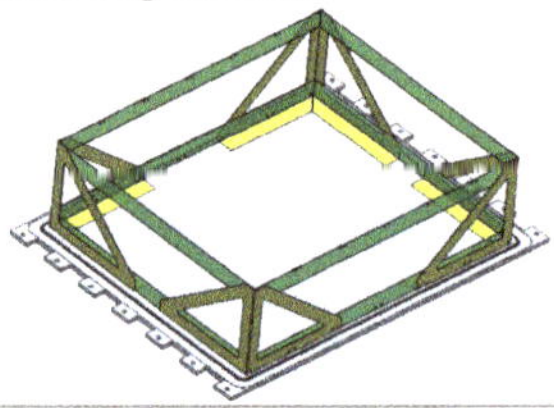
	Corner supports; top and bottom fix, 2 mm		
	Misc.	10	
	Total	32.03356	
Design option 2	SS base plate, 5 mm	42.34558	SS base plate, 5 mm; welded box side plate, 2 mm—4pc. Cells are to be fitted from the top; no access to tie up the cells together longitudinal and lateral direction due to the welded side plates. Sealing would be easier due to the welding 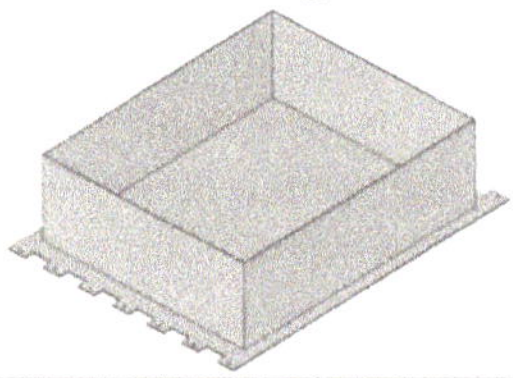
	Welded box, 2 mm		
	Misc.	10	
	Total	52.34558	
Design option 3	SS base plate, 4 mm	36.32606	SS base plate, 4 mm; welded box side plate, 2 mm—4pc. Cells are to be fitted from the top; no access to tie up the cells together longitudinal and lateral direction due to the welded side plates. Sealing would be easier due to the welding 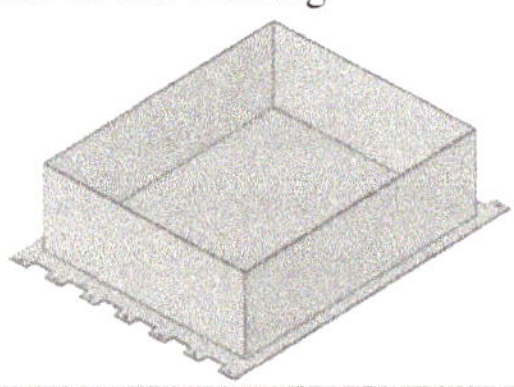
	Welded box, 2 mm		
	Misc.	10	
	Total	46.32606	

(continued)

Table 6 (continued)

Design options			
Design option 4	SS base plate, 5 mm	60.87677	SS base plate, 5 mm; U beam 7pc underneath the base plate aligned with the mounting holes; enough restraining from longitudinal, lateral and vertical direction; weight is higher than other options; can think about putting one C section by the side instead of two to decrease the weight depending on the compressive load calculation; enough space for restraining by the metal straps surroundings; U beams are supporting with the stress generation at the mounting holes 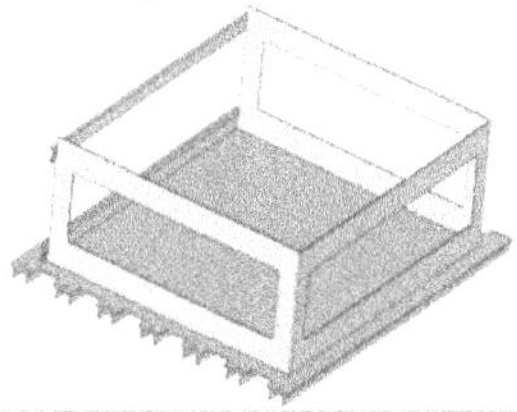
	U beam 7pc, 40 × 20 x 3 mm		
	C section, 2pc, 25 × 12 x 3 mm		
	Welded corner support, 2 mm		
	Misc.	2	
	Total	62.87677	
Design option 5	SS base plate, 4 mm	54.85725	Same design up as option 4, with only base plate thickness 4 mm 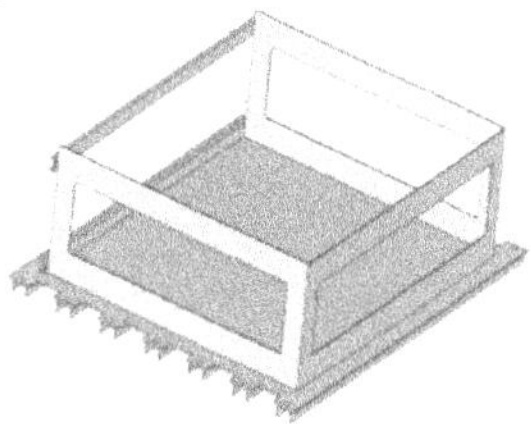
	U beam 7pc, 40 × 20 x 3 mm		
	C section, 2pc, 25 × 12 x 3 mm		
	Welded corner support, 2 mm		
	Misc.	2	
	Total	56.85725	
Design option 6	SS base plate, 5 mm	48.0425	Same restraining design set up as option 4; U beams underneath are removed from this to decrease the total weight of the battery box 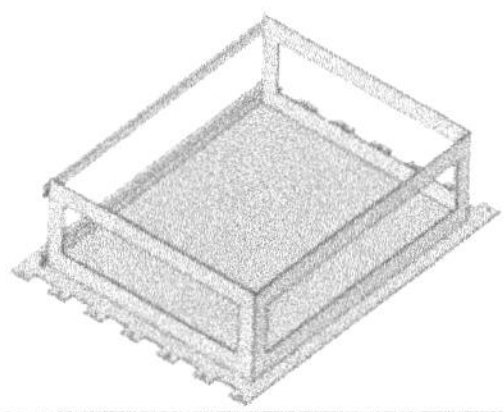
	C section, 2pc, 25 × 12 x 3 mm		
	Welded corner support, 2 mm		
	Misc.	2	
	Total	50.0425	

3.2.3 Final Battery Pack Assembly

The final pack assembly consisted of a stainless steel plate (4 mm thickness) for the base plate and channel sections (20 × 20 × 3 mm) welded underneath the base plate as reinforcements. The base plate has the toothed profile at the longitudinal

Table 7 **a** Initial assumptions used to calculate battery pack size, stiffness and strength and **b** mass calculation for battery pack designs that were considered

a	
Basic Inputs	
Total plate length, L (mm)	1008
Effective length, l (mm)	1008
Width b (mm)	769
Material	Al 6061
	AISI 316 SS
Modulus of elasticity, E (GPa)	68.9
	210
Gravity	9.81
Gravity load	1
Cell weight (kg)	1.9
Total cell in a row	22
Total cell in a column	6
U beam dimension	40 × 20 x 3
C section dimension	25 × 12 x 3
Type of support	Fixed

b			
Weight calculation of the total pack			
Items	No.	Weight (kg)	Unit weight (kg)
Total cell in a pack	132	250.8	1.9
BMS	1	2.426	2.426
Contactors	4	1.72	0.43
Current sensor	1	0.067	0.067
CM0711	1	0.6	0.6
Connectors	2	0.5	0.25
Relays	2	0.4	0.20
DC–DC converter 5 W	1	0.2	0.20
MSD	1	0.3	0.30
Com. connector, HD34-24	1	0.18	0.18
Washer, N16 Nord lock	528	0.4224	0.0008
Screws	264	0.66	0.0025
Bus bar	528	3	
Total weight of elec. accessories (kg)		261.275	
Design option 1		293.309	
Design option 2		313.621	
Design option 3		307.602	
Design option 4		324.152	
Design option 5		318.133	
Design option 6		311.318	

Fig. 6 Battery base plate and frame structure with bay walls made of SS304

end to mount the pack into the bus frame. The battery pack contains 132 CAM72 prismatic type cells in 6 rows of 22 cells (Connection: 66S2P). Longitudinally, the rows are divided into two sections (separated by 10 and 12 battery cells) by a 2 mm stainless steel sheet in the middle. Each row of cells is placed individually and restrained by a side metal plate. Restraining the battery link bars and clearance for the pressure relief vents of the cells are also considered in the design.

The design includes the electrical accessories (BMS, contactors, sensors, etc.) required for internal and external interfaces, all mounted in a position to optimise packaging, function and safe operation. Also included are Manual Service Disconnect (MSD) and terminal connectors, allowing quick and safe isolation of the battery packs during scheduled maintenance. The electrical insulation between the cells and the metal frame on each side was also considered through the inclusion of high-density polyethylene (HDPE) sheets where required.

According to the standards EN 60664-1:2007 and VDE 0110–1, clearance distance (the shortest distance between two conductive parts or between a conductive part and the bounding surface of the equipment measured through air) should be dimensioned to withstand the required impulse withstand voltage. For connections with low voltage mains, rated impulse voltage is considered the required impulse withstand voltage. However, additional clearances may be necessary to account for mechanical effects like vibration and applied forces. On the other hand, creepage distance, which is the shortest path between two conductive parts or between a conductive part and the bounding surface of the equipment,

Fig. 7 Battery pack showing top plastic layer 1 (1.5 mm) sitting on the cells with clearance holes for terminals and pressure relief vent

measured along the surface of the insulation, is defined on the basis of long-term root-mean-squared (rms) value of the working voltage.

As a general rule of thumb, the required clearance is equal to the shortest creepage distance for any application. The outer surface of the battery cell is considered as the bounding surface, as though metal foil was pressed into contact with accessible surface of insulating material (plastic HDPE or nylon, Fig. 7).

For 250 V, the recommended air clearance would be 5 mm and the creepage distance would be 8 mm. In this case, creepage distance was taken as 8 mm for the insulating material (Fig. 8).

The outer case housing (Fig. 9) was designed to seal the battery pack. The top cover of the housing allows access to the electrical accessories (contactors, current sensor, fuse), including battery controller and management system, without dismounting the whole outer case from the pack. One manual service disconnect (MSD) connected to each battery pack but mounting facilities for the MSD were made available on both sides of the battery box. Nine identical and swappable 30 kWh, 211 V (nominal) battery packs were designed for placement on the bus. To achieve the most efficient use of available space, the battery packs were set up as 3× parallel strings of 3× packs in series. A modular architecture was chosen to allow for future expansion of the battery system and to help with the weight distribution.

Battery pack testing comprised of testing battery packs individually as well as their integration into the working string of batteries to simulate the actual energy

Fig. 8 Battery pack with added insulation tape, metal cell retaining strap and additional accessories; BMS, battery controller, current sensor, contactors, fuses and other accessories were installed on a separate plate and bolted to the metal strap

storage system on-board an eBus. The battery pack was tested on charge and discharge for a period of 6 hours at a range of current capacities up to 25 A. A smooth rise and lowering of battery cell voltage was recorded, results as expected. Energy delivery was continuous with no interruptions. The endurance of the packs was established, and the analysis of the battery cell voltage and behaviour indicated an equal performance per cell. The battery cells electrical stability was verified. Three battery packs were integrated as one string to deliver energy at different levels using a small induction motor used as load. Results of test were the verification of energy delivered, instrument adjustment and software control operation.

Lastly, the majority of the packaging space was at the rear of the bus, due to the low-floor design. However, placing excessive weight on the rear axle restricts the number of passengers that can be carried in the bus. Priority was therefore given to placing batteries forward of the rear axle to help distribute the weight forward, which meant very tight packaging constraints, but increased passenger count on the bus. The high-voltage distribution unit (HVDU) is the central point for all electrical energy transfer in the bus, connecting batteries, high-voltage components and chargers to ensure power is distributed where and when it is required. This was positioned as centrally as possible to reduce excessive cable lengths and therefore reducing system energy losses, especially to the motors that require the largest

Fig. 9 Complete battery pack ready for fitment to bus

amount of power. The positioning of the remaining components was then based on proximity to functional systems, such as DC–DC converter closest to the 24 V system distribution or due to component requirements like ingress protection.

4 Summary

In this chapter, mechanical design elements affecting safety and reliability of EV battery packaging are discussed. Forces like mechanical vibration, impact energy and ambient temperature variations interact with the battery pack through different interfaces. These interactions need to be controlled for safe and reliable operation of battery pack. Restricting battery cell movement is found to be one of the successful strategies to achieve a higher degree of protection against all of them, and mechanisms that can be used for this purpose are presented. Other mechanical design solutions to increase crashworthiness and vibration isolation of the EV battery pack are also discussed. Lastly, a case study focussing on mechanical design of an eBus battery pack at Swinburne University of Technology in Australia is presented. Packaging and fitment strategy of the batteries in the eBus project was based on allowable packaging space and minimal permissible modification to the existing bus frame.

The eBus case study highlights the importance of modularity on full proofing the battery packs against future uncertainty. It can also be learnt from this case study that designing a battery pack for a high-voltage system can provide a very hazardous environment, especially if the workshop space is not equipped to deal with the required voltage. A practical design option is to produce smaller packs at a low voltage, making the work and handling of packs much safer. Using smaller individual battery packs not only improves user safety but also offers benefits in terms of prototype manufacture and testing of the packs. The increased number of packs means more complexity at a system level, which should be weighed heavily with the benefits mentioned here.

Acknowledgements Financial support from the Cooperative Research Centre for Advanced Automotive Technology (AutoCRC), Australia, for undertaking this research study is duly acknowledged. Also, support of the eBus project team at Swinburne University of Technology was useful in development of this chapter.

References

1. UNFCCC. Paris Declaration on Electro-Mobility and Climate Change and Call to Action. Electrifying Sustainable Transport. 2015. Available at: http://newsroom.unfccc.int/lpaa/transport/the-paris-declaration-on-electro-mobility-and-climate-change-and-call-to-action/. Accessed 25 Sept 2017
2. L. Ahmadi, M. Fowler, S.B. Young, R.A. Fraser, B. Gaffney, S.B. Walker, Sustain. Energy Technol. Assess. **8**, 9 (2014)
3. C. Jhu, Y. Wang, C. Wen, C. Shu, Appl. Energy **100**, 127 (2012)
4. C. Julien, Design Considerations for Lithium Batteries, in Materials for Lithium-Ion Batteries, ed. by C. Julien, Z. Stoynov (Springer, Berlin, 2000), p. 1
5. B. Diouf, R. Pode, Renew. Energy **76**, 375 (2015)
6. D. Miranda, C. Costa, S. Lanceros-Mendez, J. Electroanal. Chem. **739**, 97 (2015)
7. N. Alias, A.A. Mohamad, J. Power Sour. **274**, 237 (2015)
8. H. Han, H. Park, K.C. Kil, Y. Jeon, Y. Ko, C. Lee, M. Kim, C. Cho, K. Kim, U. Paik, Electrochim. Acta **166**, 367 (2015)
9. J. Lu, Y. Chang, B. Song, H. Xia, J. Yang, K.S. Lee, L. Lu, J. Power Sour. **271**, 604 (2014)
10. X. Sun, X. Zhang, B. Huang, H. Zhang, D. Zhang, Y. Ma, J. Power Sour. **243**, 361 (2013)
11. G. Wang, Z. Ma, G. Shao, L. Kong, W. Gao, J. Power Sour. **291**, 209 (2015)
12. M. Wu, M. Liu, G. Long, K. Wan, Z. Liang, T.S. Zhao, Appl. Energy **136**, 576 (2014)
13. B. Nykvist, M. Nilsson, Nat. Clim. Change **5**, 329 (2015)
14. E.V. Tech, EV Batteries - Falling Prices for a Clean Air Future (2015). Available at: http://myelectriccar.com.au2015. Accessed 11 Aug 2017
15. C. Arbizzani, G. Gabrielli, M. Mastragostino, J. Power Sour. **196**, 4801 (2011)
16. C.L. Campion, W. Li, B.L. Lucht, J. Electrochem. Soc. **152**, A2327 (2005)
17. D.H. Doughty, P.C. Butler, R.G. Jungst, E.P. Roth, J. Power Sour. **110**, 357 (2002)
18. C. Jhu, Y. Wang, C. Wen, C. Chiang, C. Shu, J. Therm. Anal. Calorim. **106**, 159 (2011)
19. G. Kim, A. Pesaran, R. Spotnitz, J. Power Sour. **170**, 476 (2007)
20. J. Lamb, C.J. Orendorff, L.A.M. Steele, S.W. Spangler, J. Power Sour. **283**, 517 (2015)

21. N.S. Spinner, C.R. Field, M.H. Hammond, B.A. Williams, K.M. Myers, A.L. Lubrano, S.L. Rose-Pehrsson, S.G. Tuttle, J. Power Sour. **279**, 713 (2015)
22. S. Arora, W. Shen, A. Kapoor, Designing a Robust Battery Pack for Electric Vehicles Using a Modified Parameter Diagram, SAE Technical Paper 2015-01-0041, 2015, https://doi.org/10.4271/2015-01-0041
23. Z. Rao, S. Wang, M. Wu, Z. Lin, F. Li, Energy Convers. Manage. **65**, 92 (2013)
24. N. Sato, J. Power Sour. **99**, 70 (2001)
25. J.M. Hooper, J. Marco, J. Power Sour. **245**, 510 (2014)
26. S. Arora, W. Shen, A. Kapoor, Renew. Sust. Energy Rev. **60**, 1319 (2016)
27. L.H. Saw, Y. Ye, A.A. Tay, J. Clean. Prod. **113**, 1032 (2016)
28. B. Coleman, J. Ostanek, J. Heinzel, Appl. Energy **180**, 14 (2016)
29. J.A. Jeevarajan, E.C. Darcy, Can Cell to Cell Thermal Runaway Propagation be Prevented in a Li-ion Battery Module? JSC-CN-32020 (2014)
30. J. Jeevarajan, C. Lopez, J. Orieukwu, Preventing Cell-to-Cell Thermal Runaway in Lithium-Ion Battery Modules. NASA Tech. Briefs (2014)
31. S. Al Hallaj, H. Maleki, J. Hong, J.R. Selman, J. Power Sour. **83**, 1 (1999)
32. S.C. Levy, J. Power Sour. **68**, 75 (1997)
33. P. Frey, D. Grace, *Battery Pack Exhaust Nozzle Utilizing an SMA Seal Retainer* (Tesla Motors Inc, USA, 2013)
34. W.A. Hermann, *Rigid Cell Separator for Minimizing Thermal Runaway Propagation within a Battery Pack* (Tesla Motors Inc., USA, 2012)
35. D. Andrea, Battery Management Systems for Large Lithium-Ion Battery Packs, Artech House (2010)
36. GM Announce Structural and Battery Enhancements for Chevy Volt. 2012. Available at: http://www.electric-vehiclenews.com/2012/01/gm-announce-structural-and-battery.html. Accessed 25 Aug 2017
37. A. Marino, Audi E-Tron Sportback Concept Architecture of E-mobility. 2017. Available at: https://www.driveandride.com/uk/audi-e-tron-sportback-concept/. Accessed 25 Aug 2017
38. Y. Xia, T. Wierzbicki, E. Sahraei, X. Zhang, J. Power Sour. **267**, 78 (2014)
39. J. Kukreja, T. Nguyen, T. Siegmund, W. Chen, W. Tsutsui, K. Balakrishnan, H. Liao, N. Parab, Extreme Mech. Lett. **9**, 371 (2016)
40. A. Charbonneau, M. Burgess, V. Attaluri, H.L. Gadhiya, A.P. Clarke, P.D. Rawlinson, R.D. Sumpf, B.P. Edwards, *System for Absorbing and Distributing Side Impact Energy Utilizing an Integrated Battery Pack and Side Sill Assembly* (Tesla Motors Inc., USA, 2014)
41. M. Lucas, Structural Battery Duct Assembly. U.S. Patent 8,276,696 (2012)
42. K. Watanabe, T. Abe, T. Saito, O. Shimamura, K. Hosaka, H. Sato, H. Horie, Battery Structure, Assembled Battery, and Vehicle Mounting These Thereon. U.S. Patent 8,124,276 B2 (2012)
43. S. Zhou, C.C. Husted, F.A. Benjamin, *Battery Pack Arrangements* (General Motors Co., USA, 2009)
44. M. Iwasa, S. Ogata, H. Kadota, T. Hashimura, N. Mori, *Vehicle Battery Mounting Structure* (Nissan Motor Co., USA, 2013)

Advanced Battery-Assisted Quick Charger for Electric Vehicles

Muhammad Aziz and Takuya Oda

Abstract Electric vehicles (EVs) have gained considerable attention owing to their excellent characteristics as transportation vehicles and due to their energy storage capacity. Unfortunately, this massive deployment of EVs leads to significantly high electricity demand due to their charging requirements, particularly when they are charged uncoordinatedly. In addition, the concentrated charging of EVs can potentially decrease the quality of electricity, including frequency and voltage, in addition to causing other electrical grid problems. These conditions have motivated the development of technology and policies for minimizing these negative impacts. In this chapter, an advanced quick charging system for EVs that utilizes batteries to support the simultaneous fast charging of EVs has been described, including a description of its performance under different contracted electricity capacities, ambient temperatures (seasons), and high charging demand. In addition, the charging and discharging behaviors of EVs under different ambient temperatures have been explained. Our findings suggest that charging at high ambient temperature (e.g., during summer) allows a significantly higher charging rate than charging performed at low ambient temperature (e.g., during winter). A higher charging rate leads to shorter charging time. Furthermore, the battery-assisted charging system exhibited excellent performance because it enabled optimum quick charging during simultaneous charging in addition to maintaining the contracted electricity of the charger.

M. Aziz (✉) · T. Oda
Tokyo Institute of Technology,
2-12-1 Ookayama, Meguro, Tokyo 152-8550, Japan
e-mail: maziz@ssr.titech.ac.jp

T. Oda
e-mail: oda@ssr.titech.ac.jp

G. Pistoia and B. Liaw (eds.), *Behaviour of Lithium-Ion Batteries in Electric Vehicles*, Green Energy and Technology,
https://doi.org/10.1007/978-3-319-69950-9_9

1 Introduction

The history of electric vehicles (EVs) dates back to the nineteenth century when Anyos Jedlik invented an early type of an electric motor that was further used as the power train for his developed electric car [1]. The share of EVs continued to increase until the beginning of the twentieth century, and their numbers were higher than those of internal combustion engine (ICE)-based cars [2]. Due to several challenges, such as long charging time and short travel distance, which could not be solved by EV manufacturers at that time, EVs were replaced by ICE-based cars. These two challenging factors have pushed EV developers to improve the chargeability of EVs, and the capacity and energy density of their batteries.

In general, there are four types of EVs that are currently available in the market: (i) conventional hybrid EVs (HEVs), (ii) plug-in HEVs (PHEVs), (iii) battery EVs (BEVs), and (iv) fuel cell EVs (FCEVs). An HEV combines an electric motor and an internal combustion engine; hence, it is also fitted with a battery to power the motor as well as store the electricity. The energy to drive the motor is supplied from the engine and/or regenerative breaking. However, a number of HEVs have recently been redeveloped as PHEVs because of PHEV's excellent characteristics and higher flexibility. According to the IEEE standards, a PHEV is an HEV with additional characteristics, including a larger battery storage (>4 kWh), ability to be charged from an external energy source, and ability to drive farther than 16 km [3]. In addition, a BEV is described as a vehicle that solely runs because of the electric motors, and its source of electricity is stored and converted from chemical energy in the battery. BEVs rely on external charging, and their driving range is strongly influenced by their battery capacity. In this chapter, both PHEVs and BEVs are considered as EVs because of their large battery capacity and chargeability by an external source. According to a survey by the Union of Concerned Scientists (UCS), approximately half of the drivers in the USA drive less than 60 km on weekdays [4]. Hence, a number of EVs have been designed to be able to go through weekday commuting without charging. In the future, EVs are considered to have the potential of accounting for the largest share of vehicles.

EVs have gained considerable attention over the last decade owing to their advantageous characteristics as transportation vehicles and energy carriers due to their large energy storage capacity. With regard to the total energy utilization efficiency, EVs exhibit higher energy efficiency than conventional vehicles, mainly due to their high power generation efficiency and possible regenerative breaking [5]. The massive deployment of EVs can potentially reduce the total consumption of fossil fuel; hence, greenhouse gas emissions can be reduced accordingly [6]. Moreover, from the drivers' point of view, a higher energy efficiency is believed to result in lower operating costs. However, the massive uncoordinated charging of EVs is considered dangerous in terms of electrical grid quality, which includes the degradation of power quality, larger amount of power loss, higher voltage deviation, transformer overload, and increased fault current and harmonics [7–9]. This is mainly due to the demand for a considerable amount of electricity and its fluctuation.

As an example, Paul and Aisu evaluated electricity demand in the Kanto area [Tokyo and its surroundings, connected by the Tokyo Electric Power Company (TEPCO)] [10]. They found that if 50% of currently available vehicles were converted to EVs and 50% of EVs required quick charging without any coordination, an additional 7.3 GW of power supply would be required. This amount of additional power supply is considered to be very large.

Several methods to avoid and minimize the impact of massive uncoordinated EV charging have been proposed and developed. These include coordinated charging [11], demand response [12], battery-assisted charging [13], and appropriate charger distribution [14, 15]. Additionally, an integrated vehicle to grid (V2G) can potentially avoid concentrated charging, facilitate other services [16], and improve the economic performance of EVs. Coordinated and scheduled charging has been explored by several researchers [17, 18] to avoid possible network congestion in the electrical grid and minimize both the time and cost involved in upgrading and developing the infrastructure [19]. In general, there are two types of coordinated charging: centralized and decentralized. In the former scheme, the system operator acts as a control center sending commands to each EV to set the charging time, rate, and amount [20] based on several parameters, including system capacity, minimum system loss, cost, state of charge (SoC), and node voltage profile. However, in the latter case, each EV can still independently determine its own charging pattern [21] based on system capacity and other conditions.

The integration of EVs in demand response (DR) services is also considered as a potential solution for accommodating EV charging at a lower price while minimizing the charging impact of EVs on the distribution grid and avoiding any formation or addition of peak loads during stressed conditions. DR can be active when the energy prices reach the cutoff value, which has been set by the customer [22]. In addition, V2G is thought to be considerably promising because it utilizes both the controllable charging and discharging behaviors of EVs. Several V2G services include frequency regulation, energy storage, congestion mitigation, load shifting, peak shaving, voltage control, and power loss minimization [23, 24].

However, the abovementioned technologies require further development and demonstrations to ensure the application of standards and smooth operation in a relatively massive control system. In comparison with the abovementioned strategies, battery-assisted charging is considered to be practically applicable since it is convenient to control [25]. This chapter describes the quick charging behavior and a system for EVs, particularly the recently developed battery-assisted quick charger (BAQC). Additionally, the charging levels of EVs are initially explained in terms of charging rate. Furthermore, the EVs charging behavior under different ambient temperatures (seasons) is described to observe changes in the charging rate.

2 EV Charging

EVs require recharging operations, which can be performed through external charging and regenerative braking. In relation to external charging, EV charging is strongly correlated to several parameters, such as charging equipment/standard, investment and running cost, charging rate, charger location, charging time, and electrical grid conditions. Hence, the effective selection and distribution of chargers has become crucial in accommodating these parameters. Generally, EVs have the same charging standards, which are established by car manufacturers. Hence, there is no peculiar charger feature/requirement for any single EV. The charger is designed such that it can communicate with the EV in order to ensure both safety and the flow of electricity. Furthermore, the charger measures any Earth leakage at the surrounding ground of the charger. On the other hand, at the EV side, a battery management system (BMS) is installed, which performs the entire control and monitoring of the battery, including thermal management, cell balancing, and monitoring of both overcharge and discharge. In general, the battery pack installed in the EV comprises multiple single cells with a certain operating voltage. Hence, it is required to analyze and verify whether they are operating in the range of permitted conditions in order to avoid several failures and shorter life spans.

With regard to the charging rate, the chargers or the electric vehicle supply equipment (EVSE) is categorized into several levels according to the maximum amount of electricity that can be sent to the EV battery.

(a) Level-1 charging (low-level charging)

Level-1 charging uses the onboard EV charger, while the electricity can be supplied by general household electrical sockets and power standards. It usually operates in the voltage range of 100–200 V (AC), depending on the location. This low-level charging is able to accommodate a charging rate up to approximately 4 kW. It is suitable for residential overnight charging without requiring any additional equipment. A standard passenger car requires approximately 6–8 h to reach full charge by this slow charging method.

(b) Level-2 charging (medium-level charging)

This level of EV charging has the objective of increasing the charging rate by employing a dedicated box mounted on the wall. It is able to supply power of 4–20 kW. It has a maximum voltage of approximately 400 V (AC three phase), depending on the available capacity of the local supply. Commonly, these types of chargers are deployed at certain charging facilities in residential areas or public spaces. The charging of a conventional EV by this level requires approximately 2–3 h to achieve an almost full battery charge. In addition, the charging connectors for both level-1 and level-2 chargers vary by country and manufacturer.

(c) Level-3 charging (quick charging)

Level-3 charging is considered as quick charging and is conducted in a DC system. The DC electricity is directly provided by the charger; therefore, it bypasses the onboard charger. Consequently, a considerably high charging rate (>50 kW) can be realized. Currently, there are three major standards for quick charging, which include CHAdeMO, combined charging system (CCS), and Tesla Supercharger. These standards implement several regulations related to charging specifications, interconnection, communication mode, and charging rate/power. A passenger EV is generally charged in approximately 0.5–1 h to achieve an almost full charge. This type of quick charger is usually installed in a dedicated charging station. The specifications of each charging standard are summarized in Table 1.

CHAdeMO is a trade name for quick charging and stands for "CHArge de MOve" (move by charge) in addition to being a Japanese pun of "O cha demo ikaga desuka" (let us have a tea while charging). The CHAdeMO standard is the primary DC quick charging standard and was originally developed by Japanese car manufacturers and other companies, including TEPCO, Fuji Heavy Industries, Nissan, Mitsubishi Motors, and Toyota, which are organized as the CHAdeMO Association. Additionally, the CHAdeMO standard complies with the used international standard of IEC 62196-3, which provides DC quick charging. According to its research and development road map [26], the CHAdeMO standard with higher power than currently available, which is 100 kW of continuous charging and

Table 1 Charging standards available for EVs

Specification	Combo 1	Combo 2	CHAdeMO	Tesla
AC charging				
Voltage (V)	250	400 (three phase) 230 (one phase)	–	–
Current (A)	32	63 (three phase) 70 (one phase)	–	–
Maximum output (KW)	13	44	–	–
DC charging				
Voltage (V)	600	850	500; 1000 (future)	480
Current (A)	150	200	125; 400 (future)	
Connector type	IEC 62196-3; SAE J1772	IEC 62196-3	CHAdeMO dedicated connector	Tesla dedicated connector
Power (kW)	90	170; 350 (future)	62.5; 400 (future)	120

150–200 kW (500 V; 350 A) of peak charging power, is also planned to be released. Furthermore, a higher power CHAdeMO standard is also being developed and planned to be released in 2020. It will be able to charge EVs at a charging rate of 350–400 kW (1 kV; 350–400 A). CHAdeMO chargers have the largest international coverage, with 7000, 4000, and 2000 chargers installed in Japan, Europe, and USA, respectively.

The CCS standards cover the charging standards of Combo 1 and Combo 2 and are able to accommodate both AC and DC charging, including level-1 and level-2. CCS was developed and deployed by European and US car manufacturers in 2012. The Society of Automotive Engineers (SAE) and the European Automobile Manufacturer's Association (ACEA) offered strong support to this standard, with the aim of accommodating both AC and DC charging with a single connector. CCS can provide AC and DC charging at a maximum of 44 and 200 kW, respectively. In addition, DC charging is planned to be extended up to 350 kW [27]. In total, 2500 and 1000 CCS chargers have been currently installed mainly in Europe and the USA, respectively.

On the other hand, the Tesla Supercharger employs its own charging standard, which includes multiple chargers working in parallel and can deliver up to 120 kW of DC quick charging [28]. In total, 5000 T Superchargers are currently installed in approximately 800 stations.

Other charging strategies for EVs include inductive charging, which is conducted wirelessly. Electromagnetic induction is generated by an induction coil that is charged using high-frequency AC power. The generated magnetic field is transferred to the inductive power receiver installed in the vehicle; therefore, electricity can be transferred from the power source to the vehicle. Inductive charging is based on IEC/TS 61980 standards. The adoption of inductive charging is considered to have the potential of minimizing driving range anxiety and the battery size since charging can be conducted while driving and/or while idling. However, there are several technical barriers, such as lower efficiency, slower charging rate, interoperability, and safety.

3 EV Charging Behavior: Influence of Temperature

In general, EVs employ Li-ion batteries because of their high energy density, stable electrochemical properties, longer lifetime, and low environmental impact [29, 30]. Temperature considerably influences the charge/discharge behavior of Li-ion batteries. A lower temperature results in lower rates of charging and discharging because of electrolyte limitation [31, 32] and changes in electrolyte/electrode interface properties, which include viscosity, density, electrolyte components, dielectric strength, and ion diffusion capability [33]. Furthermore, Liao et al. observed that a lower temperature leads to a significant increase in the charge transfer resistance, which can be higher than both bulk resistance and solid-state interface resistance [34].

Herein, to evaluate the influence of ambient temperature on the charging performance of EVs, charging tests were initially performed in winter and summer. Table 2 summarizes the specifications of EV and QC used during the experimental evaluation, and Table 3 summarizes the experimental conditions. The ambient temperatures referred to the weather information accessed from the database of the Japan Meteorological Agency for the corresponding time and area (Yokohama) [35]. It is worth noting that the charging test of EVs was conducted after they had been parked for a relatively long period (e.g., one night); therefore, the battery temperature was assumed to be the same as its surroundings.

Figure 1 shows the relation between charging rate, charging time, and SoC of the EV battery in winter (Fig. 1a) and summer (Fig. 1b). Generally, although the rated capacity of the charger was 50 kW, the charging rate was relatively lower, particularly during winter. Compared to charging time in winter (35 min), charging in summer resulted in a higher charging rate and, therefore, a shorter time (20 min). A significantly higher charging rate (approximately 40 kW) could be obtained during summer up to an SoC of approximately 50%. After that, the charging rate gradually decreased, following the increase in SoC, and the charging rate was 16 kW when the SoC reached 80%. On the contrary, the charging rate reached approximately 35 kW instantaneously during winter, in a very short time, and then declined when the SoC increased further. The charging rate was approximately 10 kW when the SoC was 80%.

Figure 2 shows the relation between the charging current, voltage, and time in winter and summer, which corresponds to the charging rate shown in Fig. 1. It can

Table 2 Specifications of the EV and the QC used in the experiment

Component	Property	Value
EV	Type	Nissan Leaf
	Battery type	Laminated Li-ion battery
	Total battery capacity	24 kWh
	Maximum voltage	403.2 V
	Nominal voltage	360 V
	Energy density	140 Wh kg^{-1}
	Power density	2.5 kW kg^{-1}
	Cathode	$LiMn_2O_4$ with $LiNiO_2$
	Anode	Graphite
	Cell rated capacity	33.1 Ah (0.3 C)
	Cell average voltage	3.8 V
	Cell maximum voltage	4.2 V
QC	Type	DC quick charger
	Standard	CHAdeMO
	Output voltage	DC 50–500 V
	Output current	0–125 A
	Rated power output	50 kW

Table 3 Experimental conditions for evaluating the ambient temperature effect

Season	EV No.	Ambient temperature (°C)	Starting SoC (%)	Ending SoC (%)
Winter	#1	11.4	32	80
	#2	11.0	25	
	#3	10.3	29	
Summer	#1	27.5	34	
	#2	27.1	31	
	#3	32.0	31	

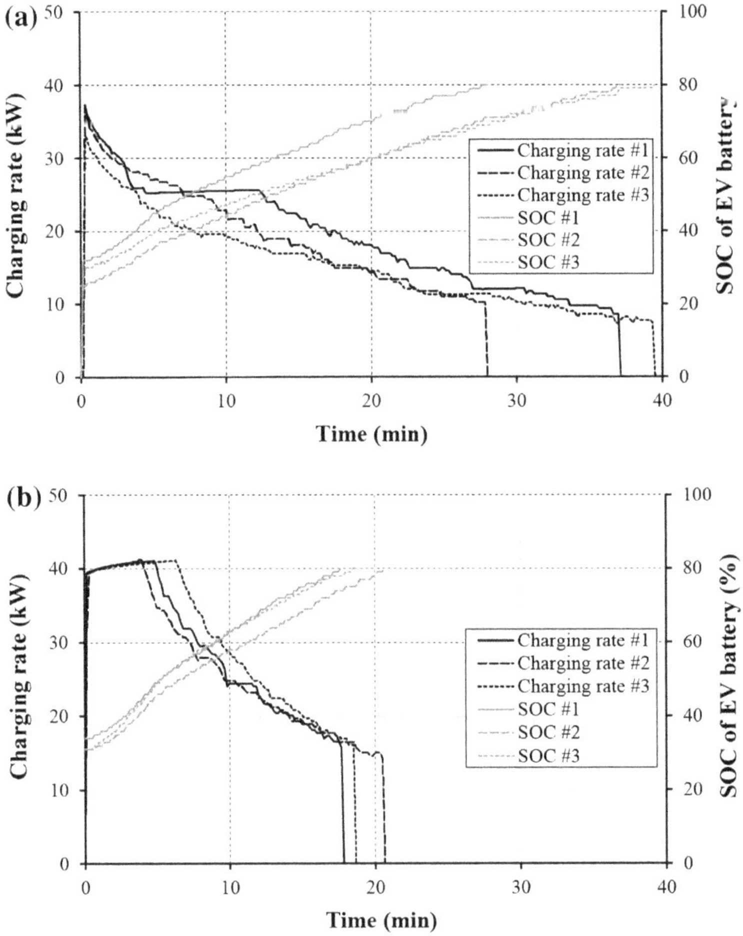

Fig. 1 Relation between the charging rate, charging time, and SoC of the EV battery: **a** winter and **b** summer (adapted from [13])

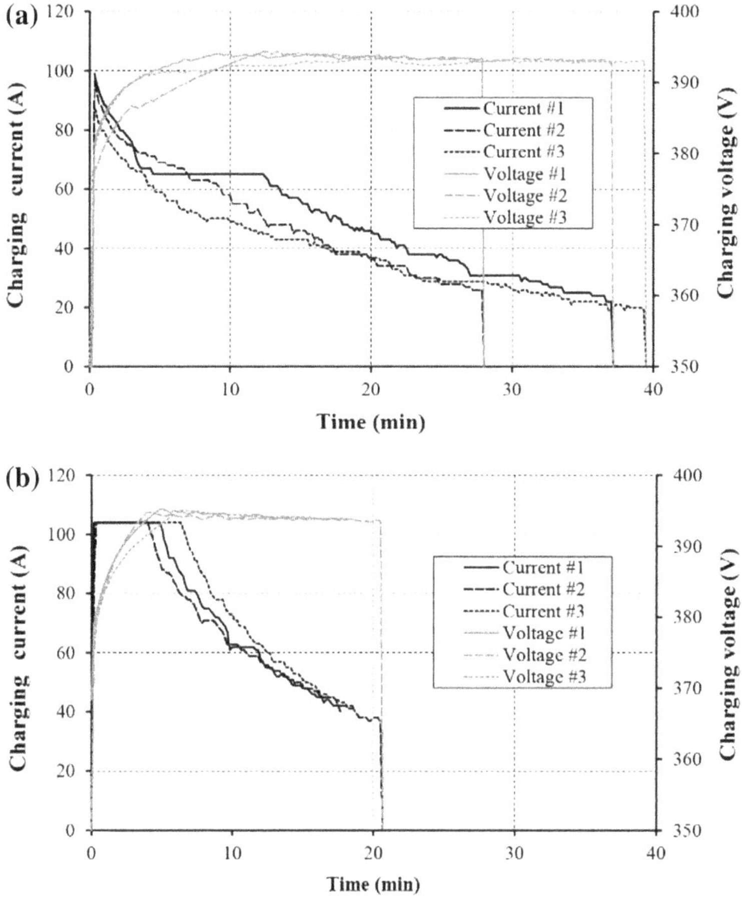

Fig. 2 Correlation between the charging current, voltage, and time in different seasons: **a** winter and **b** summer (adapted from [13])

be observed that the curves of the charging current in Fig. 2 are nearly similar to those of the charging rate in Fig. 1. Generally, Li-ion batteries are charged using a CC–CV (constant current–constant voltage) regime [36]. Charging at lower temperatures resulted in a gradual decrease in the charging current with charging time or increase in the SoC. In contrast, charging under relatively warmer conditions resulted in a higher charging current, particularly at low SoCs. A higher CC of approximately 105 A was achieved in the initial 5–10 min of charging (SoC up to approximately 50%). With regard to the charging voltage, although there was no significant difference between charging under both conditions, charging when exposed to a warmer temperature resulted in a slightly higher initial charging voltage before settling at a constant value. Hence, the CV condition was achieved faster. Temperature has a significant effect on the charging performance of EVs.

Charging in summer resulted in higher charging rates, particularly because of higher charging current and faster increase in the charging voltage. Hence, a shorter charging time could be realized.

As reported in [35, 37], a decrease in temperature leads to a slower Li^+ diffusion in the cathode and, so, to electrode polarization. Moreover, the charging transfer resistance at the electrolyte–electrode interface also increases.

4 Integrated QC System

4.1 Concept of Integrated QC System

The US Department of Energy (DOE) has issued a regulation with regard to the relatively high charging rate during quick charging, which is 10 miles of driving range per minute of charging [38]. Therefore, an EV with a battery capacity of 24 kWh (driving range of approximately 100 miles) requires 10 min of charging time at a rate of 6 C [39]. This charging rate can potentially cause significant fluctuation in the electrical grid, particularly in the case of massive simultaneous EV charging.

The installment of a stationary battery in the charger helps improve the charging performance and provides a solution to the problems related to a high electricity demand caused by EV charging [40, 41]. A BAQC that is able to control the power input received from the electrical grid (according to contracted power capacity) and the power output to the corresponding EV has been developed and evaluated. In addition, the BAQC controls the distribution of electricity inside its own system, including the stationary battery and chargers, to achieve the optimal performance by optimizing its objective function. Therefore, a BAQC is capable to satisfy both the supply side (grid load minimization through load shifting and reduction of electricity cost) and the demand side (satisfying the EV owners due to quick charging during peak hours).

The objectives of this system are: decreasing the contracted power capacity, avoiding high electricity demand during peak hours due to EV charging, reducing charging and waiting times, participating in electrical grid ancillary services, including frequency regulation and peak shaving, providing a storage for any surplus electricity by renewable energy, and providing an emergency backup for the surrounding community.

Figure 3 shows a schematic diagram of the BAQC developed by NEC Corporation in Japan. The solid and dashed lines represent the electricity and information flows, respectively. A community energy management system (CEMS) is responsible for the overall management of energy, including supply and demand across the community. CEMS optimizes the energy performance throughout the community to ensure the comfort and security of community members and minimize the environmental and social impact. Concretely, a CEMS communicates with

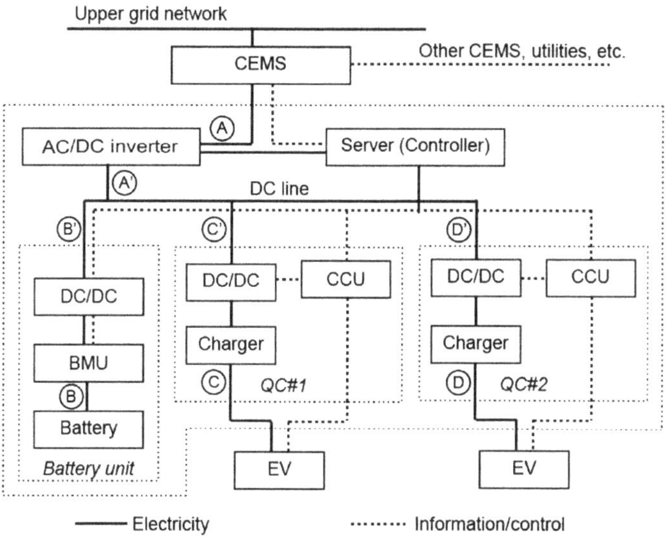

Fig. 3 Schematic diagram of the BAQC developed by NEC Corporation (adapted from [13])

other energy management systems (EMSs) under its authority, including electricity price, supply, and demand forecasting. Moreover, it also negotiates with other CEMSs or utilities to maximize the benefits of its community.

In electricity flow, there are three main modules which are connected by high-capacity DC lines: AC/DC inverter, stationary battery, and quick chargers. The AC/DC converter receives electricity from the electrical grid and converts it to high DC voltage. For this purpose, the server/controller calculates and controls the amount of electricity received/purchased from the electrical grid based on demand, electricity price, and grid condition. Moreover, the server also controls the charging and discharging behaviors of the stationary battery and the charging rate from quick chargers to the corresponding EV. In the battery unit, a bidirectional DC/DC converter and a battery management unit (BMU) are installed before the battery to control and monitor the charging and discharging of stationary battery after receiving commands from the server. Furthermore, in the quick charger modules, the DC/DC converter and the charging control unit (CCU) are also installed to provide active control during charging. The number of quick chargers could be more than one depending on the conditions.

The stationary battery is used to store electricity when there is remaining contracted capacity or when the electricity price is low (during off-peak hours). Furthermore, the stored electricity can be discharged in the case of high electricity demand for charging or during peak hours when the price of electricity is high. Basically, a stationary battery with a relatively large capacity is adopted to facilitate simultaneous charging of EVs and thus improve the quality of service.

Based on the charging and discharging conditions of the installed stationary battery and the source of electricity used for charging, the quick charging modes for the developed BAQC can be classified as follows.

1. Battery discharging

The stationary battery releases its electricity to assist the charger. Therefore, EV charging is carried out drawing electricity from both the grid and the battery. This mode is used in case of simultaneous quick charging of EVs, particularly when the price of electricity is relatively high. The flow of electricity in this mode can be expressed as follows:

$$P_{\mathrm{grid}} + P_{\mathrm{batt}} = P_{\mathrm{qc},1} + P_{\mathrm{qc},2} + P_{\mathrm{loss}} \tag{1}$$

where P_{grid}, P_{batt}, P_{qc}, and P_{loss} represent electricity purchased from the grid, electricity charged (negative value) or discharged (positive value) from a stationary battery, electricity for quick charging of EVs, and electricity loss, respectively.

2. Battery charging

In case there is still a margin in the contracted power capacity or when the price of electricity from the grid is relatively low (e.g., due to surplus electricity during the night), the stationary battery will be charged to store the electricity. The electricity flow in this mode can be expressed as Eq. (2):

$$P_{\mathrm{grid}} - P_{\mathrm{batt}} = P_{\mathrm{qc},1} + P_{\mathrm{qc},2} + P_{\mathrm{loss}} \tag{2}$$

3. Battery idling

The stationary battery can be in the idling mode under several conditions: (a) The contracted power capacity is adequate to facilitate a quick charging of EVs (low charging demand), and (b) the stationary battery is in empty condition due to the continuous quick charging of EVs (stationary battery cannot supply electricity unless being charged). In the latter case, BAQC will control the charging rate for each quick charger to maintain the contracted power capacity. Therefore, the penalty of high electricity price can be avoided. Electricity flow in this mode can be expressed as follows:

$$P_{\mathrm{grid}} = P_{\mathrm{qc},1} + P_{\mathrm{qc},2} + P_{\mathrm{loss}} \tag{3}$$

BAQC always keeps the value of P_{grid} lower than, or equal to, the contracted power capacity. In addition, P_{loss} is the power loss in the system due to several factors, including AC/DC and DC/DC conversions and electricity consumed by the system. Hence, the value of P_{loss} in each quick charging mode is different.

4.2 Pilot BAQC System

In order to evaluate the performance of the BAQC, simultaneous quick charging tests for EVs have been performed. Basically, the control systems simply followed Eqs. (1)–(3) and Table 4 lists the BAQC specifications. The DC line, which connects the modules, has a voltage of 450 V. Hence, the DC/DC converters in the stationary battery and charger modules convert the voltage from 450 V to the voltages designated for stationary battery charging and discharging. Two quick chargers, each having a maximum output power of 50 kW, were installed in the QC system. This system was located in an area of JXTG Nippon Oil & Energy Corporation, Shinkoyasu, Yokohama, Japan.

It is worth noting that the contracted power capacity is the electricity received by the AC/DC inverter (as input electricity; point A in Fig. 3). Therefore, in case no battery is installed inside BAQC, the electricity which can be transferred to EVs during charging (points C and D in Fig. 3) is less than the received electricity (point A) due to power loss during conversions and system consumption. The stationary battery of BAQC was installed together with the other controllers and converters under controlled room temperature; therefore, the performance of the stationary battery was stable and remained unchanged at different ambient temperatures.

Table 4 Specifications of the battery-assisted quick charger (BAQC)

Component	Property	Value
AC/DC inverter	Receiving voltage	200 V
	Converter output voltage	DC 450 V
	Converter output power	50 kW
DC/DC converter	Power at DC line side	50 kW
	Maximum current at the stationary battery side	150 A
	Voltage at the stationary battery side	0–400 V
Stationary battery	Type	Li-ion
	Capacity (kWh)	64.2 kWh
	Nominal voltage	364.8 V
	Maximum charging voltage	393.6 V
	Discharge cutoff voltage	336.0 V
	Maximum current in continuous discharge	176 A
	SoC threshold in charging	90%
	SoC threshold in discharging	10%
Quick charger	Number	2 units
	Standard	CHAdeMO
	Output voltage	DC 50–500 V
	Output current	0–125 A
	Rated output power	50 kW

To measure the BAQC performance, simultaneous charging tests were performed. The charging tests were conducted in three steps: (1) simultaneous charging of two EVs to measure and compare the required charging time in different seasons, (2) BAQC performance in a different contracted power capacity, and (3) simultaneous EV charging under high demand conditions. The first connected EV is prioritized with a higher charging rate, while the second EV is charged at a maximum charging rate of 5 kW until the charging rate of the first EV begins to decline. Furthermore, when the charging of the first EV is complete, the second EV would be charged according to its maximum charging rate. BAQC keeps the contracted power capacity and controls the charging rate for each EV connected to the corresponding QC.

5 Simultaneous Charging Tests

5.1 Comparison of Charging Performance

Table 5 summarizes the experimental conditions for two simultaneous charging experiments in different seasons. In addition, Fig. 4 represents the results of the simultaneous charging of two EVs during winter, with conventional QC and BAQC under a contracted power capacity of 50 kW. The electricity received from the electrical grid could be maintained at 50 kW or below. In the conventional charging system, due to limited contracted power capacity, the first connected EV was charged at a higher charging rate, while the second EV was initially charged at 5 kW. As the charging rate of the first connected EV declined, the charging rate of the second EV gradually increased, enabling the total electricity to reach its maximum contracted power capacity. As the charging rate of both EVs decreased, due to the SoC increase, the total electricity received from the grid decreased. The first and second EVs could be fully charged after 40 and 50 min, respectively.

Table 5 Conditions of two simultaneous charging experiments in conventional QC and BAQC (contracted power capacity = 50 kW)

Season	Charging system	Ambient temperature (°C)	EV No.	Starting SoC (%)	Ending SoC (%)
Winter	Conventional	10.7	#1	28	80
			#2	29	
	BAQC	11.0	#1	25	
			#2	27	
Summer	Conventional	28.4	#1	32	
			#2	35	
	BAQC	27.8	#1	33	
			#2	27	

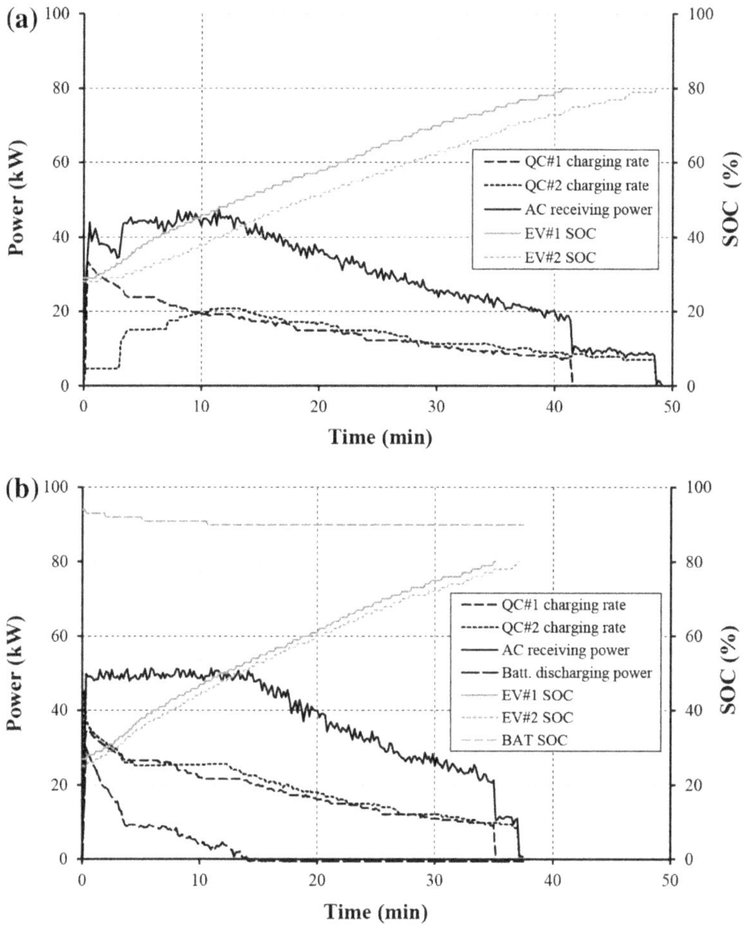

Fig. 4 Simultaneous charging test conducted during winter with a contracted power capacity of 50 kW: **a** conventional system and **b** BAQC (adapted from [13])

On the other hand, while charging with the developed BAQC, the first and second EVs nearly charged at the same rate and both EVs could complete charging almost at the same time (about 35 min). In addition, electricity from the grid could be maintained below the contracted power capacity, although the total charging rate for both EVs was larger than this power capacity. This was due to the electricity discharged from the stationary battery assisting the system. Hence, compared to a conventional charging system, the BAQC was found to provide higher quality charging with a higher rate during simultaneous charging.

Figure 5 represents the results of the simultaneous charging of two EVs performed during summer with the conventional charging system and BAQC. In the conventional system, due to the limited contracted power capacity, the first connected EV had a higher charging rate, while the second EV had a charging rate of

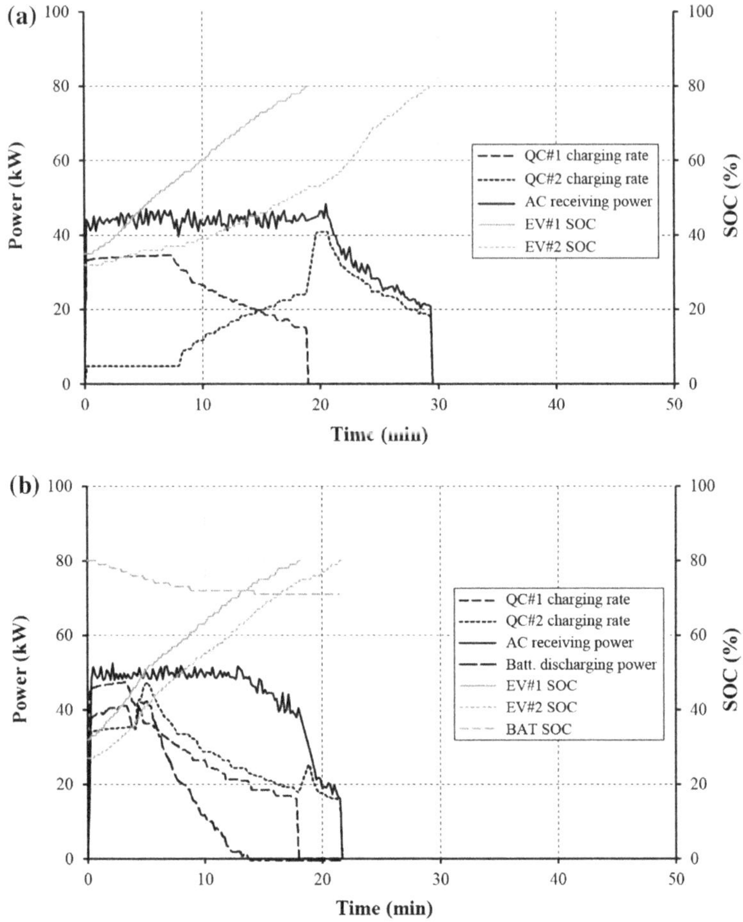

Fig. 5 Simultaneous charging experiments during summer with a contracted power capacity of 50 kW: **a** conventional system and **b** BAQC (adapted from [13])

5 kW until the charging rate of the first EV started to decline. The first and second EVs finished their charging after approximately 20 and 30 min, respectively.

On the other hand, charging with BAQC, a similar tendency to the case in winter was observed. Both EVs could be charged almost at the same charging rate while maintaining the contracted power capacity. Both EVs were charged in a relatively shorter period of approximately 20 min. When the stationary battery was fully discharged, the capacity of the two EVs was the same or smaller than the contracted power capacity from the grid.

Generally, it could be observed that the developed BAQC could improve charging quality, especially during simultaneous charging of multiple EVs. Moreover, from the point of view of the electrical grid, the application and deployment of BAQC could minimize the stress of the grid due to the high demand for EV charging.

5.2 *Influence of Contracted Power Capacity*

In order to evaluate the performance of BAQC in terms of different contracted power capacities, charging tests with different amounts of received power were conducted. In addition to the charging tests explained in Sect. 5.1 (contracted power capacity = 50 kW), power capacities of 30 and 15 kW were evaluated. Table 6 summarizes the experimental conditions under each contracted power capacity in winter and summer.

Figures 6 and 7 show the experimental results for the simultaneous charging of two EVs in the above conditions. For an additional comparison, the results obtained with a contracted power capacity of 50 kW in winter and summer can be seen in Figs. 4b and 5b, respectively. With regard to the simultaneous charging of two EVs, there is no significant difference in terms of charging time in each season. As the contracted power capacity decreased, the total charging rate for the two QCs also declined. As the contracted power capacity decreased, the second connected EV was initially charged with a lower charging rate. This rate increased as the charging rate of the first EV declined due to the increase in the SoC. As a result, there was insignificant difference between the contracted power capacities of 50 and 30 kW with regard to the required charging time in both seasons. It appears that although the second connected EV initially received lower power, it increased to its maximum charging rate as the charging rate of the first connected EV declined. The charging time became increased as the contracted power capacity was lowered to 15 kW. In this case, the required charging times were ca. 38 and 25 min in winter and summer, respectively.

In addition, the discharged electricity from the stationary battery was higher in the case of the lower contracted power capacity. Change in the SoC of the stationary battery was greater after the decrease in the contracted power capacity. Numerically, the SoC decreased by approximately 5, 10, and 20% during the simultaneous charging of the two EVs in the case of the contracted power capacities of 50, 30, and 15 kW, respectively. In this system, since the SoC thresholds for

Table 6 Experimental conditions for the simultaneous charging of two EVs with a BAQC under different contracted power capacities

Contracted power capacity (kW)	Season	Ambient temperature (°C)	EV No.	Starting SoC (%)	Ending SoC (%)
30	Winter	10.7	#1	28	80
			#2	29	
	Summer	11.0	#1	25	
			#2	27	
15	Winter	28.4	#1	32	
			#2	35	
	Summer	27.8	#1	33	
			#2	27	

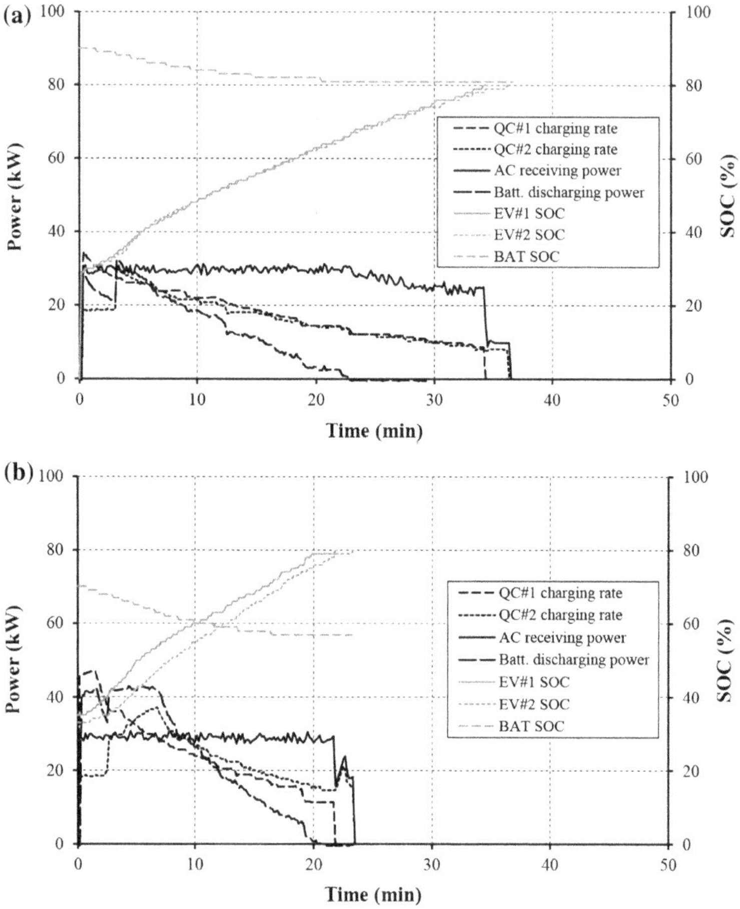

Fig. 6 Simultaneous charging test of two EVs using a BAQC with a contracted power capacity of 30 kW: **a** winter and **b** summer (adapted from [41])

charging and discharging were 90 and 10%, respectively, the total SoC change that could be utilized to assist simultaneous charging was 80%. Therefore, it can be roughly assumed that the stationary battery can continuously assist approximately 32, 16, and 8 EVs during simultaneous charging with contracted power capacities of 50, 30, and 15 kW, respectively.

5.3 Tests Performed During High Charging Demand

In order to evaluate the performance of the developed system during high charging demand, simultaneous charging of eight EVs was performed. The EVs were

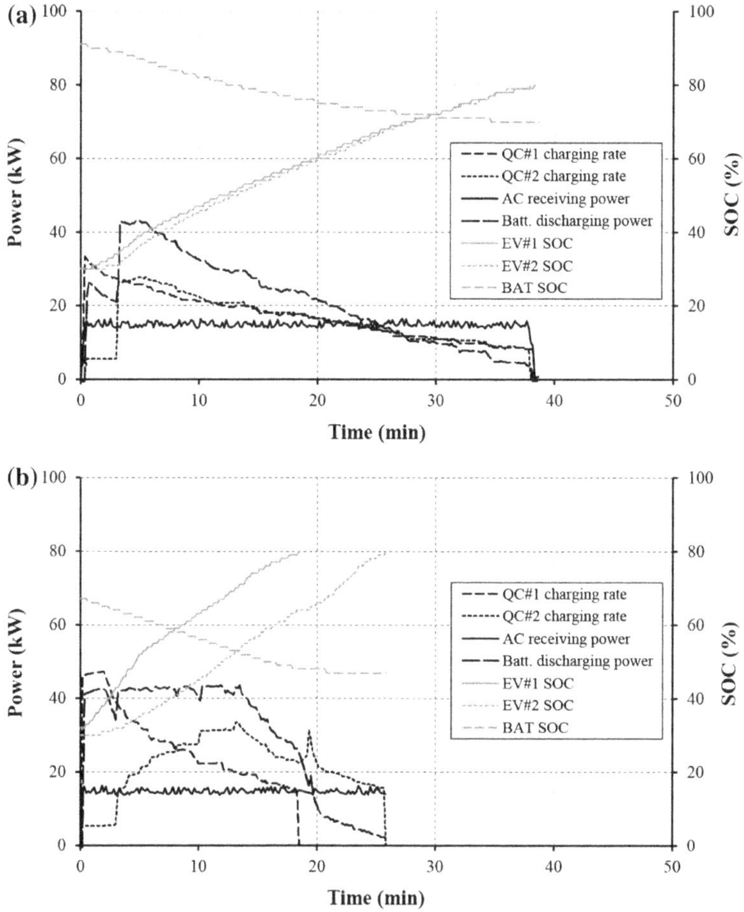

Fig. 7 Simultaneous charging test of two EVs using a BAQC with a contracted power capacity of 15 kW: **a** winter and **b** summer (adapted from [41])

initially prepared and parked near the BAQC system as queuing EVs waiting for their turn to charge. The experimental conditions are listed in Table 7. As one EV finished its charging (SoC of 80%), the next EV (SoC of approximately 30%) would replace it in the same charger.

Figure 8 shows the results of the experiments performed in winter and summer under a contracted power capacity of 30 kW. In general, the simultaneous charging of eight EVs during summer can be performed faster compared to that during winter due to the higher charging rate. Although the charging time required for simultaneous charging in summer was significantly less, the SoC of the stationary battery declined substantially. This was due to the high discharging rate of the stationary battery to assist the chargers. The stationary battery could not be charged during this kind of simultaneous charging as the marginal electricity was unavailable. On the

Table 7 Experimental conditions for simultaneous charging test of eight EVs using a BAQC under different contracted power capacities

Contracted power capacity (kW)	Season	Ambient temperature (°C)	EV No.	Starting SoC (%)	Ending SoC (%)
30	Winter	10.7	#1	29	80
			#2	34	
			#3	33	
			#4	33	
			#5	27	
			#6	30	
			#7	29	
			#8	33	
	Summer	29.0	#1	31	
			#2	20	
			#3	25	
			#4	29	
			#5	31	
			#6	36	
			#7	27	
			#8	32	
15	Summer	27.8	#1	30	
			#2	30	
			#3	30	
			#4	30	
			#5	30	
			#6	30	
			#7	27	
			#8	29	

other hand, during winter, the discharging rate of the stationary battery was significantly lower because of the slower charging rate of EVs. The total charging rate of the two chargers might be lower than the contracted power capacity, leading to marginal electricity, which could be used to charge the stationary battery (Fig. 8a). Therefore, the SoC of the stationary battery did not substantially decline in comparison with the SoC in summer.

Figure 9 shows the simultaneous charging test of eight EVs during summer under a contracted power capacity of 15 kW. In comparison with simultaneous charging with a higher contracted power capacity (30 kW), almost no significant change could be observed in the EV charging rate, except for the last connected EV. Unfortunately, the discharging rate of the stationary battery was considerably high and led to a faster decline of its SoC. As shown in Fig. 9, the SoC of the stationary battery dropped considerably fast and reached 10% during the charging of the last two EVs. Consequently, the last connected EV was charged using the electricity received from the electrical grid without any support from the stationary battery.

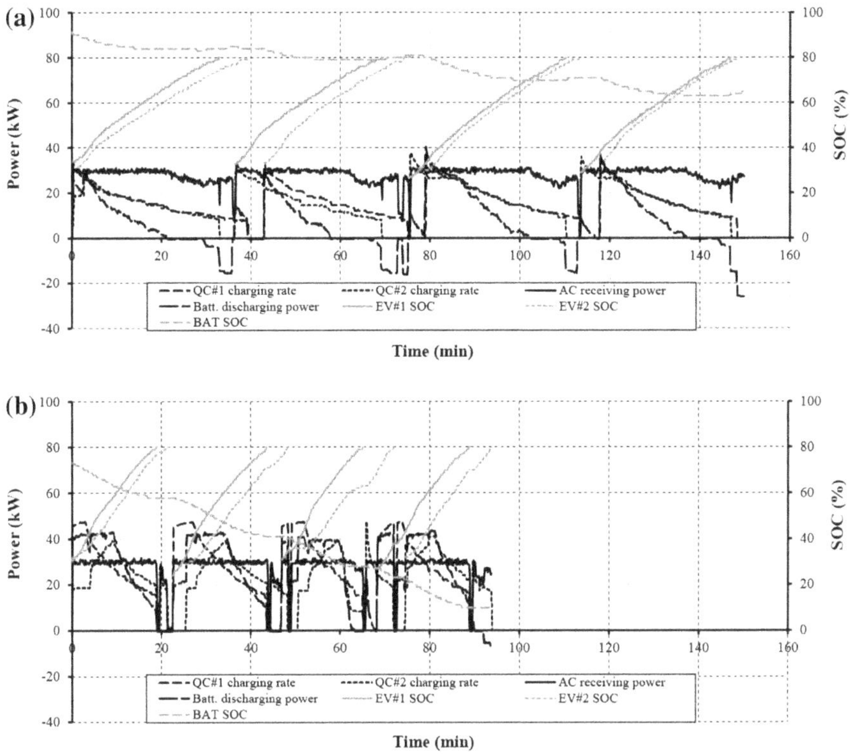

Fig. 8 Simultaneous charging test of eight EVs using a BAQC under a contracted power capacity of 30 kW: **a** winter and **b** summer (adapted from [13])

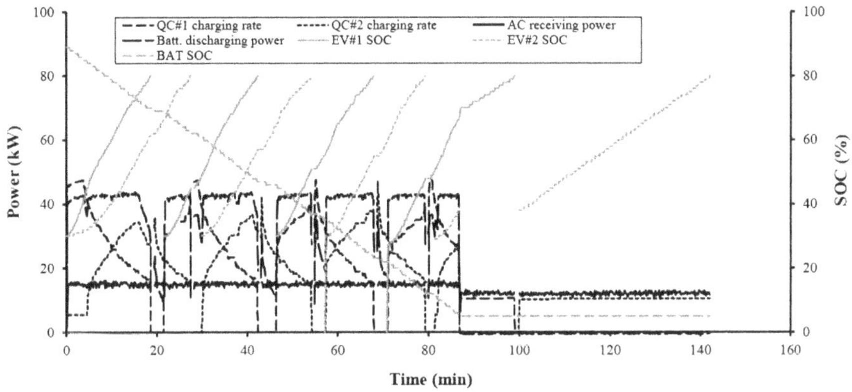

Fig. 9 Simultaneous charging test of eight EVs using a BAQC under a contracted power capacity of 15 kW during summer (adapted from [13])

Hence, since the contracted power capacity was considerably low, the last connected EV was not charged until the charging of the EV connected previously was complete. The stationary battery could not be charged during simultaneous charging due to the unavailability of the marginal electricity and the high charging rate of EVs.

The experimental results indicate that the application of the BAQC can significantly improve the charging behavior of the QCs, particularly during simultaneous EV charging. It appears that the balance between the charging rate of EVs, contracted power capacity, and stationary battery SoC is a considerably important issue that needs to be cautiously addressed. In addition, EVs charging demand should be initially forecast.

6 Conclusions

A BAQC-based QC has been developed, and its performance in simultaneous charging was assessed. First, the charging behavior of EV in different seasons, winter and summer, was analyzed to clarify the influence of temperature on the charging rate. The battery temperature has a strong effect on the battery's charging behavior: It was seen that the charging rate during summer was higher than that during winter.

In the simultaneous charging experiments, the application of the BAQC clearly improved the performance of EV chargers. Charging could be performed in a shorter time while maintaining the contracted power capacity. In the future, since the demand for EV charging is predicted to increase, the electrical grid stress due to charging demand and its fluctuation will increase accordingly. Hence, the adoption of BAQC is believed to be able to reduce this stress and maintain the quality of grid electricity.

Furthermore, since BAQC has a large quantity of stationary batteries, it can also participate in ancillary service programs for the electrical grid when the demand for quick charging is relatively low. In this case, a bidirectional AC/DC inverter is required to provide both services: up (from the battery to the electrical grid) and down (from the electrical grid to the battery). This participation can improve the economic performance of BAQC in addition to improving electrical grid reliability.

Acknowledgements The authors express their sincere thanks to the JXTG Nippon Oil & Energy Corporation and the NEC Corporation for providing support and assistance during experimentation and data collection.

References

1. W. Zhu, Y. He, Eur. J. Oper. Res. **258**, 165 (2017)
2. G.J. Suppes, T.S. Storvick, *Sustainable Power Technologies and Infrastructure* (Academic Press, Amsterdam, 2016)

3. K.J. Dyke, N. Schofield, M. Barnes, IEEE Trans. Ind. Electron. **57**, 3917 (2010)
4. Union of Concerned Scientists (UCS), BEV and PHEV driving range. Available from: http://www.ucsusa.org/clean-vehicles/electric-vehicles/bev-phev-range-electric-car. Accessed 25 Sept 17
5. C. Qiu, G. Wang, Energy Convers. Manage. **119**, 389 (2016)
6. M. Aziz, T. Oda, T. Mitani, Y. Watanabe, T. Kashiwagi, Energies **8**, 3720 (2015)
7. M. Aziz, T. Oda, A. Morihara, T. Murakami, N. Momose, Utilization of EVs and their used batteries in factory load leveling, in *2014 IEEE PES Innovative Smart Grid Technologies Conference (ISGT), Washington, USA* (2014)
8. R.J. Hamidi, H. Livani, Electric Power Syst. Res. **143**, 522 (2017)
9. M. Aziz, T. Oda, T. Energy Procedia **75**, 1938 (2015)
10. T.K. Paul, H. Aisu, Management of quick charging of electric vehicles using power from grid and storage batteries, in *2012 IEEE International Electric Vehicle Conference (IEVC), Greenville, USA* (2012)
11. M.F. Shaaban, A.A. Eajal, E.F. El-Saadany, Renew. Energy **82**, 92 (2015)
12. N. Nezamoddini, Y. Wang, Energy **116**, 836 (2016)
13. M. Aziz, T. Oda, M. Ito, Energy **100**, 82 (2016)
14. T. Oda, M. Aziz, T. Mitani, Y. Watanabe, T. Kashiwagi, Electr. Eng. Jpn. **198**, 11 (2017)
15. T. Oda, M. Aziz, T. Mitani, Y. Watanabe, T. Kashiwagi, J. Jpn. Soc. Energy Res. **37**, 7 (2016). (in Japanese)
16. M. Aziz, T. Oda, Load levelling utilizing electric vehicles and their used batteries, in *Modeling and Simulation for Electric Vehicle Applications*, ed. by M.A. Fakhfakh (InTech, Rijeka, 2016)
17. S. Deilami, A. Masoum, P.S. Moses, M.A.S. Masoum, IEEE Trans. Smart Grid **2**, 456 (2011)
18. E. Sortomme, M.M. Hindi, S.D.J. MacPherson, S.S. Venkata, IEEE Trans. Smart Grid **2**, 198 (2011)
19. M. Moghbel, M.A.S. Masoum, A. Fereidouni, Intell. Ind. Syst. **1**, 141 (2015)
20. K. Clement-Nyns, E. Haesen, J. Driesen, IEEE Trans. Power Syst. **25**, 371 (2010)
21. Z. Ma, D.S. Callaway, I.A. Hiskens, IEEE Trans. Control Syst. Technol. **99**, 1 (2011)
22. K.N. Kumar, K.J. Tseng, Energy **111**, 190 (2016)
23. C. Peng, J. Zou, L. Lian, Renew. Sustain. Energy Rev. **68**, 147 (2017)
24. M. Aziz, Charging/discharging behaviors and integration of electric vehicle to small-scale energy management system, in *6th International Conference on Smart Cities and Green ICT Systems (SMARTGREENS 2017), Porto, Portugal* (2017)
25. M. Aziz, Advanced charging system for plug-in hybrid electric and electric vehicles, in *Hybrid Electric Vehicles*, ed. by T. Donatio (InTech, Rijeka, 2017), p. 63
26. CHAdeMO Association, High Power CHAdeMO. Available from: https://www.chademo.com/technology/high-power/. Accessed 25 Sept 17
27. CharIN e. V. Combined charging system specification. Available from: http://www.charinev.org/ccs-at-a-glance/ccs-specification/. Accessed 25 Sept 17
28. Tesla Supercharger. Available from: https://www.tesla.com/supercharger. Accessed 25 Sept 2017
29. Y.T. Shih, C.H. Wu, F.Y. Hung, T.S. Lui, L.H. Chen, Surf. Coat. Technol. **215**, 79 (2013)
30. J.W. Yao, F. Wu, X.P. Qiu, N. Li, Y.F. Su, Electrochim. Acta **56**, 5587 (2011)
31. L.F. Xiao, Y.L. Cao, X.P. Ai, H.X. Yang, Electrochim. Acta **49**, 4857 (2004)
32. P. Ping, Q. Wang, P. Huang, J. Sun, C. Chen, Appl. Energy **129**, 261 (2014)
33. A.N. Jansen, D.W. Dees, D.P. Abraham, K. Amine, G.L. Henriksen, J. Power Sources **174**, 373 (2007)
34. L. Liao, P. Zuo, Y. Ma, X. Chen, Y. An, Y. Gao, G. Yin, Electrochim. Acta **60**, 269 (2012)
35. Japan Meteorological Agency. Available from: http://www.jma.go.jp/jma/index.html. Accessed 25 Sept 17
36. Y. Seyama, T. Shimozono, K. Nishiyama, H. Nakamaura, T. Sonoda, GS News Tech. Rep. **62**, 76 (2003)
37. H. Zheng, H. Zhang, Y. Fu, T. Abe, Z. Ogumi, J. Phys. Chem. B **109**, 13676 (2005)

38. D. Howell, P. Faguy, B. Cunningham, US DOE Vehicle Battery R&D: Progress Update, US Department of Energy (2011). Available from: http://www.hydrogen.energy.gov/pdfs/htac_nov2011_howell.pdf. Accessed 25 Sept 17
39. R. Chandrasekaran, J. Power Sources **271**, 622 (2014)
40. Y. Ishii, T. Shizuno, J. Kogure, N. Itabashi, H. Watanabe, T. Nyu, NEC Tech. J. **7**, 24 (2012)
41. M. Aziz, T. Oda, Simultaneous quick-charging system for electric vehicles, in *9th International Conference on Applied Energy, Cardiff, UK* (2017)

Charging Optimization Methods for Lithium-Ion Batteries

Jiuchun Jiang

Abstract Traditional charging technology uses external battery parameters, e.g., terminal voltage and current, as the control target, and only controlling external parameters does not give information on internal characteristics of the battery, and thus, the effects of different charging currents and cutoff voltages on battery degradation are not clear. In this chapter, the electrochemical reaction mechanisms and external characteristics of the battery during charging process are studied, and the mechanisms of battery charging performance and characteristics of charging polarization are revealed. By researching the electrochemical reaction law and potential distribution characteristics of the battery during the charging process, a novel electric model based on the Butler–Volmer equation was employed to outline the unique phenomena induced by changing rates for high-power lithium batteries. The robustness of the developed model under varying loading conditions, including galvanostatic test and Federal Urban Dynamic Schedule (FUDS) test, is evaluated and compared against experimental data. The analysis of polarization voltage features at different charging rates indicates that polarization voltage is high on both ends of the SOC range but low in the middle SOC range, and the shape of the polarization voltage curve is like a bowl. In the middle SOC range, an approximate linear relationship exists between the steady-state polarization voltage and the charging rate. The two time constants (TCs) representing polarization voltage change are in 10- and 1000-s orders of magnitude, respectively, which corresponds to three charging reaction processes. The dynamic polarization voltage exhibits a lagged effect and an overshoot effect when the charge current is changed. Depending on the polarization voltage characteristics, setting battery polarization voltage and charging cutoff voltage as the constraint conditions, the calculation method for the maximum charge current of a Li-ion battery based on the battery polarization time constant is established, which can help engineers design a practical charging strategy. An optimal charging strategy is devised to balance charging

J. Jiang (✉)
School of Electrical Engineering,
Beijing Jiaotong University, Beijing 100044, China
e-mail: jcjiang@bjtu.edu.cn

G. Pistoia and B. Liaw (eds.), *Behaviour of Lithium-Ion Batteries in Electric Vehicles*, Green Energy and Technology,
https://doi.org/10.1007/978-3-319-69950-9_10

time and temperature rise, with polarization constraints fulfilled. The charging target function is constructed by setting limits to the charging temperature rise and shortening the charging time as the optimization target. The optimal charging current curve is determined by the genetic algorithm (GA) under the constraint of the maximum charge current and limited by polarization voltage. The experimental results indicate that the developed charging protocol can reduce charging time remarkably with reasonable temperature rise, highlighting its advantages over conventional CC–CV charging methods. Aging experiments further verify that the developed charging protocol has a similar capacity retention ratio, compared to that of 0.5C CC–CV charging after 700 cycles. By effectively combining the external characteristics and the internal electrochemical reaction during the charging process, the optimized charging strategy with polarization voltage as the control target results in a fast charging process without damage to the battery life.

1 Introduction

Li-ion batteries are widely used in electrical devices and energy storage systems because of their high energy density, good cycle-life performance, and low self-discharge rate [1–6]. However, the charging strategy for Li-ion batteries has become a bottleneck for their wider application, due to the slow charging speed and uncertainty effects on battery life. The charging process is closely related to a battery's state of health [7, 8]. The Li-ion power batteries used in electric vehicles (EVs) are made up of many layers of cells and have a large capacity, usually tens of ampere-hours. The charging optimization technology for Li-ion power batteries, however, is a challenge.

Numerous charging methods have been reported in the literature, with various objectives such as increasing charging speed, enhancing charging performance, and maximizing battery life. In Ref. [9], Vo et al. proposed a charging strategy based on an integration of the Taguchi method and state of charge (SOC) estimation to search for an optimal charging current profile. Guo et al. [10] proposed an optimum charging technique for Li-ion batteries using a universal voltage protocol, which has the potential to improve charging efficiency and cycle life. Hu et al. [11] developed a dual-objective optimal charging strategy based on equivalent circuit models, whereby charging time and charging energy loss are optimally traded off. Zhang et al. [12] used a dynamic programming optimization algorithm to find a suboptimal charging protocol under a certain balance between charging time and charging loss. Hussein et al. [13] proposed an online tracking algorithm to allocate and track the optimal charging frequency for common batteries in real time under any conditions. Ji et al. [14] presented a framework for optimizing Li-ion battery charging subject to side-reaction constraints, which can help avoid damage phenomena.

Battery charging optimization methods can be mainly categorized as improved charging current waveform-based methods [9–13, 15–17], battery model-based methods [11, 14, 18–23], polarization-based methods [24, 25], and enhanced battery material-based methods [26]. Improved charging current waveform-based methods are generally simple to control and implement. However, they are often heuristic and lack the theoretical foundations for choosing the battery charging current in an optimal manner. Battery model-based methods can predict charging current by employing, e.g., a lumped equivalent circuit model, an ac-impedance model, or an electrochemical model. They combine external electrical behavior with internal reaction mechanisms, searching for the optimal charging current. Diverse modeling and identification methods are developed to acquire parameters with high precision [27–31]. Nevertheless, the effects of battery temperature and health fade on model parameters need to be further investigated. Polarization-based methods provide acceptable charging current with constraints on battery polarization voltage. Polarization modeling and its quantitative effects on battery life are worth further examination.

Several battery models have been reported to meet critical requirements of diversified circumstances over the past decades. Commonly used battery models fall into three categories [1, 2, 8, 32], i.e., electrochemical models, analytical models, and electrical circuit models. Electrochemical models accurately characterize material properties and the reaction mechanism inside the battery, serving as a basis for the optimal design of battery systems [32–35]. However, electrochemical models contain complicated nonlinear differential equations with many unknown variables, which not only increase the model complexity but also are difficult to be employed in power control systems. Analytical models usually are simplified forms of electrochemical models and remain too complicated for accurate prediction of dynamic performance during battery runtime [36]. Electrical circuit models can capture battery current–voltage (*I–V*) characteristics through a combination of electrical components, such as voltage sources, resistors, and capacitors [37–40]. These models have simpler structures and fewer unknown variables than the other two kinds of models and also can be easily incorporated into control models of battery-powered systems. Low et al. [41] presented an improved model comprising two resistance–capacitance (RC) parallel networks, which gave a sufficiently accurate prediction of behavior for lithium iron phosphate (LFP) batteries. The second-order RC model has been proved to be universal and has been widely used for polarization modeling.

To satisfy various power and energy demands of different applications, batteries can be classified into two groups—high-power designs and high-energy designs. Although the second-order RC model can apply in most conditions, it is sometimes unsuitable when applied for high-power batteries. So, the second-order RC model with new structures or new mechanisms needs to be improved.

There are two unique phenomena induced by rate variation, namely the rate capacity effect and the recovery effect. The rate capacity effect refers to the change of battery capacity due to changing rates, which is more visible at high rates. After a battery discharges at a specific rate, it is still able to discharge for a period at a lower rate, which is known as the recovery effect. To improve the model feasibility for batteries with high-power designs, researchers have developed many enhanced electrical circuit models. The SOC estimation of different rates was achieved by using a normalization method, which was based on the definition of rate factors [42]. More comprehensively, Zhang et al. [43] proposed an organic combination of an electrical circuit model and the Rakhmatov diffusion model, which was sufficient to capture the recovery effect. Nevertheless, this improved model was difficult to configure for its complicated structures. To enhance the model's adaptability to high rates and suitability for system simulation, a hybrid battery model was presented in [44], which utilized a kinetic model to represent the rate capacity effect instead of the highly coupled diffusion model.

Notably, the aforementioned methods rarely discussed the change of model parameters depending on the quantity and direction of current, although they were accurate enough from the perspective of quantitative results. Lam et al. [45] proposed an empirical formula to describe current dependence of parameters using curve fitting. However, the popularization of this method is deficient because of a lack of theoretical derivations and the fact that the model validation was only actualized at a rate lower than 2C, this being the rate at which the battery is fully discharged in half an hour after being fully charged at room temperature. Apart from the general circuit model, an improved nonlinear battery model was presented in [46], which utilized the well-known electrochemical kinetics equation, namely the Butler–Volmer (BV) equation, to outline the nonlinear electrode behavior of the battery.

Temperature has a significant impact on Li-ion battery performance and lifetime [47–49]. Battery activity increases as temperature is elevated. Nevertheless, if temperature increases over allowable limits, the stability of the battery cathode lattice structure worsens, which not only accelerates battery degradation, but also results in battery safety hazards. Accelerated aging in thermal cycling for lithium manganese oxide (LMO) batteries at 40 °C was observed, which is mainly caused by active material losses in both electrodes, generalized loss of lithium inventory, and inhibited kinetics [4]. During rapid charging, the temperature gradient will inevitably increase since average charging current is enlarged, leading to lifetime decrease if the battery is operated in unreasonable thermal excursions. However, maximum charging currents differ at various regions of the SOC, which makes it possible to balance the charging speed and temperature rise. In our previous study [25], an acceptable charging current curve was pursued in accordance with Li-ion battery polarization voltage behavior. The charging curve could prevent polarization from being out of range and thus was conducive to increasing charging capacity and charging speed.

In this chapter, a polarization-based charging boundary curve is developed as a reference and is based on the second-order RC model. Based on this curve, by changing the charging current, the charging speed can be improved and polarization of the battery can be reduced. Considering there will be a large calculation error when the model is applied to the high-power battery, an improved BV equation-based electrical model is employed to capture unique phenomena induced by changing rates for the high-power battery. Finally, a polarization-based charging time and temperature rise optimization strategy for Li-ion batteries is developed to equilibrate charging speed and lifetime. This strategy uses the GA method to search for the optimal charging current trajectories, considering the temperature rise constraint and charging time.

2 Acceptable Charging Current Based on Polarization Voltage Model

2.1 Modeling of Charging Polarization Voltage

Battery polarization mainly includes two components, namely electrochemical polarization and concentration polarization. Electrochemical polarization occurs once the electric field of the battery has been established, whereas concentration polarization, because of Li ions transfer in the electrolyte phase when it is dominated by the diffusion effect, requires a longer time to be built completely [50]. Polarization voltage is one of the characteristic parameters in the battery. It involves many factors and complex variations and is one of the key factors affecting battery performance.

The equivalent circuit model, neural network model, and simplified electrochemical model are the most widely used models for battery simulation. The equivalent circuit model is based on the battery working principle that describes the performance characteristics of the battery with a circuit network, and it is suitable for many types of batteries. The RC network model is the most representative. From the above characteristic analysis, the establishment of the polarization voltage, with exponential function change rule, is consistent with the structure characteristics of the circuit model. Polarization voltage consists of concentration polarization voltage and electrochemical polarization voltage, so it is reasonable to use the second-order RC model to simulate the two parts of polarization effects, respectively. The equivalent circuit model is illustrated in Fig. 1, where R_Ω is the ohmic resistance, R_{P1} is the charge-transfer resistance, and C_{P1} is the electric double-layer capacitance of the electrode interface. R_{P2} is the diffusion resistance of charge in the electrodes and the electrolyte, and C_{P2} is the concentration diffusion capacitance. V_o is the battery terminal voltage, I is the charging current, and OCV represents the open-circuit voltage. V_P is the polarization voltage of the battery.

Based on Kirchhoff's law, the model can be described by the following equation:

$$\begin{cases} V_{P1}(t)/R_{P1} = I(t) - C_{P1} \times \mathrm{d}V_{P1}(t)/\mathrm{d}t \\ V_{P2}(t)/R_{P2} = I(t) - C_{P2} \times \mathrm{d}V_{P2}(t)/\mathrm{d}t \\ V_O(t) = \mathrm{OCV}(t) + I(t) \times R_\Omega + V_{P1}(t) + V_{P2}(t) \end{cases} \tag{1}$$

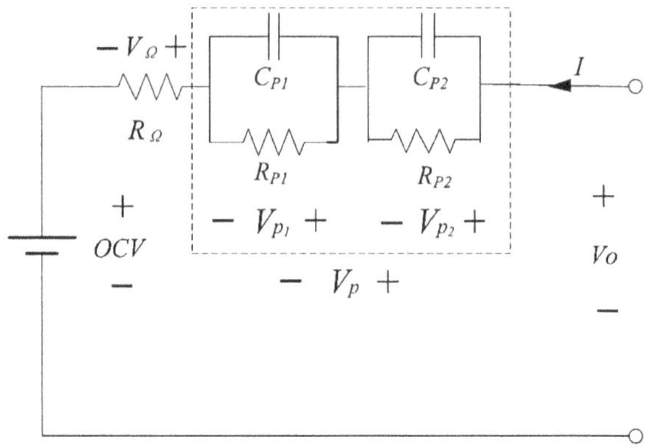

Fig. 1 Equivalent circuit model of the battery

The solution of Eq. (1) is given by:

$$\begin{cases} V_{P1}(t) = V_{P1}(0) \times \mathrm{e}^{-\frac{t}{R_{P1}C_{P1}}} + \mathrm{e}^{-\frac{t}{R_{P1}C_{P1}}} / C_{P1} \times \int_0^t I(t) \times \mathrm{e}^{\frac{t}{R_{P1}C_{P1}}} \mathrm{d}t \\ V_{P1}(t) = V_{P1}(0) \times \mathrm{e}^{-\frac{t}{R_{P1}C_{P1}}} + \mathrm{e}^{-\frac{t}{R_{P1}C_{P1}}} / C_{P1} \times \int_0^t I(t) \times \mathrm{e}^{\frac{t}{R_{P1}C_{P1}}} \mathrm{d}t \\ V_O(t) = \mathrm{OCV}(t) + I(t) \times R_\Omega + V_{P1}(t) + V_{P2}(t) \end{cases} \tag{2}$$

The initial polarization voltage $V_{P1}(0)$, $V_{P2}(0)$ was equal to zero after the battery was kept in an open-circuit condition for a long time. The charge current I can be regarded as constant with the time interval, and Eq. (2) can be further simplified. The polarization voltage V_P is the sum of V_{P1} and V_{P2}. The simplified equation can be expressed by:

$$\begin{cases} V_{P1}(t) = I \times R_{P1} - I \times R_{P1} \times \mathrm{e}^{-\frac{t}{R_{P1}C_{P1}}} \\ V_{P2}(t) = I \times R_{P2} - I \times R_{P2} \times \mathrm{e}^{-\frac{t}{R_{P2}C_{P2}}} \\ V_O(t) = \mathrm{OCV}(t) + I \times R_\Omega + V_{P1}(t) + V_{P2}(t) \\ V_P(t) = I \times (R_{P1} + R_{P2}) - I \times R_{P1} \times \mathrm{e}^{-\frac{t}{R_{P1}C_{P1}}} - I \times R_{P2} \times \mathrm{e}^{-\frac{t}{R_{P2}C_{P2}}} \end{cases} \tag{3}$$

The model parameters of R_Ω, R_{P1}, R_{P2}, C_{P1}, and C_{P2} need to be identified to simulate the charge polarization voltage dynamics. The ohmic resistance R_Ω is significantly influenced by battery temperature. R_Ω can be regarded as constant at a specified temperature during the charging/discharging process. The dc internal resistance can be achieved by measuring the battery terminal voltage response at a certain charge current since it performs pure resistance behavior. As for the R_{P1}, R_{P2}, C_{P1}, and C_{P2} identifications, some rules can be found from the V_P formula, in which the constant part equals the sum of the coefficients of two exponential functions. The constant is divided into two parts—the high-frequency components expressing charge-transfer polarization and the low-frequency components describing mass-transport polarization. According to time constant $\tau = RC$, the parameters of R_{P1}, R_{P2}, C_{P1}, and C_{P2} can be identified based on the nonlinear least squares method.

Table 1 Constant current test data-based model parameters as a function of SOC

SOC/%	10	20	30	40	50	60	70
$R_{\Omega}/\text{m}\Omega$	1.38	1.41	1.41	1.38	1.36	1.38	1.41
$R_{P1}/\text{m}\Omega$	0.31	0.31	0.29	0.31	0.37	0.35	0.31
τ_{P1}/s	25.19	76.74	122.23	140.28	84.02	158.29	139.59
$R_{P2}/\text{m}\Omega$	0.72	0.72	0.68	0.72	0.88	0.83	0.72
τ_{P2}/s	255.8	150.1	122.2	140.3	268.6	158.2	139.5

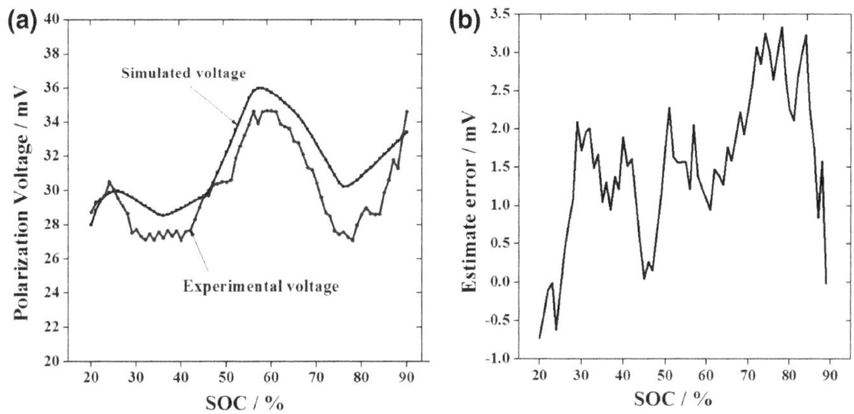

Fig. 2 **a** Simulated and measured polarization voltage of the battery for constant charging current; **b** simulation error for the battery charge polarization model

The model parameters extracted from the data of a constant current (CC) test are listed in Table 1.

The Simulink model was established for simulating battery charge polarization. The initial SOC was set to 20%, and the parameters were varying with SOC increase during the simulation. The simulated and experimental polarization voltage is illustrated in Fig. 2a, and the estimate error is shown in Fig. 2b, respectively. It is found that the developed model with identified parameters can effectively simulate the charge polarization dynamics of the battery, and the maximum estimation error is controlled within 3.5 mV, satisfying the accuracy requirement of the battery usage in EVs.

2.2 *Characteristics of Charging Polarization Voltage*

Polarization voltage can be calculated from Eq. (4), according to the aforementioned model

$$V_P(\mathrm{SOC}) = V_O - \mathrm{OCV}(\mathrm{SOC}) - IR_\Omega \tag{4}$$

Battery terminal voltage and charging current can be easily measured in real time. The polarization voltage is small at a very low current. The values of charging/discharging polarization voltage are approximately equal to each other, but the signs are opposite in the steady state. Ohmic resistance voltages are same. Therefore, the OCV curve can be obtained by calculating the average values of a full-charge curve and a full-discharge curve at low current, such as 0.05C. The relational expressions of the model are as follows:

$$\begin{aligned} V_O^{\mathrm{ch}} &= \mathrm{OCV} + V_\Omega + V_P \\ V_O^{\mathrm{dis}} &= \mathrm{OCV} - V_\Omega - V_P \end{aligned} \tag{5}$$

OCV can be calculated by

$$\mathrm{OCV} = \left(V_O^{\mathrm{ch}} + V_O^{\mathrm{dis}}\right)/2 \tag{6}$$

Figure 3 shows the calculated OCV–SOC curve. In [51, 52], the authors estimate the SOC based on the Thevenin model or an improved Thevenin model. The offline identification results of the ohmic resistance parameters illustrate that the ohmic resistance at various SOCs is nearly constant, independent of the battery load current. Therefore, the ohmic resistance at a certain temperature is regarded here as a constant. The ohmic resistance is the ratio of voltage changes to current changes, which can be described as

$$R_\Omega = \Delta V_O / \Delta I \tag{7}$$

According to Eq. (4), a V_P–SOC curve is available.

Charging polarization voltage profiles at various constant current–constant voltage (CC–CV) charging rates are shown in Fig. 4. The initial polarization voltage is large and then declines slowly, with the local minimum value at around

Fig. 3 OCV versus SOC curve

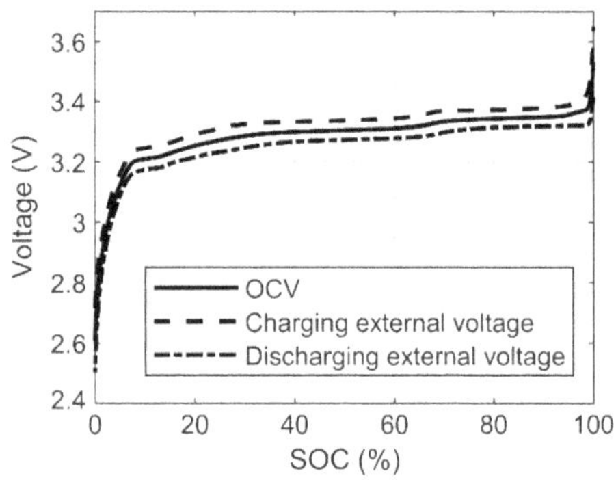

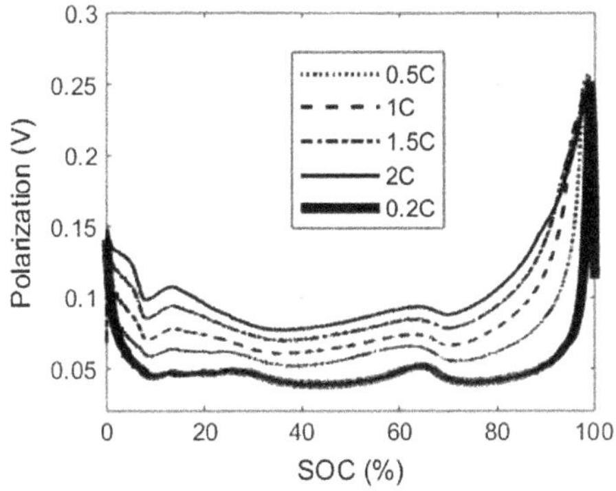

Fig. 4 Polarization voltage profiles versus SOC at various charge rates

the 10% SOC point. The polarization voltage then becomes relatively stable, but its amplitude still fluctuates slightly. When SOC reaches about 80%, the polarization voltage starts to increase sharply.

The LFP battery is chosen as an example to describe the charging progress in order to reveal the variation of polarization voltage. The polarization voltage variation within the whole SOC range is closely related to the internal charging reaction process. As shown in Fig. 5, the external charger drives the flow of electrons to interact with the electrode materials. Before charging, the positive cathode material is in the Li-rich LFP phase. After absorbing enough energy, Li^+ becomes active and exceeds the limit of the activation barrier. Li^+ will move along with the direction, which requires a minimum energy of the olivine-like structure. Nakayama et al. [53] report that the nucleation of the LFP phase occurs first on the surface of $LiFePO_4$ when Li^+ is removed. This process needs relatively more energy. Hence, polarization voltage is higher in the initial SOC region. Once the LFP phase is formed on the particle surface, during the following charging process, the sequential path of Li ions extracted from particles is first through the core of the

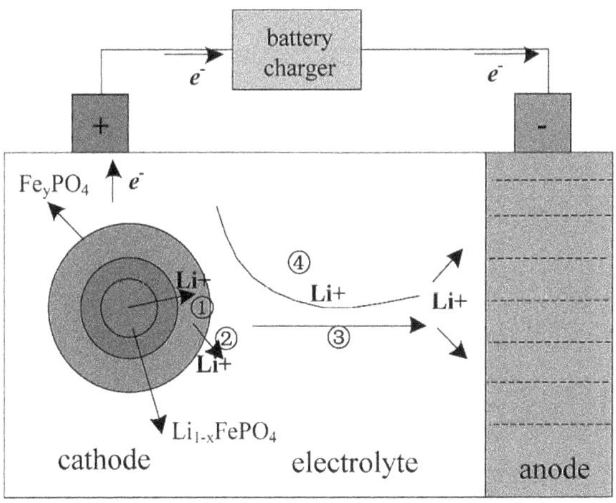

Fig. 5 Charging process diagram: ① transfer of Li ions in the solid-state phase; ② Li ions moving into the electrolyte from the electrode material; ③ diffusion of Li ions; ④ concentration gradients of Li ions in the electrolyte

Li-rich LFP phase, then the two-phase coexistent interface, and finally the shell of the Li-poor FP phase. The Li ions' flow is then driven by the concentration potential in the electrolyte. Finally, Li ions are inserted into the graphite anode. Because FP nucleation has been finished in the low SOC region, the whole process does not need excess energy to overcome the interface energy barrier. The polarization voltage is therefore relatively low and stable in the middle SOC range.

As charging continues, the shell grows, the core is consumed, and the two-phase coexistent interface moves toward the interior. The charging process is completed when the cathode material particles become the single FP phase. In the last charging stage, the travel path of Li ions through the Li-poor phase is long; in addition, with the Li-ion concentration increasing in the negative materials, it is difficult for Li ions to insert. The polarization voltage, therefore, shows obvious growth in the high SOC region.

The correlation between polarization voltage and charging rate is illustrated in Fig. 6, which can be divided into three parts. The first part corresponds to the nonlinear regions containing 0–5% SOC and 85–100% SOC. The second part corresponds to the approximately linear regions containing 5–25% SOC and 65–85% SOC, and their slopes are different. The third part is also characterized by linear regions, and their slopes are nearly the same. The slopes at 5 and 85% SOC are larger than those at other SOC points, which suggests that the polarization at these two SOC points is sensitive to the charging rate. The polarization voltage is significantly increased when the charging rate increases, and vice versa.

The polarization-rate coefficient (PRC) κ is defined, which refers to the polarization voltage increment while the charging rate is raised per 1C. It describes the degree to which the charging current affects polarization voltage. In the 5–85% SOC range, the polarization has variable slopes: At the ends, the PRC is greater than those measured at intermediate percentages.

Linear fitting was conducted on the curves in Fig. 6 to calculate κ. Its relationship with SOC is shown in Fig. 7 where κ is smaller in the middle SOC regions than that at both ends of the whole range. In the initial and final charging stages, Li-ion concentration is relatively high in one of the two electrodes, and the diffusion

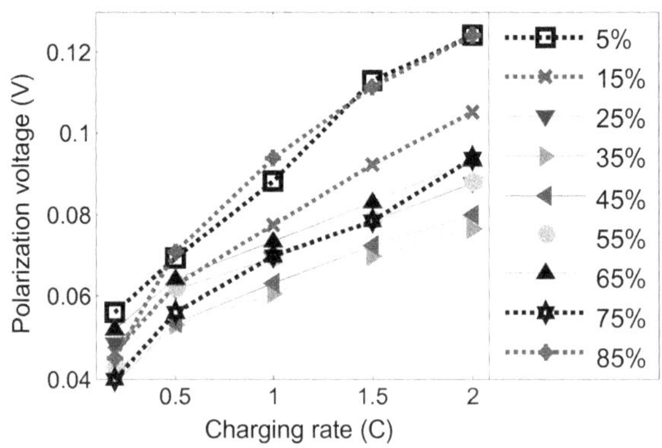

Fig. 6 Polarization voltage versus charging rate at various SOCs

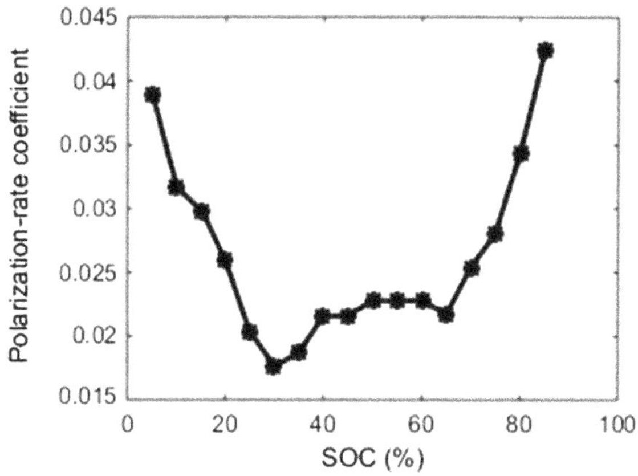

Fig. 7 Polarization-rate coefficient versus SOC

coefficient is smaller. The solid-state phase diffusion polarization dominates the total polarization, and the polarization rate is high. In the middle SOC region of charging, Li ions distribute well in both electrodes. The solid-state phase diffusion polarization and concentration polarization play a major role. The diffusion coefficient is larger as Li ions have enough free space to move easily and the polarization is relatively small.

A series of experiments were conducted to investigate the effects of initial SOC on PRC. The procedure can be summarized as follows: The battery was first charged from a fully discharged state ($SOC_0 = 0\%$) at a current of 0.5C until the terminal voltage was up to 3.65 V. The battery was then kept in the open-circuit state for 1 h. The battery was discharged at a current of 0.5C until 80% nominal capacity was consumed, and then, it was kept in an open-circuit state for 1 h to eliminate the polarization of the battery. The battery was charged from initial $SOC_0 = 20\%$ at the same current until the charge limit voltage was obtained. Similarly, having been kept in an open-circuit state for 1 h, the battery was discharged at 0.5C until 50% of nominal capacity was consumed. Leaving the battery in the open-circuit state for 1 h, the battery was then charged from $SOC_0 = 50\%$ to the charge limit voltage. The charge polarization characteristics at different initial SOCs are reported in Fig. 8 wherein the dynamic overshot values of the polarization voltage are affected by the initial SOC value, which is higher at $SOC_0 = 50\%$ than that at $SOC_0 = 20\%$. However, the steady value of the polarization voltage at various initial SOCs is approximately the same, which suggests that the steady value of the polarization voltage is independent of initial SOC.

Equation (2) shows that when the battery stops charging suddenly, polarization voltage declines as an exponential function, containing two time constants $\tau 1$ and $\tau 2$. The polarization charging time constant (PCTC) can be determined through experiments directly and is approximately equal to the polarization standing time constant (PSTC). The voltage variation of a battery charged to a certain SOC was monitored for 2 h. It was found that the change of terminal voltage was around 1 mV from 3600 to 7200 s (the voltage ranging from 3.767 to 3.768 V), suggesting

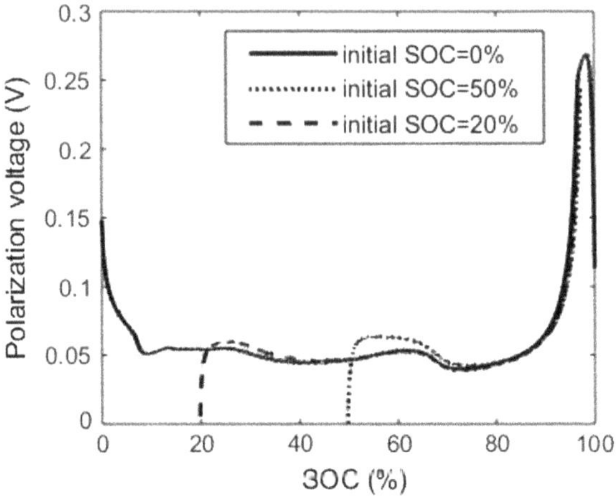

Fig. 8 Charge polarization characteristics at different initial SOCs

that the terminal voltage was getting steady after 1 h. Therefore, the terminal voltage change for 1 h is approximately regarded as the steady value of the polarization voltage. Figure 9 shows the changing polarization voltage curve of the battery charged to 15% SOC with CC and then resting for 1 h.

Figure 10 compares polarization voltages under variable current charging and 1/3C CC–CV charging. The variable current charging profile is: Charge the battery at 1/3C at the starting stage to 20% SOC, then charge at 1.2C during the middle stage of 20–80% SOC, and 1/3C charge in the last stage. Figure 10 shows that the battery polarization voltage overshoot appears as the charging current is increasing. By contrast, the polarization voltage hysteresis can be observed while the charge current decreases. It is noticeable that the hysteresis in this study resulted from current change and disappeared as the battery polarization became steady. The overshoot amplitudes at 20% SOC and 80% SOC points are not the same, since in the current increasing and decreasing process reaction kinetics is not symmetrical, which results in asymmetric charging/discharging polarization [54, 55].

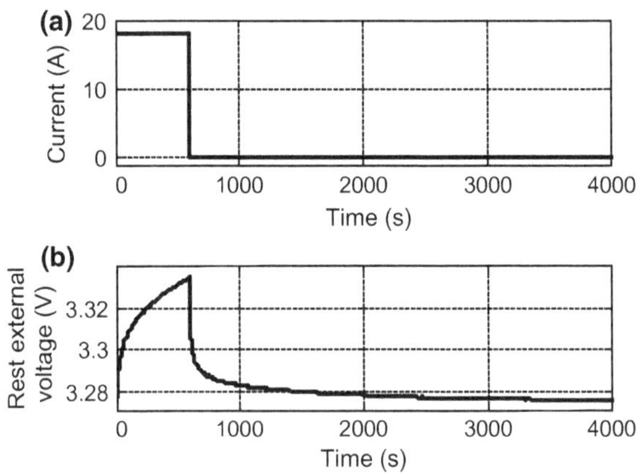

Fig. 9 Voltage and current curve change profiles: **a** current curve and **b** polarization voltage curve

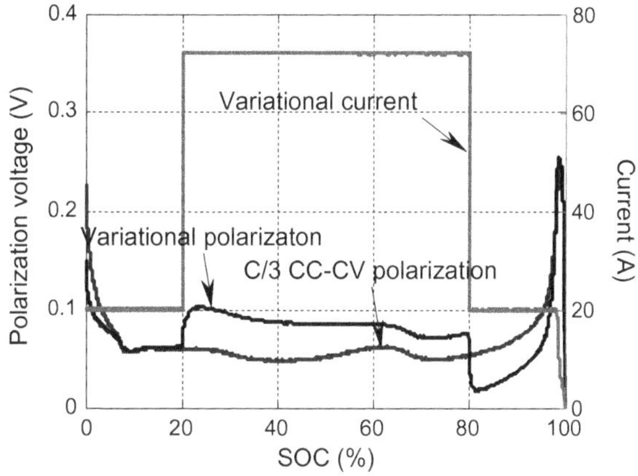

Fig. 10 Polarization step response experiment results

The polarization process includes at least three aspects: electronic and ionic conduction inside the electrode, charge-transfer reaction at the electrode and electrolyte interface, and a Li-ion transport process such as concentration gradient-driven diffusion. The process can be divided into two types. One includes the first and second fast processes, and the other includes the third slow process [53]. The number of electrons flowing through the electrode per second changes rapidly when charging current changes, but due to the time constant, these three processes cannot respond quickly. The polarization hysteresis effect is produced. The overshoot effect is a result of the excess energy cost in the initial period. However, according to Eq. (2), the overdamping response has no overshoot effect. This phenomenon indicates that the equivalent circuit model shown in Fig. 1 cannot describe the mechanism of the overshoot effect.

2.3 *Evaluation of Acceptable Charging Current*

The polarization acceptable charging current (PACC) is proposed based on the polarization time factor: The CC charging time length is equal to the larger polarization time factor at a specified SOC point. If the external battery voltage exactly gets the preset cutoff voltage at the end of charging, then the charging current has an acceptable value at this SOC point.

The PACC relational expression is as follows:

$$V_{\mathrm{OCV}}(\mathrm{SOC}_0 + I\tau/C) + IR + V_P(I) = V_{\mathrm{cutoff}} \tag{8}$$

The iterative calculation of PACC is shown in Fig. 11a. The map is divided into two parts by PACC: Region I is the acceptable charging area under the boundary curve, and region II is the unsuitable charging area. The battery can be charged in region I, which scarcely affects the battery. The current will cause polarization to increase too much in region II, so that the battery cannot be fully charged. Here, the larger time factor was selected as the charging time, because after such a long time, the polarization voltage tends to be stable and jumps over the dynamic change stage of polarization voltage. According to the battery tests, parameters are set as follows: The cutoff voltage is 3.65 V; the ohmic internal resistance is 1.37 mΩ; the charging time is 1000 s. The result of acceptable charging current at different SOCs is shown in Fig. 11. It can be seen that the simulation error is within 10%.

The maximum charging current is set as 2C as recommended by the battery manufacturer. In addition, the battery polarization is large at the initial charging stage, in which small current charging is applied. The optimized charge strategy diagram is illustrated in Fig. 12. The battery is first charged at C/3, ranging from

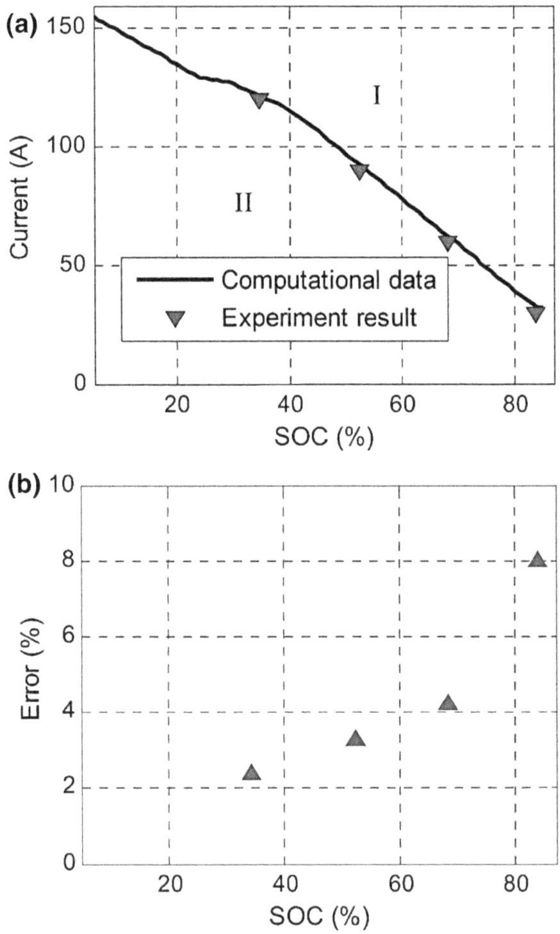

Fig. 11 **a** Acceptable charging current compared with experiment results and **b** calculated error

0 to 20% SOC. The current of 2C is applied after 20% SOC until it intersects with the PACC. Then, every 5% SOC, the charging current is reduced according to the boundary curve. The charging current will be reduced to C/3 at around 80% SOC, followed by a C/3 CC–CV process.

Table 2 compares the proposed charging mode and the 1C and 0.5C CC–CV charging modes. It can be observed in Table 2 that the total charging time of experiment I is medium. CV charging time is almost equal, around 10 min. The CV charging capacity of the three experiments increases in sequence. The initial CV charging currents of experiments I to III are C/3, 0.5C, and 1C, respectively. The polarization increased as the charge current increased. The actual equivalent charging rate of the proposed charging mode is medium between experiments II and III. Table 2 also shows that, compared with the 0.5C CC–CV charging mode (experiment II), the charging time and equivalent charging rate of the proposed charging mode (experiment I) are improved by 18 and 21.3%, respectively. The charging time and charging capacity at the CV charging stage are less than 0.5C CC–CV charging mode, having higher charging efficiency and requiring only CC charging in practice. In addition, the average polarization voltage of both charging modes is nearly the same. It is suggested that the proposed charging mode can well balance charging time and charging polarization compared to the 0.5C CC–CV charging mode. Furthermore, the SOC ranging from 20 to 80% is chosen as the fast charging region in the proposed charging mode since it is the commonly used interval of batteries particularly for plug-in hybrid EVs. Compared with 1C CC–CV charging mode (experiment III), the charging time is approximately identical from 20 to 80% SOC, and the average polarization voltage is much less than the 1C CC–CV charging mode. From a temperature rise point of view, the charging current of

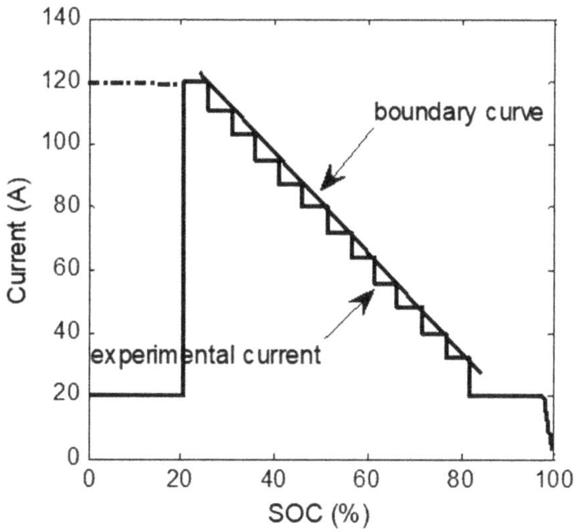

Fig. 12 Optimized charging strategy diagram

Table 2 Comparison of proposed charging mode and CC–CV charging mode

Experiment	Charging mode	Total charging time/min	CV charging time/min	CV charging capacity/Ah	Equivalent rate/C	Average polarization voltage/mV	20–80% SOC charging time/min
I	Proposed	105	7.2	1.082	0.57	79	33.6
II	0.5C CC–CV	128	9	3.023	0.47	80	72
III	1C CC–CV	69	10.2	4.514	0.87	102	36

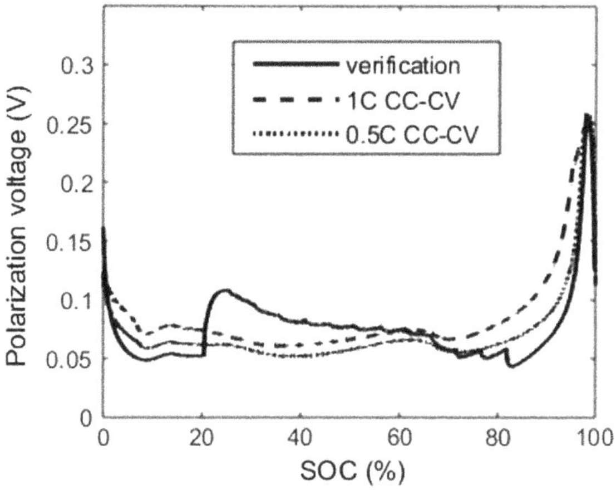

Fig. 13 Polarization voltage profiles

the proposed charging mode is gradually decreasing, and the temperature rise is slowing down, which can effectively prevent the battery from thermal runaway. Therefore, the proposed charging mode can balance charging speed, charging polarization, and temperature rise.

Figure 13 compares polarization voltage profiles for the three experiments. In 0–20% SOC and 80–100% SOC, the polarization of experiment I is the lowest due to the low charging current. With charging current increasing suddenly, the polarization rises to the maximum at 20% SOC point. The polarization is going down as the current is falling. The minimum polarization of experiment I happens at 80% SOC.

Battery temperature distribution was obtained by thermal infrared imaging for different charging modes. It was found that 0.5C CC–CV charging mode has the minimum temperature range (5 K), and the 1C CC–CV picture has the maximum temperature rise (9.7 K), whereas the proposed charging mode is between the two and, so, temperature rise is acceptable.

All the experiments were done with two batteries that were randomly selected from 10 batteries made from the same production batch. The results and methods have generality and representativeness and can be reproduced.

For modular Li-ion batteries, equalization [56] and impedance measurement [57] are inevitable challenges. The final acceptable charging current curve should take temperature rise, battery life, and charge efficiency into account for getting reasonable charging current values. In addition, an acceptable charging current curve may not be the practically applied current. Taking multiple constraints into account, the acceptable charging current requires further investigation.

3 Optimization of Polarization Voltage Model in High C-Rate Application

3.1 Model Optimization Based on the Butler–Volmer Equation

Although the second-order RC model is widely used in the study of polarization voltage, there are performance differences when applying it to differently designed batteries, especially high-power batteries. To rigorously investigate unique battery performance caused by different designs, two batteries produced by the same manufacturer but with different designs were selected and analyzed using the second-order RC model, where the two RC parallel networks with different TCs are employed to describe the polarization voltage.

Figure 14 shows the different polarization distributions of the two batteries from the experiment. The battery runtime decreased regularly in an inverse proportion way when current flowing through the battery increased because of Peukert's law, resulting in a changing tendency shown as ①. As a consequence of current growth, these two totally distinct polarizations will be more serious. Polarization with a

Fig. 14 Polarization distributions of two types of batteries: **a** a high-energy battery; **b** a high-power battery

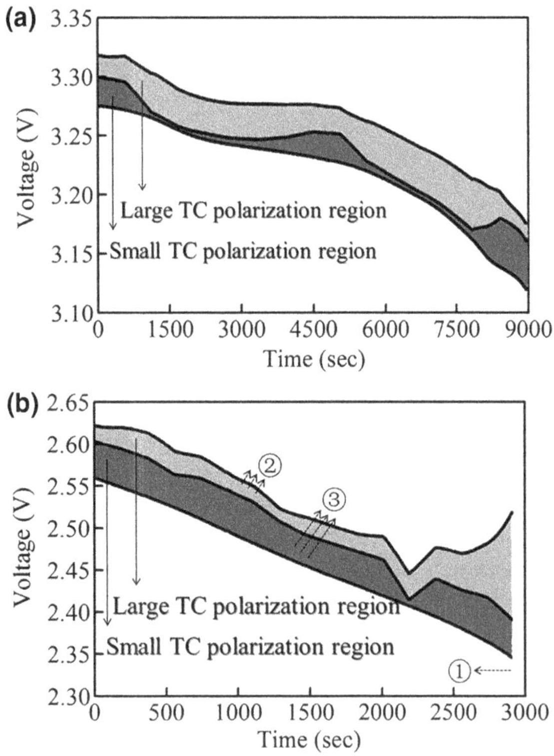

smaller TC, in particular, will be much more likely in an extremely critical condition owing to a power battery's short recovery time. The variation tendencies of these two polarizations are marked as ② and ③, respectively.

It can be concluded that polarization with a smaller TC will be affected heavily as the current increases. Moreover, it should be quite appropriate to regard this short polarization as electrochemical polarization since ion diffusion is more inclined to generate polarization with a larger TC. Therefore, methods capable of characterizing electrochemical polarization induced by changing rates are required.

The Butler–Volmer equation is one of the most fundamental relationships in electrochemical kinetics. It describes how the electrical current on an electrode depends on the electrode potential, when the electrode reaction is controlled by electrical charge transfer at the electrode. Accordingly, it presents a quantitative form of electrochemical polarization.

After the kinetics of ion transfer is completely established, the BV equation is proposed to model an ideal overpotential $\eta(t)$ coming from electrochemical polarization under the steady state, as follows:

$$J(t) = J_0\left[\exp\left(\frac{\alpha F}{RT}\eta(t)\right) - \exp\left(\frac{-(1-\alpha)F}{RT}\eta(t)\right)\right] \tag{9}$$

Considering there are few approaches to measuring current density externally, the BV equation can be written as Eq. (11) according to

$$I(t) = J(T) \times S \tag{10}$$

$$\frac{1}{J_0 \times S}I(t) = \left[\exp\left(\frac{\alpha F}{RT}\eta(t)\right) - \exp\left(\frac{-(1-\alpha)F}{RT}\eta(t)\right)\right] \tag{11}$$

However, it seems impossible to get the overpotential caused by current variation because Eq. (11) cannot be solved without any assumptions. Fortunately, the transfer coefficient rarely changes suddenly and generally is argued as $\alpha \approx 0.5$ [33–35]. Consequently, the relationship is expressed as

$$\eta(t) = \frac{2RT}{F}\ln\left(\frac{1}{2J_0 S}I(t) + \sqrt{\left(\frac{1}{2J_0 S}I(t)\right)^2 + 1}\right) \tag{12}$$

Despite the fact that there are still a variety of electrochemical variables that cannot be obtained externally, the corresponding change of overpotential excited by changing current can be acquired. Meanwhile, J_0 and S are closely related to the SOC, reflecting the complexity during the polarization establishment.

Whereas other variables available are fixed, the increase of current will result in a sharp deviation from the equilibrium state. On the contrary, a battery gradually becomes steady and reaches equilibrium eventually when the current approaches zero. More meaningfully, the battery overpotential at a specific SOC, which is only induced by changing currents, is affirmatively accessible in accordance with Eq. (12). It suggests the battery has completely reached a nonequilibrium state after deviating from equilibrium, which particularly requires some time.

The structure of the Thevenin model has been selected as a reference for the developed model owing to these reasons: (1) The Thevenin model is mainly used as the basic part of the improved models developed by other authors [1, 58]; thus, it permits comparison within diverse models. (2) The lithium titanate battery (LTO) has been selected mainly to study the electrochemical polarization. The Thevenin model has been proved preferable for LiNMC batteries in [30], and appropriate for LFP batteries after adding a hysteresis component. Particularly, hysteresis voltages of lithium titanate batteries are within 6 mV and will reduce or even vanish in high-rate applications. Hence, hysteresis is ignored (3). The Thevenin model is more feasible and suitable for high-power batteries than the second-order RC model, given that dynamic high-rate characteristics are affected more severely by polarization with a smaller TC.

Figure 15 displays the developed electrical model. It takes the same electrical circuit as the Thevenin model but uniquely embeds a simplified form of the BV equation to predict parametric variations, particularly the polarization voltage drops caused by changing currents. Consequently, the developed model is sufficient for characterizing comprehensive battery behaviors. Moreover, its framework is so minimalistic that less computation and lower storage costs than other enhanced models mentioned in [42–45] are needed.

Fig. 15 Schematic of the developed electrical model

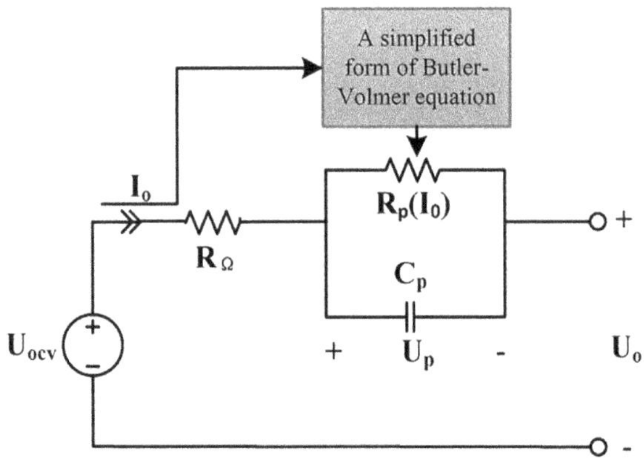

Assuming a battery discharges with a current of I_1 during $t_0 < t < t_1$, followed by rest in $t_1 < t < t_2$, the developed model is described as

$$\mathrm{SOC}(t) = \mathrm{SOC}(t_0) - \frac{1}{C_{\max}} \int_{t_0}^{t} I_0(t)\mathrm{d}t \tag{13}$$

$$U_P(t) = \begin{cases} R_P I_0(t) \cdot \left[1 - \mathrm{e}^{-\frac{t-t_0}{\tau}}\right], & t_0 < t < t_1 \\ U_P(t_1) \cdot \left(\mathrm{e}^{-\frac{t-t_1}{\tau}}\right), & t_1 < t < t_2 \end{cases} \tag{14}$$

$$U_O(t) = U_{\mathrm{OCV}}(t) - U_P(t) - R_\Omega I_O(t) \tag{15}$$

where $\tau = R_P C_P$. During the constant discharging process, the SOC of the battery reduces linearly according to Eq. (13). In addition, the initial value of SOC is necessary for SOC estimation when a battery provides the requisite power for electric motors. This value of SOC is always calculated from the specific OCV–SOC curve of the battery. Given the circuit parameters are likely to change with SOC, all of these parameters are functions of SOC. Equation (16) expresses SOC dependence of U_{OCV}, R_Ω, C_P, where a_1, a_2, b_1, b_2, c_1, and c_2 are obtained from circuit parameters of typical SOC points by the linear interpolation method.

$$\begin{cases} U_O[\mathrm{SOC}(t)] = a_1 \mathrm{SOC}(t) + a_2 \\ R_\Omega[\mathrm{SOC}(t)] = b_1 \mathrm{SOC}(t) + b_2 \\ U_O[\mathrm{SOC}(t)] = c_1 \mathrm{SOC}(t) + c_2 \end{cases} \tag{16}$$

When the discharge current suddenly changes from I_1 to I_2, the time interval can be considered as $\Delta t < \varepsilon_1$. Hence, we can easily get Eq. (17) and infer that $\Delta\mathrm{SOC} \approx 0$.

$$\Delta\mathrm{SOC} \le \max\{I_1 \Delta t, I_2 \Delta t\} \le \max\{I_1 \varepsilon_1, I_2 \varepsilon_2\} \tag{17}$$

As a lumped representation of resistive forces, internal resistance R_Ω is precisely correlated with the battery's SOC but clearly independent of current amplitude [59, 60]. However, battery polarization, particularly polarization resistance, is heavily affected by current variation [25]. Based on Eqs. (12) and (14), the steady-state value of battery polarization at a specific SOC can be expressed as

$$\begin{aligned} U_{P\max} &= R_P I_O(t) \cdot \left[1 - \mathrm{e}^{-\frac{t-t_0}{\tau}}\right]\Big|_{t \cdot \tau} \\ &= R_P I_O(t) = f_1(\mathrm{SOC}) \cdot \ln\left(f_2(\mathrm{SOC}) \cdot I_o(t) + \sqrt{(f_2(\mathrm{SOC}) \cdot I_o(t))^2 + 1}\right) \end{aligned} \tag{18}$$

where $f_1(\mathrm{SOC})$ and $f_2(\mathrm{SOC})$ are coefficients of the simplified form of the BV equation. More importantly, these two coefficients are also functions of the current direction. According to Eq. (14), battery polarization can be divided into two categories: polarization settle when the battery current is nonzero and polarization recovery when the battery is rested. As a dominant part of polarization settle, polarization resistance is derived from Eq. (18). On the other hand, polarization resistance of polarization recovery is irrelevant to current amplitude, since no external excitation is applied to the battery. Therefore, polarization resistance is analytically calculated as Eq. (19):

$$R_P = \begin{cases} \dfrac{\left[f_1(\mathrm{SOC})\cdot \ln\left(f_2(\mathrm{SOC})\cdot I_o(t) + \sqrt{\left(f_2(\mathrm{SOC})\cdot I_o(t)\right)^2 + 1}\right)\right]}{I_o(t)}, & I_o(t) \neq 0 \\ R_{PC}, & I_o(t) = 0 \end{cases} \quad (19)$$

where R_{PC} represents polarization resistance measured at the current of 1C. Clearly, model accuracy is more sensitive to R_P rather than C_P, given C_P is solely dominated by the τ and the fact that recovery time remains short for high-power batteries. To reduce the storage space required, C_P is assumed to be independent of the current amplitude.

3.2 Parameter Extraction of the Optimized Model

As an indispensable component of the proposed model, OCV as a function of SOC can be determined after a battery has rested for a long period. The hysteresis mentioned in [61, 62] specifically describes the OCV difference for a certain SOC, which is caused by current directions. Nevertheless, the OCV–SOC curves of three selected lithium titanate batteries, shown in Fig. 16, indicate that there is an extremely high consistency between two parts of an OCV–SOC curve, namely the hysteresis voltages of lithium titanate batteries are within 6 mV as shown in Table 3. Li et al. [63] pointed out that hysteresis easily appeared when a battery

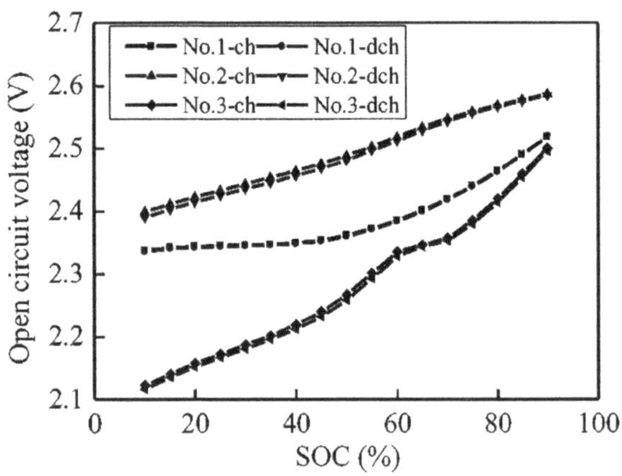

Fig. 16 OCV–SOC curves of selected lithium titanate batteries

Table 3 Hysteresis voltage of tested batteries

Battery number	Hysteresis voltage (average value of measured points) (mV)
1	1.0
2	6.0
3	6.0

required a fairly long time to be entirely relaxed. More specifically, it was confirmed that lithium titanate oxide (LTO) batteries also showed OCV hysteresis and high rates affected the macroscopic processes in a way that hysteresis even vanished for Li-ion batteries [64]. Since high-power lithium titanate batteries for high-rate applications are evaluated and the actual hysteresis voltages of 1C measured current are already within 6 mV, hysteresis voltages will decrease or even disappear when battery current increases. Hence, hysteresis is ignored to simplify the model without a noticeable effect on accuracy.

Figure 17 demonstrates a typical voltage response curve for a positive pulse sequence (PPS). The internal resistance of the battery is calculated from Eq. (20) by using the instantaneous voltage rise from t_3 to $t_4 = t_3 + \Delta t$, where Δt is equal to the sampling period, namely 1 s. Then, the polarization voltage can be obtained, according to Eq. (15) and U_{ocv} measured at a fixed SOC, providing data for identification of the RC parallel network. Using Eq. (21), polarization resistance and polarization capacitor, where polarization resistance refers to that measured at the current of 1C, particularly, are determined from the least squares curve fitting tool in MATLAB. It is important to note that two groups of parameters are used to capture characteristic differences between charging and discharging behaviors of the battery.

$$R_{\Omega} = \frac{U_O(t_4) - U_O(t_3)}{I(t_4) - I(t_3)} = \frac{U_O(t_3 + \Delta t) - U_O(t_3)}{I_{PC}} \tag{20}$$

$$\begin{cases} U_P(t) = A \cdot \left[1 - \mathrm{e}^{-\frac{t}{\tau}}\right] \\ R_P = \frac{A}{I_{PC}} \\ C_P = \frac{\tau}{R_P} \end{cases} \tag{21}$$

In this part, Battery 1 is chosen to illustrate abnormal battery performances of typical rates. However, polarization voltages of typical rates, which are displayed in Fig. 18, maintain a continuously growing trend, despite the fact that the speed is clearly not increasing as quickly as before.

When the battery was tested during charging or discharging, the polarization voltage of the middle SOC range clearly achieved a steady-state value. Given the fact that the internal resistance is independent of current amplitude and the data of rate charging tests, the battery polarization voltage of varied charging rates can be acquired according to Eq. (15) after obtaining the OCV value and resistance at 1C measured current. Similarly, the battery polarization voltage at various discharging

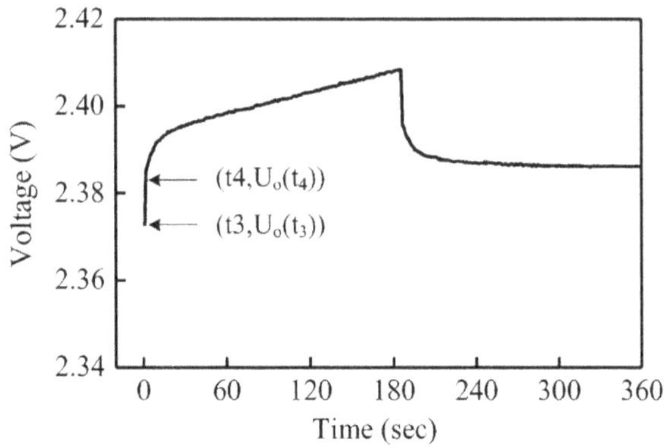

Fig. 17 Typical voltage response curve for a PPS

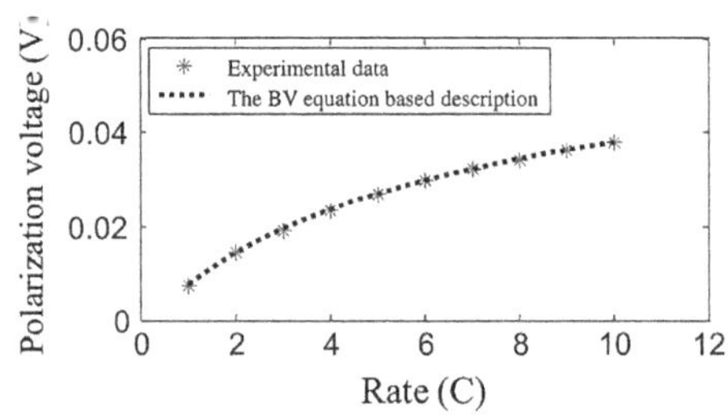

Fig. 18 Polarization voltages of typical rates

rates can be obtained with the same method. In addition, rate charging and discharging, particularly in high-rate applications, will not require too much time compared with the proposed identification methodology in [46]. In order to outline battery performances, particularly polarization voltage relationships with rates, and to enable the model feasibility for system-level simulation, two coefficients of Eq. (19), which are closely related to the SOC of the battery, are determined from the least squares curve fitting tool. In particular, the battery enters the CV charging region earlier when the rate rises, complicating the model description. In addition, the current obviously decreases after the battery enters the CV charging region, resulting in much less polarization inside the battery. Therefore, coefficients of Eq. (19) to capture charging behavior are refined calculated in 10–75% SOC, whereas polarization resistances at the current of 1C are adopted in other SOC intervals.

3.3 Model Verification

To verify the developed model, the simulation results of the galvanostatic test at a high rate and the FUDS test, which were carried out on three types of LTO batteries, are, respectively, compared with the experimental data. The developed model was implemented in MATLAB/Simulink, ensuring the model's promising popularization in high-powered applications.

The experimental setup for the battery tests consisted of an Arbin BT2000 tester, a thermal chamber providing controllable temperature, a computer for monitoring and data storage, and tested batteries. The voltage and current were measured and recorded per second during the whole test. In this work, we prefer to evaluate battery characteristics depending on changing rates rather than other factors that will be discussed in our subsequent works. Hence, the thermal chamber is maintained at 25 °C all the time.

Three types of lithium titanate batteries were selected, and their shapes were totally different. The first two batteries were from leading manufacturers in China, and the third one was produced by a well-known Japanese company. Their major parameters are shown in Table 4. Using LTO as its anode instead of graphite, LTO batteries are more inclined to high-rate applications in EVs than other lithium batteries with graphite anode, resulting in an unusual reference rate for the capacity test presented in Table 4. Moreover, Takami et al. [65] also mentioned that 1C was chosen as the reference rate for the LTO battery capacity test.

The test procedure was performed to adequately evidence the comprehensive behaviors of these batteries. It began with the capacity test, which was measured by discharging the battery down to the lower cutoff voltage with a current of 1C after CC–CV charging to the lower cutoff current. This test will not be finished until the difference of results between two adjacent cycles is within 0.1 Ah. The OCV dependency on directions of current using a PPS 20 times was further examined, followed by a negative pulse sequence (NPS) repeated 20 times. A PPS contains 1C charge for 180 s to get 5% of the battery SOC, and rest of 1 h on purpose of acquiring magnitude of OCV precisely. Similarly, a NPS has the same structure as PPS but its current is negative. For the internal resistance test, a rich pulse sequence was applied, including several steps at each specific SOC: (1) 1C charge for 10 s; (2) rest for 40 s; (3) 1C discharge for 10 s; (4) repeat steps (1) to (3) after increasing 1C to 2C, 3C, 4C, 5C, 6C, 7C, 8C, 9C, 10C, respectively. Then, charging and discharging tests were carried out with selected typical rates to investigate the relationship between the detailed parameters and rates. The batteries were also excited by the FUDS profile, providing the data set for model verification.

Table 4 Power batteries used for tests

Battery number	Anode material	Nominal capacity (Ah)	Nominal voltage (V)	Cutoff voltages (V)	Cutoff currents (C)	Reference rate for capacity test and cycling given by manufacturers (C)	Cycle life at 25 °C (times)
1	LTO	8.5	2.3	2.8/1.5	0.1	1	10,000
2	LTO	16	2.5	2.8/1.5	0.1	1	15,000
3	LTO	20	2.3	2.7/1.5	0.05	1	6000

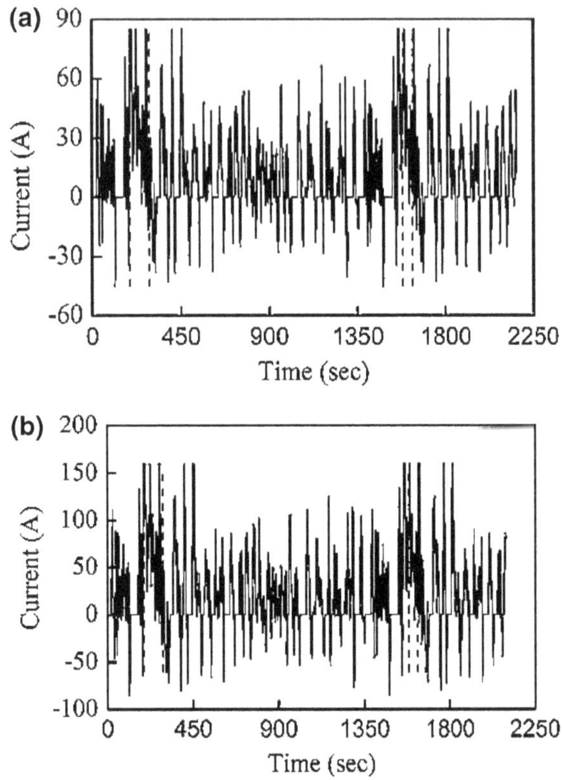

Fig. 19 Current profiles for FUDS test: **a** Battery 1; **b** Battery 2 and Battery 3

The typical current profile of the FUDS tests is shown in Fig. 19. The regions between the two adjacent dotted lines are chosen as the representative parts for subsequent discussions.

Figure 20 compares simulation results of the galvanostatic discharge test at a high rate with experimental data. There is a close accordance between the terminal voltage extracted from the experimental data and the simulation results of the developed electrical model. The maximum voltage errors for the galvanostatic tests of the three batteries were found to be similar at the end of discharge and are all within 1.5%. The voltage error of Battery 1 for the galvanostatic test at 10C (85 A) was less than 0.5% in 240 s (23.3–90% SOC), and the voltage error was less than 1.5% during the discharge process. The same distribution of voltage errors for the discharge process at 10C (160 A) was found for Battery 2. Although the simulation result of the Battery 3 is slightly different from that of the prior two batteries, the voltage errors of the whole discharging process are still within 1.5%. Therefore, the developed model can precisely describe the galvanostatic performances of batteries.

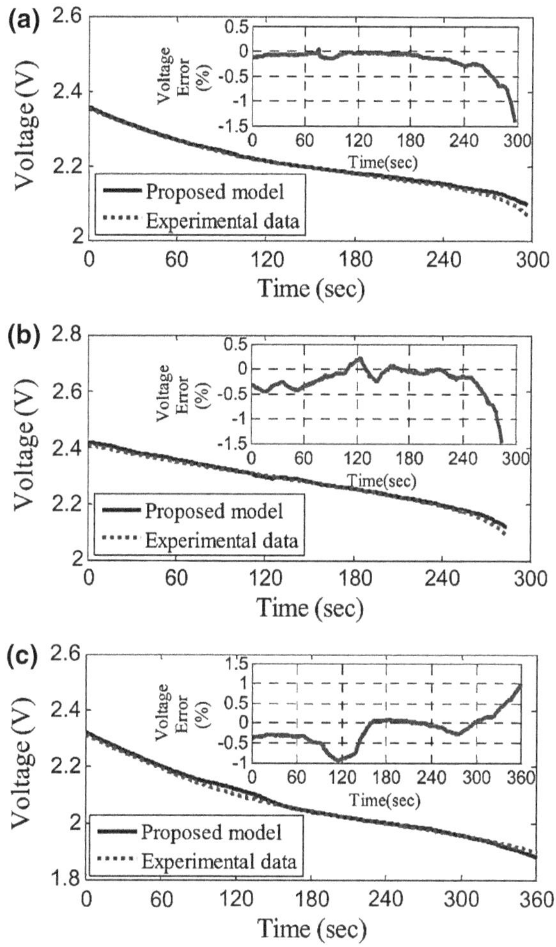

Fig. 20 Comparison between simulation and experimental data of the galvanostatic discharge test: **a** Battery 1; **b** Battery 2; **c** Battery 3

Furthermore, the simulation results of the galvanostatic charging test at a high rate and experimental data are displayed in Fig. 21, confirming that the developed electrical model fits well voltage profiles obtained from experimental data. It can be concluded that the maximum voltage errors for the galvanostatic charge tests of the three batteries are all less than 2% and occur at the end of the charge process. Although the voltage errors of Battery 1 at 8C (68 A) increased in the last 60 s, they are less than 1% during the first 240 s. Moreover, the voltage errors of Battery 2 at 10C (160 A) and Battery 3 at 8C (160 A) during the whole discharging process are within 1%. Therefore, the developed model can precisely capture both the galvanostatic discharge and charge behavior of lithium titanate batteries.

The developed model with FUDS profile was further verified. The comparison between experimental data and terminal voltage obtained from battery simulation, shown in Fig. 22, indicates that the simulation results match the experimental data well.

Fig. 21 Comparison between simulation and experimental data of the galvanostatic charge test: **a** Battery 1; **b** Battery 2; **c** Battery 3

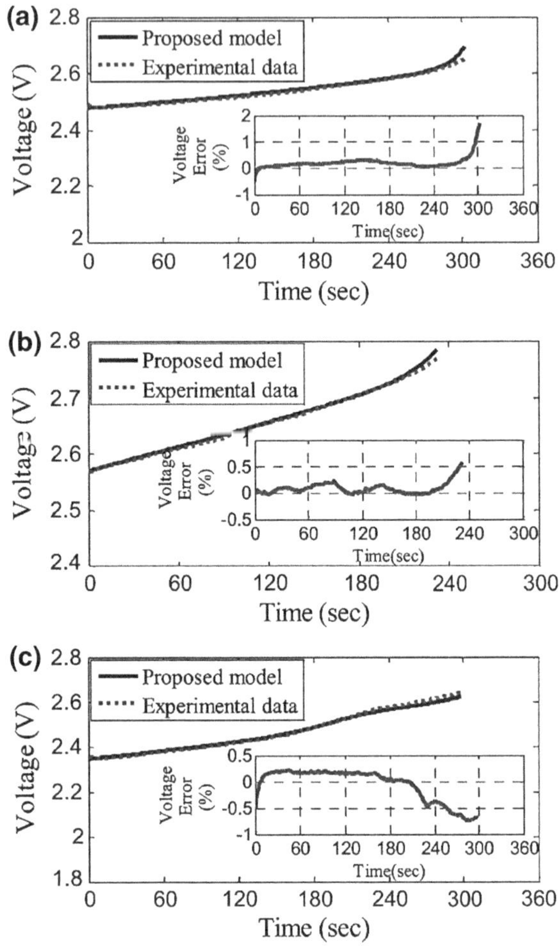

The maximum voltage errors of Battery 1 and Battery 2 are less than 1.5%, while the maximum voltage error of the Battery 3, at the end of discharge, is a little larger than that of the other two batteries but still within 2%.

To better illustrate the BV equation's advantage, both the Thevenin model and the developed model are used to describe the dynamic performances of Battery 1 in the galvanostatic test at 10C and the FUDS test. Figure 23 exhibits voltage errors of these two models when the battery discharges at a current of 10C. During the discharging process, the voltage error of the Thevenin model continuously rises and eventually exceeds 80 mV. Instead, the developed model shows a small voltage error, which is lower than 30 mV. The BV equation produces two entirely different trends of model errors. Hence, the BV equation is capable of meaningfully characterizing comprehensive battery behaviors and can enhance electrical model applicability for high-rate operations.

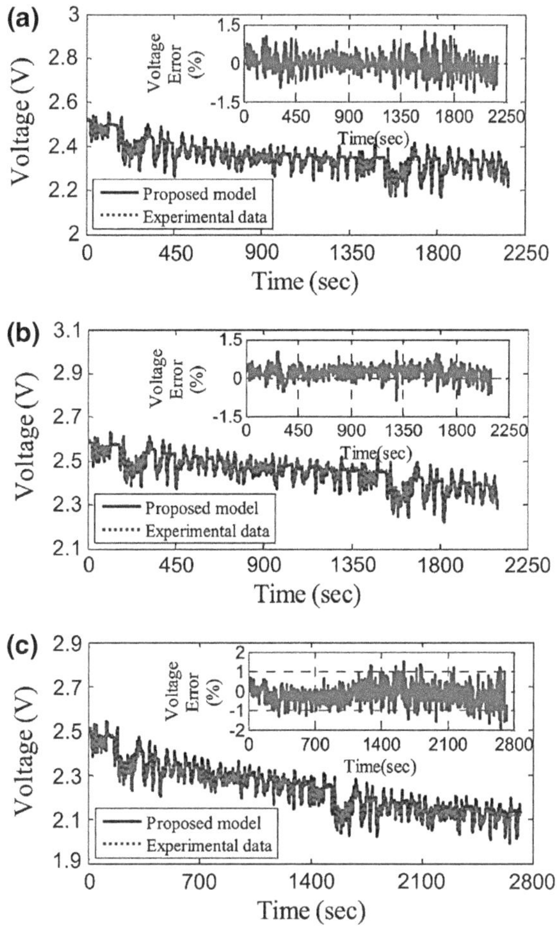

Fig. 22 Comparison between simulation and experimental data of the FUDS test: **a** Battery 1; **b** Battery 2; **c** Battery 3

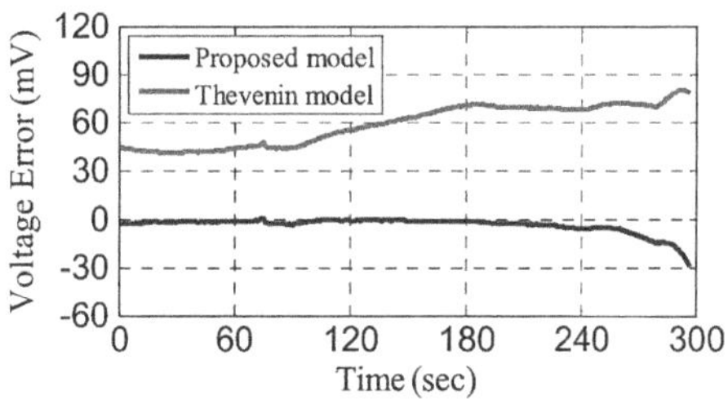

Fig. 23 Comparison between the developed model (lower curve) and the Thevenin model (upper curve) for the galvanostatic test of Battery 1

In this part, the regions chosen in the battery FUDS test are selected and labeled as representative part I and II for our subsequent discussions due to three reasons:

(1) The regions include three different trails that deviate from a high rate, nearly covering all available trajectories after the current has reached a high value, such as 10C (85 A) for Battery 1.
(2) Corresponding to the three typical performances of a battery, these trails are sufficient to compose all possible high-rate battery behaviors. At the time interval of [190, 200 s] of representative part I, the current slightly varies near the 10C rate, simulating the application in which the battery continuously provides power to power equipment or EVs at a substantially high rate. The second typical performance appears when the battery discharges at [200, 240 s], outlining a massive current shift from 10C (85 A) but still larger than zero. This current shift accurately simulates dynamic discharging of a battery. Moreover, the third typical performance, which happens at [275, 285 s], displays a sudden change of the current direction. The sudden change at the beginning of 10C (85 A) obviously indicates an abrupt charging after battery discharging.
(3) It becomes clearer how the model description is affected by the BV equation using these regions instead of the whole FUDS test.

Figure 24 expresses current pulses and voltage errors of these two representative parts, based on analyses of battery characteristics by using the two models mentioned above. The Thevenin model retains a value that is larger than 20 mV when the battery discharges at [190, 200 s] of representative part I. However, the voltage

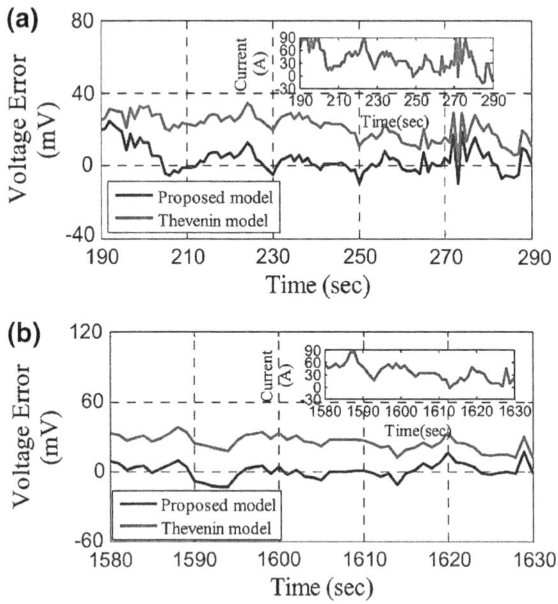

Fig. 24 Comparison between the developed model (lower curve) and the Thevenin model (upper curve) for the FUDS test: **a** representative part I; **b** representative part II

error of the developed model falls rapidly to 5 mV at the same time interval. As for discharging at [200, 240 s], the developed model is clearly superior to the Thevenin model with a voltage error that is lower than 15 mV. Similarly, there is the same comparison when the battery current direction changes at [275, 285 s]. It is worth pointing out there is little difference between the two representative parts. Therefore, the BV equation accurately outlines the battery characteristics of three different trails deviated from a high rate, especially when the current varies slightly near the high rate.

The accuracy and reliability of the developed model are verified by the close agreement between the simulation results and the experimental data of the three groups of lithium titanate batteries. Based on the discussions on error distributions of the two models, the advantage of the developed model compared with the Thevenin model has also been demonstrated.

4 Charging Optimization Based on Temperature Rise and Charge Time

4.1 *Enhanced Thermal Behavior Model*

A battery's surface is assumed to have a uniform temperature when the charge current is within some limits, so that the battery can be regarded as a single cell for thermal modeling. The thermal behavior model can be expressed by the following equation [66]:

$$C_{\text{cell}} \frac{\mathrm{d}T_{\text{cell}}}{\mathrm{d}t} = Q_S + Q_O - Q_B \tag{22}$$

where C_{cell} represents the heat capacity of the battery, T_{cell} expresses battery temperature, Q_S is the reversible reaction heat by entropy change ΔS, Q_O is the energy loss by the overpotential during charge and discharge cycles, and Q_B is the transferred heat. It should be noted that the average heat capacity is adopted in the study since it differs among the battery's active material components. The battery's temperature change is governed by both endothermic transition and exothermic transition. The battery temperature is gradually rising while the absorption of heat is greater than the release of heat due to heat accumulation.

The reversible reaction heat by entropy change can be described by

$$Q_S = T_{\text{cell}} \Delta S \frac{I}{nF} \tag{23}$$

$$\Delta S = -\frac{\partial \Delta G}{\partial T_{\text{cell}}} = nF \frac{\partial E}{\partial T_{\text{cell}}} \tag{24}$$

where F is the Faraday constant, I is the charge current, n is the number of moles of electrons transferred in the cell reaction ($n = 1$ for a Li-ion battery), ΔG is the Gibbs free energy change for the cell reaction, and E is the cell equilibrium, which can also be replaced with close-to-equilibrium OCV in approximate treatment. The differential $\partial E/\partial T_{\text{cell}}$ is negative, and Q_S is endothermic during charging.

The transferred heat is expressed by

$$Q_B = hA(T_{\text{cell}} - T_{\text{amb}}) \tag{25}$$

where h is the heat transfer coefficient, A is the total surface of cell, and T_{amb} is the ambient temperature.

Cell overpotentials comprise mass-transport polarization overpotential, charge-transfer overpotential at positive and negative electrodes, and ohmic overpotential. For overpotential heat Q_O is usually described as the square of current multiplied by overpotential resistance $Q_O = I^2R_\eta$, in which the overpotential resistance R_η is regarded as constant [29], and the R_η at SOC = 0% is always applied. However, the polarization voltage has a linear relationship to the charge current, rather than a proportional relationship, and its slope is a function of SOC, which has been reported in our prior paper [19]. The simulation error of temperature using the conventional calculation method will increase as the current increases. Considering the influence of SOC on polarization overpotential and ohmic overpotential, the overpotential heat can be rearranged as follows:

$$Q_O = Q_P + Q_J \tag{26}$$

$$Q_P = IU_P \tag{27}$$

$$Q_J = I^2R_\Omega \tag{28}$$

where Q_P represents the energy loss by polarizations, U_P is the cell polarization voltage expressed as $U_P = kI + b$ [23], and Q_J represents the energy loss by ohmic resistance.

Discretizing Eqs. (22–28), by using the sampling time interval of 1 s, the battery temperature at time k can be expressed by

$$T_{\text{cell},k} = T_{\text{cell},k-1} + \frac{I^2R_\Omega + IU_P + T_{\text{cell},k-1}I\frac{\partial E}{\partial T_{\text{cell}}} - Ah\left(T_{\text{cell},k-1} - T_{\text{amb}}\right)}{mC_{\text{cell}}} \tag{29}$$

where m is the mass of the battery. The initial temperature is 25 °C in this study.

Making use of the battery parameters shown in Table 5, the simulated and experimental results for battery temperature during charging are illustrated in Fig. 25, where R_Ω represents the simulated results with the DC resistance to calculate the overpotential heat, and $kI + b$ expresses the simulated charging temperature based on the improved thermal model. The simulation results of both

Table 5 Sample battery parameters

Capacity (Ah)	Weight (kg)	Total surface area (cm^2)	Average heat capacity (J/g k)	Charge limit voltage (V)	Discharge limit voltage (V)
25	0.899	753	0.9932	4.2	3.0

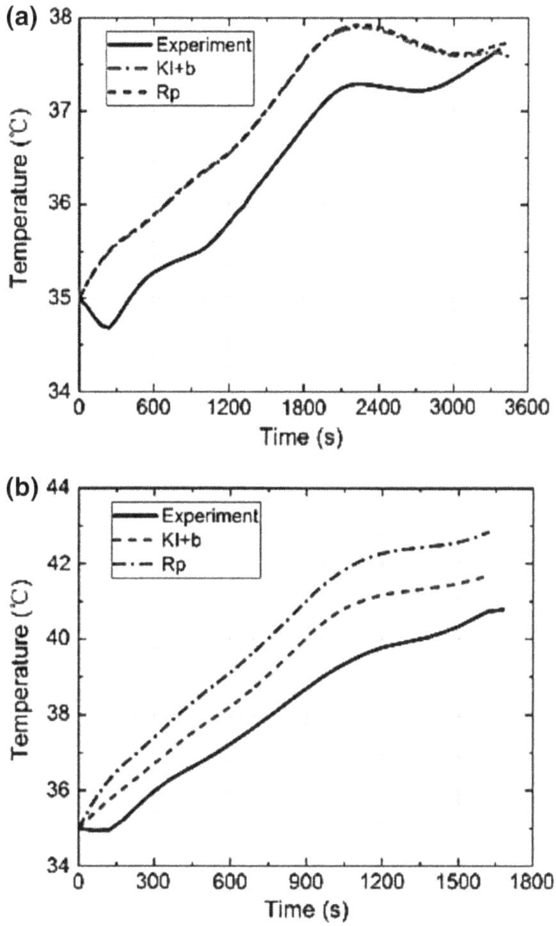

Fig. 25 Simulated and experimental temperatures during battery charge: **a** 1C and **b** 2C

methods are almost identical at the charge current of 1C, and the maximum error is approximately 0.9 °C.

The simulation error with the $kI + b$ method is within 1.7 °C at the charge current of 2C. The accuracy is remarkably improved compared to the DC resistance calculation method with the maximum error of 2.7 °C. This is because the polarization resistance is not constant, but decreases to some extent as the charge current increases. The estimation error with the DC resistance calculation method is small at the low charge current. Nevertheless, the constant DC resistance cannot effectively express battery polarization at high charge currents, resulting in larger errors. It is clear that the $kI + b$ method can better describe the relationship between polarization voltage and current, demonstrating its high fidelity of thermal simulation, particularly in the case of fast charging.

4.2 Formulation of Battery Charging Optimization

The performance and lifetime of a battery will be significantly affected if the current flowing into the battery is beyond the maximum charge current value, which can even lead to safety issues. However, from the aspect of electrochemistry, the maximum charging current is difficult to obtain in real time because it requires battery chemical parameters and detailed knowledge of battery material properties normally unavailable to BMS designers. The polarization voltage describes the deviation of the battery terminal voltage from its equilibrium, which can be observed based on lumped equivalent circuit models. The side reactions and lithium deposition may occur during charging when the polarization of the battery is too large, noticeably affecting battery performance and life. The polarization corresponds to an acceptable battery charging current to a large extent. The polarization constraint is therefore employed to calculate the maximum charging current in this work. The charging currents at various SOC values are optimized within the charging boundary curve. The acceptable charge current is high at the low side of battery SOC and declines with augmented SOC, based on polarization properties [25]. The maximum charge current is determined by battery kinetics limits. The temperature rise also needs to be considered to extend battery lifetime when designing charging patterns. In consideration of battery charge polarization and temperature rise constraints, the optimized charging strategy can be summarized as follows. First, taking the acceptable charge current as the optimal charge current limit, the battery is charged with high current at the initial charging stage to speed up the charging process. Smaller charge current is then employed at the end of charging to decrease battery polarizations and to procure more charge capacities. Meanwhile, the temperature rise is calculated as a constraint, preventing thermal runaway and ensuring charge safety. Because the ohmic resistance, polarization resistance, and entropy change coefficient of the battery vary with SOC, the temperature rise rate differs among various SOC regions. Second, considering temperature rise characteristics at different SOC regions, the charge current can be augmented in SOC regions with lower temperature rise. The charge current and temperature rise can consequently be optimized for the whole SOC range, balancing battery lifetime and charging speed.

The whole SOC ranging from 0 to 100% is divided into N segments in the developed charging strategy, and the charge current for each step needs to be optimized based on the charging time and temperature rise constraints. The SOC change for each segment is defined as ΔSOC, and the relationship between N and SOC is expressed as $N = \frac{1}{\Delta \mathrm{SOC}}$. The charging time for the kth step can be written as

$$t_k = \frac{Q_k}{I_k} = \frac{\Delta \mathrm{SOC} \cdot Q}{I_k} \tag{30}$$

where I_k is the charging current for the kth step, Q_k represents the charged capacity, and Q is the maximum available capacity of the battery, charged at a current of 0.05C. Then, the total charging time can be calculated by

$$t = \sum_{k=1}^{N-1} t_k + t_N = \sum_{k=1}^{N-1} t_k + \frac{Q_N}{I_N} \tag{31}$$

and the total charged capacity can be derived by

$$Q_{\mathrm{ch}} = \Delta \mathrm{SOC} \cdot Q \cdot (N-1) + Q_N \tag{32}$$

The charging time and charged capacity for the Nth step are calculated alone, since the Q_N may not reach the theoretical value $\Delta \mathrm{SOC} \cdot Q$ resulting from the charging current higher than 0.05C, and the total charged capacity is consequently less than the maximum available capacity. Supposing the charging voltage limit is 4.2 V, the open-circuit voltage (OCV) at the end of charging for the Nth step can be described as

$$\mathrm{OCV}_N = 4.2 - U_{\Omega,N} - U_{P,N} = 4.2 - (I_N R + k I_N + b) \tag{33}$$

Based on the SOC-OCV mapping curve, the SOC at the end of charging for the Nth step can be expressed by

$$\mathrm{SOC}_N = f^{-1}(4.2 - I_N R - k I_N - b) \tag{34}$$

The charged capacity Q_N is given by

$$Q_N = Q \cdot [\mathrm{SOC}_N - \Delta \mathrm{SOC} \cdot (N-1)] \tag{35}$$

The charging time t_N can be obtained based on the formula $t_N = Q_N / I_N$. Using the developed thermal model, the charge temperature rise for the kth step is

$$\begin{aligned} m C_{\mathrm{cell}} \left(T_k^{\mathrm{end}} - T_k^{st} \right) = & \int_0^{t_k} I_k^2 R \mathrm{d}t + \int_0^{t_k} I_k U_P \mathrm{d}t + \int_0^{t_k} T_{\mathrm{cell}} I_k \frac{\partial E}{\partial T_{\mathrm{cell}}} \mathrm{d}t \\ & - \int_0^{t_k} A h (T_{\mathrm{cell}} - T_{\mathrm{amb}}) \mathrm{d}t \end{aligned} \tag{36}$$

$$\Delta T_{\max} = T_{\max} - T_1^{st} \tag{37}$$

where $T_{\max}$, is the maximum temperature for the entire charging process.

The total charging time and temperature rise are two parameters to be optimized based on the developed charging strategy. The objective is to speed up the charging process and to reduce the resultant temperature rise, hence aiming at a balance between such two parameters. The fitness function is consequently portrayed by the following equation:

$$\begin{cases} F(t, \Delta T) = \alpha\left[\frac{40}{t_{\max} - t_{0.05C}}(t - t_{0.05C}) + 60\right] + \beta\left[100 + \frac{40(\Delta T - \Delta T_{0.05C})}{\Delta T_{0.05C} - \Delta T_{\max}}\right] \\ \alpha + \beta = 1 \end{cases} \tag{38}$$

where α and β are weighting coefficients of charging time and temperature rise, respectively. Moreover, $t_{\max}$ represents the total charging time when the battery is charged with the acceptable current, whereas $t_{0.05C}$ is the charging time at the current of 0.05C. Here, $\Delta T_{0.05C}$ is the battery temperature rise when charged at the current of 0.05C. The linear weighting method is used to calculate the fitness values.

GA is an adaptive optimization technique, including selection, crossover, and mutation, that is inspired by the principles of natural selection and survival of the fittest from biological evolution. GA is an efficient method for solving both constrained and unconstrained optimization problems and requires a genetic representation of the solution, known as the population, and a fitness function used to evaluate the solution. To use GA to find the optimal solution, a population of solutions is first randomly initialized and then, through repetitive applications of the selection, crossover, and mutation operators, produces the next generation, which make the population evolve toward an optimal solution. This heuristic (also sometimes called a metaheuristic) is routinely used to generate useful solutions to solve engineering problems [67, 68]. The flowchart of optimal charge current estimation using the GA method is shown in Fig. 26, and the specific procedure is summarized as follows. An initial population is first produced as the basic currents. In the process, m column vectors with N charging currents are randomly generated based on the acceptable charge current limits. Second, fitness values are calculated, in which the charging time and temperature rise are regarded as the optimization objectives. Third, the selection operator, recombination operator, and mutation operator are carried out in the population to optimize the parameters. Finally, the procedure ends when the preset termination criterion is met. Actually, each SOC segment has m optional charge currents, thus will have m^N charging current permutations for the whole charging process. Each permutation has a fitness value, and the charging current permutation with the highest score will be finally selected.

4.3 Optimal Charging Current Calculation

The simulated charging current at various weighting coefficients is illustrated in Fig. 27, where α and β represent weighting coefficients of charging time and temperature rise, respectively. It is shown that the maximum fitness value of every

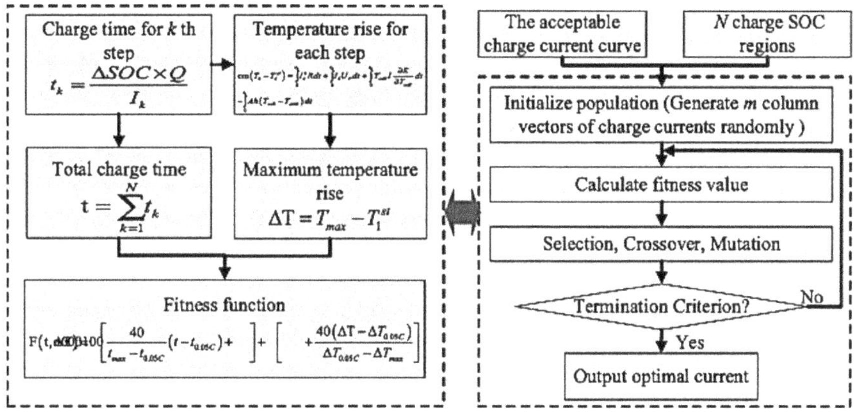

Fig. 26 Flowchart of charge current estimation using the GA method

generation approximately tends to be stable when the reproduction generations of the GA approach approximately reach 40, indicating that the fitness function is convergent. The overall variation tendency of the optimal charging current at various weighting coefficients is decreasing gradually as the acceptable charge current is declining with SOC increase. However, in some SOC regions, the charging current increases result from smaller resistance.

According to the simulation results, when $\alpha:\beta = 3:7$, the total charging time is 119 min, and the temperature rise is 1.9 °C; when $\alpha:\beta = 7:3$, the charging time of the optimal current is 64 min, and the temperature rise is 3.8 °C. Considering various weighting coefficients, the optimization results sufficiently reveal interactions between charging time and temperature rises. Charging time will decrease with a high value of α and a small value of β. The relationship among charge time, temperature rise, and weighting values is shown in Fig. 28. Charging time exhibits a monotonically increasing function of the $\alpha:\beta$ ratio, while temperature rise is monotonically decreasing, without having a local optimal solution. To speed up the charging process and to minimize battery life degradation, the ratio of α to β, 7:3, is finally picked.

4.4 Validation of Optimized Charging Method

The experiments involved the developed charging protocol, a CC–CV charging whose current rate is an average in the former (average charge rate CC–CV), and a 1/3C CC–CV charging. The average charge rate is calculated by dividing the charge capacity by charging time. The experiment outcome is displayed in Table 6. The developed charging solution is 5 min faster than the average charge rate in CC–CV charging and has almost the same total charge capacity and the temperature rise. This is because more attention is paid to improving charging speed in our protocol. Specifically, CV charging time in our protocol is almost 60% less than that of the

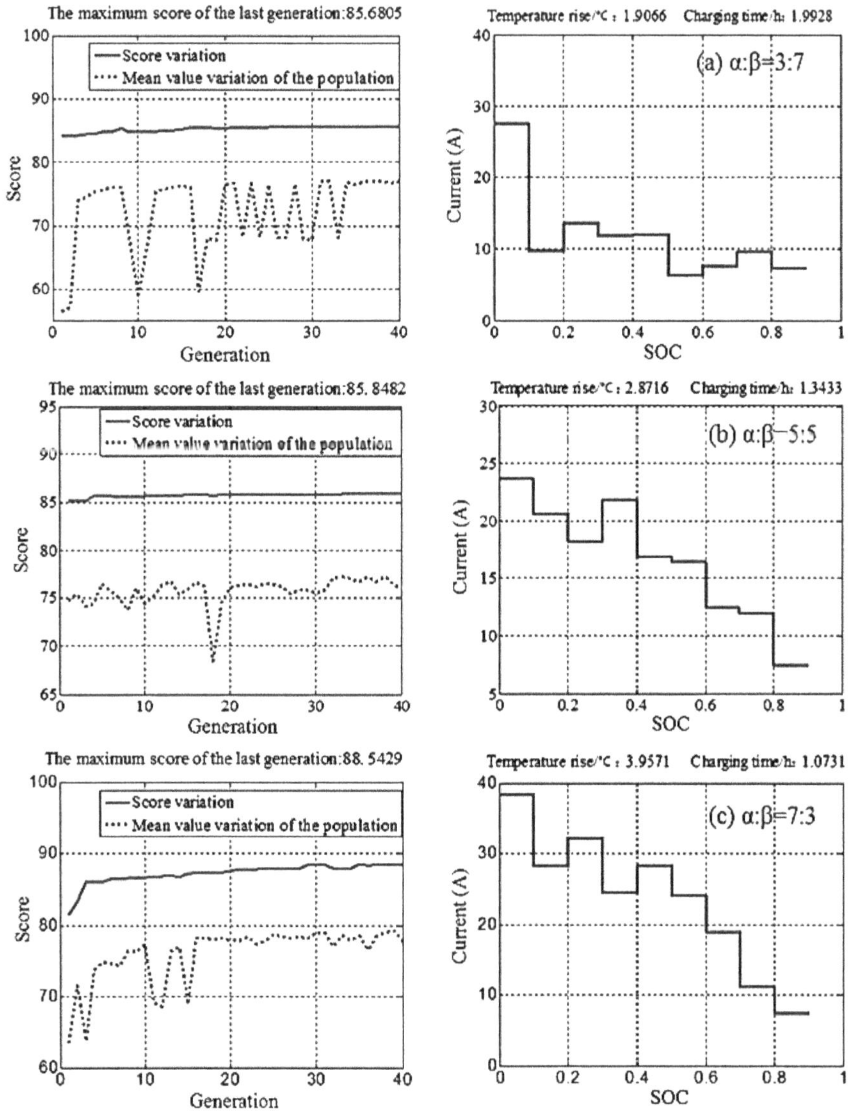

Fig. 27 Optimization results of charging current at various weighting coefficients: **a** $\alpha:\beta = 3:7$, **b** $\alpha:\beta = 5:5$, and **c** $\alpha:\beta = 7:3$

average charge rate CC–CV charging. The capacity during the CV part in the developed charging protocol only accounts for 2.4% of the total charge capacity, whereas that in the average charge rate CC–CV charging is 7%. Smaller charge current is then employed at the end of charging in the developed charging protocol to decrease battery polarization, so that its polarization at the end of charging is small, as confirmed in Fig. 29. Because such a lower polarization level results in

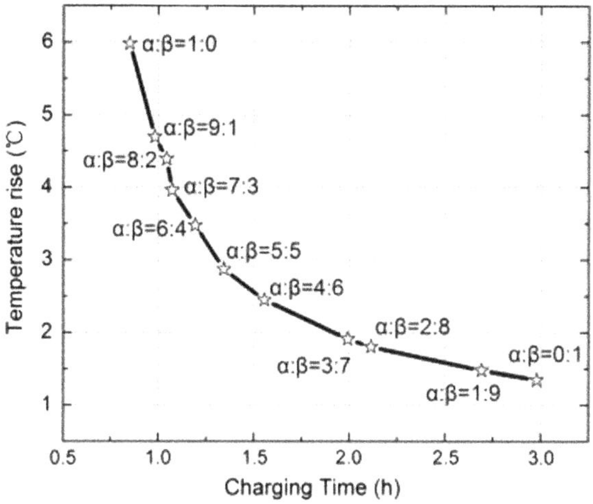

Fig. 28 Evolution of charging time and temperature rise, as the ratio of α to β decreases

Table 6 Experimental comparisons of the developed and typical charging patterns (α: β = 7:3)

Charging pattern	Total charge time/s	Total charge capacity/Ah	Charge time to 4.2 V/s	Charge capacity to 4.2 V/Ah	Temperature rise Δ°C
Proposed	3938	19.912	3520	19.428	3.8
Average charge rate	4263	20.084	3219	18.305	3.7
1/3C CC–CV	10,089	20.661	8997	18.742	1

Fig. 29 Polarization voltage characteristics at different charging strategies. Upper curve: optimized charging; middle curve: average rate charging; lower curve: 1/3C rate charging

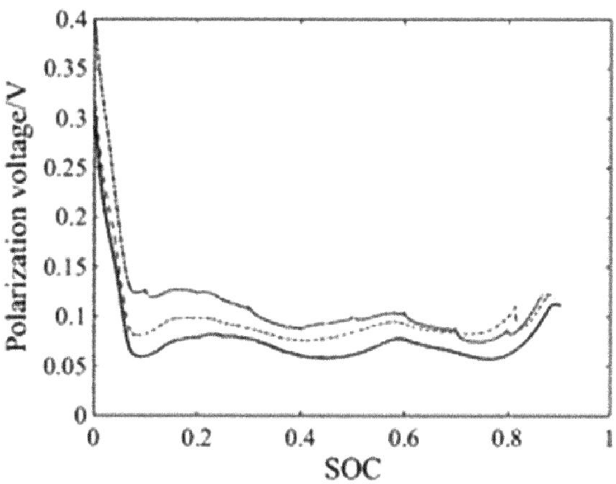

more charge capacity until 4.2 V, and less CV charge time and capacity, the CV charging can be neglected in the developed charging protocol to reduce the total charging time. Compared to the 1/3C CC–CV charging, the charge time of the developed solution when $\alpha:\beta = 7:3$ is reduced by about 60%, and the temperature rise is comparable, underlining the efficacy of our charging scheme.

These experimental results demonstrate that charge time and temperature rise can be optimized in the whole SOC region using the developed method. The polarization is decreased during the charging process, and more capacity can be charged, desirably balancing charge capacity and charge time. With low temperature rise and polarization level, the developed charging protocol can alleviate battery degradation as well.

References

1. A. Seaman, T.-S. Dao, J. McPhee, J. Power Sources **256**, 410 (2014)
2. M. Einhorn, F.V. Conte, C. Kral, J. Fleig, IEEE Trans. Power Electr. **28**, 1429 (2013)
3. X. Hu, R. Xiong, E. Bo, IEEE Trans. Ind. Inform. **10**, 1948 (2014)
4. Z. Ma, J. Jiang, W. Shi, W. Zhang, C.C. Mi, J. Power Sources **274**, 29 (2015)
5. H. Rahimi-Eichi, U. Ojha, F. Baronti, M. Chow, IEEE Ind. Electron. M **7**, 4 (2013)
6. N.M.L. Tan, T. Abe, H. Akagi, IEEE Trans. Power Electr. **27**, 1237 (2012)
7. H. Lin, T. Liang, S. Chen, IEEE Trans. Ind. Electron. **9**, 679 (2013)
8. M. Shahriari, M. Farrokhi, IEEE Trans. Ind. Electron **60**, 191 (2013)
9. T.T. Vo, X.P. Chen, W.X. Shen, A. Kapoor, J. Power Sources **271**, 413 (2015)
10. Z. Guo, B.Y. Liaw, X.P. Qiu, L. Gao, C.S. Zhang, J. Power Sources **274**, 957 (2015)
11. X.S. Hu, S. Li, H. Peng, F.C. Sun, J. Power Sources **239**, 449 (2013)
12. S. Zhang, C.N. Zhang, R. Xiong, W. Zhou, Energies **7**, 6783 (2014)
13. A.A. Hussein, A.A. Fardoun, S.S. Stephen, IEEE Trans. Sustain. Energy **7**, 32 (2016)
14. L. Ji, L. Guang, H.K. Fathy, J. Dyn. Syst. Meas. Control **138**, 021009 (2016)
15. Y.H. Liu, C.H. Hsieh, Y.F. Luo, IEEE Trans. Energy Conver. **26**, 654 (2011)
16. Y.H. Liu, Y.F. Luo, IEEE Trans. Ind. Electron **57**, 3963 (2010)
17. D. Anseán, M. González, J.C. Viera et al., J. Power Sources **239**, 9 (2013)
18. N. Legrand, B. Knosp, P. Desprez et al., J. Power Sources **245**, 208 (2014)
19. L.R. Chen, S.L. Wu, D.T. Shieh et al., IEEE Trans. Ind. Electron. **60**, 88 (2013)
20. S.J. Huang, B.G. Huang, F.S. Pai, IEEE Trans. Power Electr. **28**, 1555 (2013)
21. J.Y. Yan, G.Q. Xu, H.H. Qian et al., Energies **4**, 1178 (2011)
22. P. Sourav, A. Sohel, J. Power Sources **313**, 164 (2016)
23. Abdollahi, X. Han, G.V. Avvari, N. Raghunathan et al., J. Power Sources **303**, 388 (2016)
24. J.C. Jiang, C.P. Zhang, J.P. Wen et al., IEEE Trans. Veh. Technol. **62**, 3000 (2013)
25. J.C. Jiang, Q.J. Liu, C.P. Zhang et al., IEEE Trans. Ind. Electron. **61**, 6844 (2014)
26. K.M. Kim, Y.G. Lee, K.Y. Kang et al., RSC Adv. **2**, 3844 (2012)
27. R. Xiong, F. Sun, Z. Chen, H. He, Appl. Energy **113**, 463 (2014)
28. F. Sun, R. Xiong, J. Power Sources **274**, 582 (2015)
29. F. Sun, R. Xiong, H. He, Appl. Energy **162**, 1399 (2016)
30. X.S. Hu, S.B. Li, H. Peng, J. Power Sources **198**, 359 (2012)
31. C.P. Zhang, L.Y. Wang, X. Li, W. Chen, G.G. Yin, J.C. Jiang, IEEE Trans. Ind. Electron. **62**, 4948 (2015)
32. M. Urbain, M. Hinaje, S. Rael, B. Davat, P. Desprez, IEEE Trans. Energy Conver. **25**, 862 (2010)

33. H.V.M. Hamelers, A.T. Heijne, N. Stein, R.A. Rozendal, C.J.N. Buisman, Bioresource Technol. **102**, 381 (2011)
34. R.F. Mann, J.C. Amphlett, B.A. Peppley, C.P. Thurgood, J. Power Sources **161**, 775 (2006)
35. E. Prada et al., J. Electrochem. Soc. **159**, A1508 (2012)
36. P. Rong, M. Pedram, IEEE Trans. Very Large Scale Integr. (VLSI) Syst. **14**, 441 (2006)
37. M. Greenleaf, H. Li, J.P. Zheng, IEEE Trans. Sustain. Energy **4**, 1065 (2013)
38. N. Watrin, R. Roche, H. Ostermann, B. Blunier, A. Miraoui, IEEE Trans. Veh. **61**, 3420 (2012)
39. X. Hu, F. Sun, Y. Zou, Simul. Model. Pract. Theory **34**, 1 (2013)
40. Blanco et al., IEEE Trans. Veh. Technol. **63**, 3592 (2014)
41. W.Y. Low, J.A. Aziz, N.R.N. Idris, R. Saidur, J. Power Sources **221**, 201 (2013)
42. L. Gao, S. Liu, R.A. Dougal, IEEE Trans. Compon. Packag. Technol. **25**, 495 (2002)
43. J. Zhang, S. Ci, H. Sharif, M. Alahmad, An enhanced circuit-based model for single-cell battery, in *Proceedings of 25th Annual IEEE APEC* (2010), p. 672
44. T. Kim, W. Qiao, IEEE Trans. Energy Conver. **26**, 1172 (2011)
45. L. Lam, P. Bauer, E. Kelder, A practical circuit-based model for Li-Ion battery cells in electric vehicle applications, in *Proceedings of 33rd IEEE INTELEC*, vol 1 (2011)
46. L.W. Juang, P.J. Kollmeyer, T.M. Jahns, R.D. Lorenz, IEEE Trans. Ind. Appl. **49**, 1480 (2013)
47. L.X. Liao, P.J. Zuo, Y.L. Ma et al., Electrochim. Acta **60**, 269 (2012)
48. M. Dubarry, C. Truchot, B.Y. Liaw et al., J. Electrochem. Soc. **160**, A191 (2013)
49. F. Leng, C.M. Tan, M. Pecht, Sci. Rep. **5**, 1 (2015)
50. S. Liu, J. Jiang, W. Shi et al., IEEE Trans. Ind. Electron. **62**, 7557 (2015)
51. H. He, R. Xiong, X. Zhang, F. Sun, J. Fan, IEEE Trans. Veh. Technol. **60**, 1461 (2011)
52. H.W. He, R. Xiong, H.Q. Guo, Appl. Energy **89**, 413 (2012)
53. M. Nakayama, K. Iizuka, H. Shiiba, S. Baba, M. Nogami, J. Ceram. Soc. Jpn. **119**, 692 (2011)
54. V. Srinivasan, J. Newman, Electrochem. Solid-State Lett. **9**, A110 (2006)
55. H.C. Shin et al., Electrochem. Commun. **10**, 536 (2008)
56. F. Baronti et al., IEEE Trans. Ind. Informat. **10**, 1003 (2014)
57. P. Munoz-Condes et al., IEEE Trans. Ind. Electron. **60**, 191 (2013)
58. M. Einhorn, F.V. Conte, C. Kral, J. Fleig, IEEE Trans. Power Electron. **28**, 1429 (2013)
59. M. Chen, G. Rincon-Mora, IEEE Trans. Energy Convers. **21**, 504 (2006)
60. K.M. Tsang, L. Sun, W.L. Chan, Energy Convers. Manag. **51**, 2857 (2010)
61. S. Bangaru, R. Alugonda, P. Palacharla, Modeling and simulation of Lithium-Ion battery with hysteresis for industrial applications, in *Proceedings of ICEETS*, vol 771 (2013)
62. H.A.-H. Hussein, N. Kutkut, I. Batarseh, A hysteresis model for a Lithium battery cell with improved transient response, in *Proceedings of 26th Annual IEEE APEC*, vol 1790 (2011)
63. Z. Li, X. Han, L. Lu, M. Ouyang, Chin. J. Mech. Eng. **47**, 115 (2011)
64. M.A. Roscher, O. Bohlen, J. Vetter, Int. J. Electrochem. 2011, Art ID. 984320 (2011)
65. N. Takami et al., J. Power Sources **244**, 469 (2013)
66. K. Onda, T. Ohshima, M. Nakayama et al., J. Power Sources **158**, 535 (2006)
67. K.S. Sajan, V. Kumar, B. Tyagi, Int. J. Electr. Power Energy Syst. **73**, 200 (2015)
68. R. Gholami, M. Shahabi, M.R. Haghifam, Int. J. Electr. Power Energy Syst. **71**, 335 (2015)

State of Charge and State of Health Estimation Over the Battery Lifespan

Abbas Fotouhi, Karsten Propp, Daniel J. Auger and Stefano Longo

Abstract The battery management system (BMS) plays a critical role in battery packs especially for the lithium-ion battery chemistry. Protecting the cells from overcharge and overdischarge, controlling the temperature at the desired level, prolonging the life of the battery pack, guaranteeing the safety and indicating the available power and energy of the battery are the key functionalities of a BMS. In this chapter, two important concepts of a BMS are discussed: (i) battery state-of-charge (SoC) and (ii) battery state-of-health (SoH). Battery SoC and SoH are variables which should be determined precisely in order to use the battery optimally and safely. Batteries are time-varying systems that behave very differently at various states. In other words, the internal states of a battery tell us what we should expect from it. Depending on the battery chemistry, various techniques have been developed in the literature for SoC and SoH estimation. This covers a wide range from simple integration of current over time (i.e. coulomb counting) to advanced estimation techniques such as Kalman filter. In this study, almost all the existing battery SoC and SoH estimation approaches are reviewed and proper references are cited for further studies in each category.

A. Fotouhi (✉) · K. Propp · D. J. Auger · S. Longo
Advanced Vehicle Engineering Centre, School of Aerospace,
Transport and Manufacturing, Cranfield University,
MK43 0AL Cranfield, UK
e-mail: a.fotouhi@cranfield.ac.uk; abfotouhi@gmail.com

K. Propp
e-mail: k.propp@cranfield.ac.uk

D. J. Auger
e-mail: d.j.auger@cranfield.ac.uk

S. Longo
e-mail: s.longo@cranfield.ac.uk

G. Pistoia and B. Liaw (eds.), *Behaviour of Lithium-Ion Batteries in Electric Vehicles*, Green Energy and Technology,
https://doi.org/10.1007/978-3-319-69950-9_11

1 Introduction

In the near future, batteries will play a significant role in the transportation sector regarding vehicle electrification. Lots of research activities are going on, trying to improve different aspects of battery technology such as more cycling life, more energy and power densities, more safety, less cost, less environmental effects after recycling and so on. Part of this development process is focused on developing efficient battery management systems (BMSs). A BMS is responsible for different tasks such as protecting cells from damage, prolonging life of the battery and maintaining the battery in a state in which it can meet the requirements of the application for which it was designed. In one sentence, BMS is for 'optimal and safe use of a battery by monitoring and control'. The monitoring process consists of real-time measurements of variables like current, voltage and temperature, which are used to calculate some 'unmeasurable' variables like battery state-of-charge (SoC) and state-of-health (SoH).

In simple terms, SoC is defined as an indicator of the remaining battery energy. There are many techniques in the literature developed for battery SoC estimation, and a number of them are explained in this chapter. Here we introduce different approaches, from simple 'bookkeeping' techniques, like the integration of the current signal over time, to methods that exploit some behavioural challenges during the usage range. While these methods theoretically work independently, they can be combined or used together to increase the estimation precision.

The simplest method of battery SoC calculation, called 'coulomb counting', works based on integration of the battery current over time to calculate the amount of energy flowing through the battery. This method is generally precise in short term, but the integration leads to difficulties for long-term usage. The addition of small measurement errors or biases over time can lead to large errors in this technique especially in real applications without full cycles, using less accurate sensors with more noise. Another issue can be the need for a precise initial condition, as this information might not be given at any point in practice. Because of these practical limitations, more advanced battery SoC estimation techniques have been developed in the literature as discussed in this chapter.

On the other hand, battery SoH is defined as a quantitative indicator of healthiness of the battery and is determined based on the battery end-of-life definition. However, there is not a single definition of battery end-of-life that is accepted by all scientists in this area. Consequently, different definitions exist in the literature such as 'calendar life' which presents the age of a battery as months or years. So, battery end-of-life is confirmed based on a time period. However, we know that the life of a battery is affected by different scenarios of usage. So, another definition of battery life has been proposed called 'cycling life'. In this second definition, the age of a battery is presented in the sense of battery use depending on the charge/discharge profile. So, a battery life is calculated based on the number of cycles. Such a definition can be used in practice when the load condition is consistent and repeating. Both definitions of SoH could not be used in practice without challenge. Consequently, more advanced techniques were developed to get a better estimation of battery age in real applications as discussed in this chapter.

2 Battery State-of-Charge (SoC)

Naturally, for all types of energy storage systems, information about the capacity and available energy inside the system is vital. For classic cars with combustion engine and petrol tank, this is relatively easy since the amount of petrol in the tank is measurable at any given moment and the capacity of the tank itself is constant. This leads to precise range predictions which are also needed in an electric car, especially due to the smaller range and less well-established charging network. However, for chemical energy storage systems like batteries, the situation can be more complex. Firstly, there is usually no direct measurement available showing the remaining energy, or state-of-charge, and secondly, the capacity of a battery varies with temperature, discharge currents and age. These difficulties stimulated an intense research activity focusing on accurate SoC estimations for different battery types and applications. Main motivations for accuracy, next to advantages for the user due to accurately predicted operation times for battery-powered devices, are the properties of some battery types. While the SoC estimation for lead–acid batteries can be relatively simple [1], direct measurement of physical properties of the battery is possible, due to the battery's tolerance for overcharge and overdischarge to some extent. However, this is not the case for other battery types. Li-ion batteries, for example, are sensitive to overcharge and overdischarge. So, the SoC must be known when dynamic currents are applied, as in electric vehicles, to guarantee the safety and an optimal lifetime [2].

As shown in Fig. 1, less uncertainty in the SoC estimation allows battery operations closer to its limits without compromising safety or reliability. For these reasons, many SoC estimation methods were developed for Li-ion batteries with accuracies up to 1% [3]. Here, the estimation effort, and therefore the precision, also depends on the use. While most simple consumer electronics use basic methods,

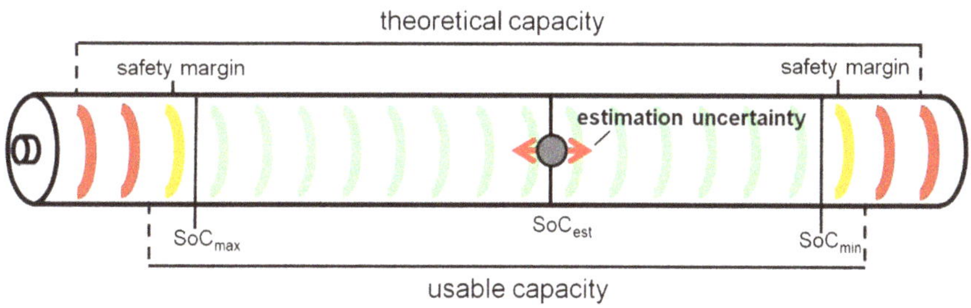

Fig. 1 Schematic of theoretical and usable capacity of a cell; the usable capacity also depends on the estimation uncertainty

the SoC estimation for electric vehicles is usually more complex since it has to be robust against a wide range of user conditions and highly dynamic current profiles.

2.1 Cell Capacity Definition

Based on the chemical reactions, the material for the individual cell components and the operational parameters (e.g. temperature), each type of cell has a voltage window for safe operation. In practice, it is important to know how much current can be exchanged between the maximum and minimum cell's potentials. Therefore, a cell's capacity is simply defined as

$$Q_n = \int_{t_{100\%}}^{t_{0\%}} I d\tau \tag{1}$$

where $t_{100\%}$ is the time necessary for a full charge and $t_{0\%}$ corresponds to full discharge.

As mentioned before, the actual capacity of the cell varies under different conditions. To accommodate for these, common experimental techniques use a standardised slow discharge at C/10 [4], in a controlled environment on a fresh cell for determining its capacity. The result is then called nominal or rated capacity, emphasising that it is just valid under these conditions. Furthermore, cell manufacturers usually supply more information for other discharge conditions in their technical cell descriptions. With the given data for capacity over temperature or power density, the application engineer has the opportunity to narrow potential cell candidates for the intended purpose. The effect of cell ageing is usually not considered in detail. The manufacturer only guarantees for a number of full cycles with the nominal capacity.

2.2 SoC Definition

SoC of a cell is defined as a fraction of the nominal capacity, which is available for the discharge process. While 1 (100%) indicates a fully charged battery, 0 (0%) represents a fully discharged cell. In mathematical terms, the definition is

$$SoC_t = \frac{Q_t}{Q_n} \tag{2}$$

where Q_t is the remaining capacity of the cell. A cell is defined fully charged when the charging current does not change within 2 h at a constant voltage and temperature (DIN 43539); this is rarely the case in practical applications [4]. Here the cell might not be fully charged from an electrochemical point of view, but is seen as fully charged in practice due to safety thresholds or charge time limitations. This shows that a precise SoC, based on the amount of chemical reactants, is very difficult to obtain. Nevertheless, the introduced standard definitions for cell capacity and SoC work well in practice and efficient methods have been developed to handle the complex variations of batteries. Similar to the nominal capacity, the main focus is on the current flow over time. Here the variations of battery voltage and therefore the variations of available power over the discharge range are not considered further. For the SoC estimation, however, these 'behavioural' changes are very useful as described in the following parts. Thereby we concentrate on explaining the principles of the main and commonly used methods. For a complete review of SoC estimation methods, the reader should refer to [1, 5].

3 Battery State-of-Health (SoH)

Unlike SoC definition which is agreed by almost all researchers in this area, battery SoH has been defined in different ways in the literature. SoH is an indicator showing the remaining life of the battery; however, the problem is that there is not a single definition of battery end-of-life (EoL) that is widely accepted. In order to clarify this, some common expressions in the literature are explained below.

3.1 Calendar Life

In this definition, the age of a battery, and its EoL, is presented in terms of months or years. We agree that like other devices, the life of a battery is affected by different scenarios of usage. Of course, the life of a battery may end before the calendar's prediction; therefore, another definition of battery life has been proposed called 'cycling life'.

3.2 Cycling Life

In this definition, the battery life is calculated from the number of cycles the battery can sustain in given conditions. Such a definition can be used in practice when the load conditions are consistent and reproducible. However, an accurate cycle counting can be difficult or impossible in some applications like automotive, where the battery is not cycled on a regular basis due to the variable driving condition [6]. On the other hand, the rate of charge/discharge can significantly affect the number of usable cycles. In addition, all similar cells do not necessarily show the same behaviour and the number of usable cycles may change from cell to cell.

3.3 Battery SoH Definition Based on Capacity Fade

Regarding the practical limitations of battery cycle counting, it has been concluded that some other quantitative indices are needed to reflect battery ageing. For example, battery capacity fade has been used in a wide range of studies as an index reflecting battery ageing. Lithium-ion battery degradation begins just after the battery manufacturing due to the electrochemical reactions taking place inside the battery. This process leads to degradation of the battery's active material, and consequently, battery's internal resistance increases which means more battery internal losses and capacity fade. A good review about the major ageing processes of Li-ion cells is given in [7]. An estimation of a battery's capacity can give us useful information about the age of the battery. This would be possible by comparing the battery capacity (C_{batt}) to its initial value when the battery is brand new (C_{init}). In a common definition, battery end-of-life is when the capacity falls to 80% of its initial value. SoH can be formulated as follows:

$$\text{SoH} = 1 - \frac{C_{\text{init}} - C_{\text{batt}}}{0.2C_{\text{init}}}, \quad 0.8C_{\text{init}} \leq C_{\text{batt}} \leq C_{\text{init}} \tag{3}$$

where SoH can change between 0 and 1 in which zero SoH means battery EoL ($C_{\text{batt}} = 0.8\ C_{\text{init}}$). The coefficient 0.2 in the denominator comes from: $C_{\text{init}}-0.8\ C_{\text{init}} = 0.2\ C_{\text{init}}$.

3.4 Battery SoH Definition Based on Power Fade

Another definition of battery SoH is based on 'power fade' instead of 'capacity fade'. This refers to how the battery power capability is decreased by the ageing process. This different definition is also presented as battery state-of-power (SoP) in the literature [8–10]. The power that a battery can deliver directly depends on battery internal resistance. The battery ageing leads to an increase in internal ohmic resistance in almost all types of battery. Consequently, this parameter can be used as an indicator of battery SoH. The higher the internal resistance becomes, the less power is available from the battery. The reason is due to the higher battery terminal voltage drop caused by the higher internal resistance. Assuming a simple battery model consisting of a resistor (R_{O}) in series with a voltage source (V_{OC}), battery terminal voltage (V_t) is obtained as follows showing the direct effect of ohmic resistance on terminal voltage drop ($R_{\text{O}}I_L$).

$$V_t = V_{\text{OC}} - R_{\text{O}}.I_L \tag{4}$$

There are a number of studies in the literature in which the effect of a battery ageing on its internal resistance is assessed. For example in [11], battery end-of-life is considered to happen when the internal ohmic resistance becomes twice its initial value. Using such a definition, battery SoH can be formulated as follows:

$$\mathrm{SoH} = 1 - \frac{R_{\mathrm{batt}} - R_{\mathrm{init}}}{R_{\mathrm{init}}}, \quad R_{\mathrm{init}} \leq R_{\mathrm{batt}} \leq 2R_{\mathrm{init}} \tag{5}$$

where R_{init} is the battery's initial resistance and R_{batt} is the battery's resistance at a given time. Again, SoH can change between 0 and 1 showing end and beginning of battery life, respectively.

In another study [12], battery end-of-life is defined when maximum power is dropped to 70% of its initial value. Again this formulation can be rewritten in terms of resistance ratio as follows:

$$\frac{P_{b,\mathrm{max}}}{P_{\mathrm{init}}} = \frac{R_{\mathrm{init}}}{R_{\mathrm{batt}}} \tag{6}$$

where P_{init} and R_{init} are battery's initial maximum power and initial ohmic resistance, whereas $P_{b,\mathrm{max}}$ and R_{batt} are battery's maximum power and ohmic resistance after a number of cycles.

It has been demonstrated in the literature that battery's polarisation resistance is not as effective as the ohmic resistance for reflecting Li-ion battery ageing.

Figure 2 demonstrates a comparison between the ohmic resistance and the polarisation resistance variations over cycling for a Li-ion battery [13]. This result shows higher sensitivity of the ohmic resistance against battery ageing comparing to the polarisation resistance.

Regarding the various definitions of battery SoH, different techniques have been developed in the literature for battery SoH estimation. Usually, one or a combination of the above-mentioned definitions is used along with a suitable measurement and/or estimation technique. A big difference between SoC and SoH estimation is the matter of 'time'. SoC might change from 100 to 0% in few hours; however, SoH might change in this range in a decade. Various SoH estimation techniques have been proposed in the literature considering one or more battery

Fig. 2 Li-ion battery ohmic and polarisation resistance variations over cycling (replotted based on a figure in [13])

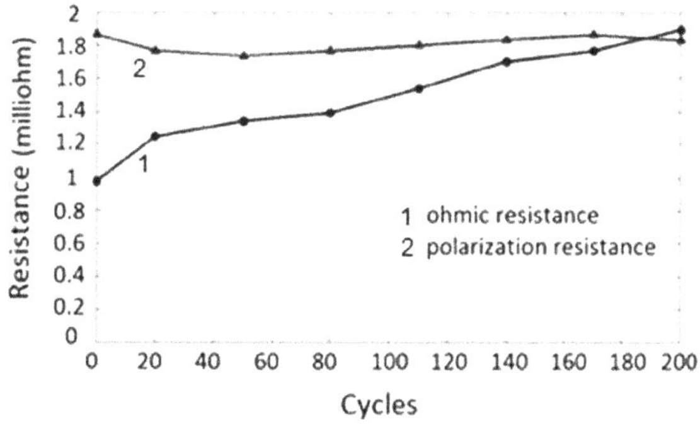

parameters, changing along the battery ageing, to get an estimation of battery SoH. In the following sections, some of these techniques are presented.

4 Coulomb Counting

The simplest method of battery SoC calculation is coulomb counting. Its basic principle is integration of current measurements over time to calculate the amount of energy flowing through the cell. The mathematical description given in [1] is as follows:

$$\mathrm{SoC} = \mathrm{SoC}_0 + \frac{1}{C_N}\int_{t_0}^{t}(I_{\mathrm{batt}} - I_{\mathrm{loss}})d\tau \tag{7}$$

where C_N is the nominal capacity, I_{batt} the measured current and I_{loss} a factor accounting for charge and discharge efficiency. While this method is generally precise in the short term, the integration leads to difficulties for long-term usage without a laboratory environment. The addition of small measurement errors or biases over time can lead to large errors. In laboratory tests, using full cycles, a controlled environment and high-precision current sensors, this error is reasonably small. However, in real applications without full cycles, less accurate sensors and more intense sensor noise, the drift leads to imprecisions that limit the usefulness of the method significantly. Another issue can be the need for a precise initial condition, as this information might not be accessible at any point.

In a group of studies, additional features are added to the original coulomb counting technique to make it suitable for real applications (enhanced coulomb counting). For example, 'current density' can be considered as an effective factor in coulomb counting. The Peukert law [14] formulates the discharge capacity as a function of current density as follows:

$$C_P = i^{\mathrm{PC}}.t \tag{8}$$

where C_P is the Peukert capacity, t is the time and PC is the Peukert coefficient. This formulation is valid under certain conditions, with the cell discharged with almost constant current. In real applications, the average current during a time step can be used instead. In order to obtain the Peukert coefficient, first a reference condition should be defined in which the cell is discharged with a certain current rate under a controlled temperature. The capacity of the cell under this condition is considered as the nominal capacity. Then experimental tests are performed at different discharge rates to find the relationship between real capacity and discharge rate. This relationship is mathematically modelled by using the Peukert coefficient obtained from experimental data [15]. In addition to current density, 'temperature' can remarkably affect the coulomb counting accuracy. This factor is also considered by defining modification coefficients covering the influence of temperature. Again experimental tests are performed to tune the modification coefficients [16].

As mentioned before, Peukert's law is obtained during discharge; however, in many applications, a cell might be charged in between as well. A good example is the regenerative braking in automotive application. To consider the charging effect in the coulomb counting technique as well, 'coulombic efficiency' is used [17]. Coulombic efficiency is defined as a modification coefficient (λ) multiplied by the charging current building a new variable called effective charging current (I_{eff}):

$$I_{\text{eff}}(\text{SoC}, i, T) = i.\lambda(\text{SoC}, i, T) \tag{9}$$

As seen in the above equation, coulombic efficiency is a function of SoC, current and temperature. Each of these dependencies should be investigated separately. Having the coulombic efficiency, we can modify the charging part of the coulomb counting method as well by replacing the current by the effective current.

Finally, an 'enhanced coulomb counting' technique would be a rate- and temperature-dependent coulomb counting technique that works based on a large number of lookup tables (or polynomials) to cover all possible operating conditions. Even such enhanced coulomb counting methods have some limitations, such as the need of initial SoC values and accurate current sensor readings, which restrict their practical use.

5 Open-Circuit Voltage SoC Estimation Method

Another relatively simple and reliable method of SoC estimation uses the relationship between OCV and SoC (Fig. 3). This method can get a reasonable estimation of the SoC without initial conditions or the need of long-term measurements just by letting the battery rest. Performing a piecewise linear interpolation, mathematical expression of OCV as a function of SoC can be presented as follows:

$$V_{\text{OC}}(t) = a.\text{SoC}(t) + b \tag{10}$$

where the coefficients, 'a' and 'b', should be obtained in the form of lookup tables at different SoC and temperature levels. The difficulty for this method is that it cannot be used for high or dynamic currents, since it can only be used when the

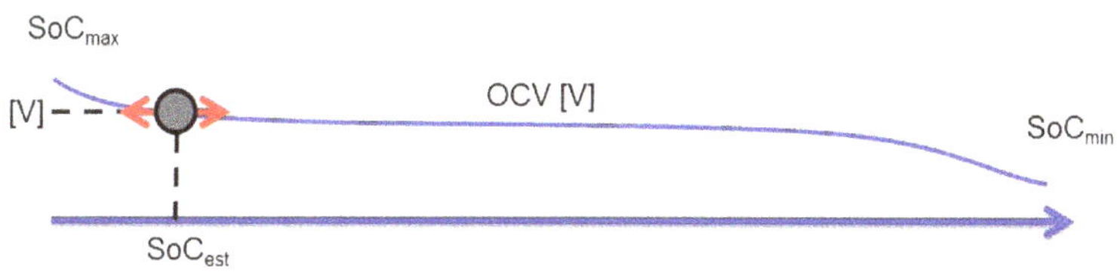

Fig. 3 Battery SoC estimation based on OCV

battery is in, or close to, a resting condition. Therefore, the OCV method itself is limited to applications with long breaks or small and constant currents. Due to their monotonic relationship between OCV and SoC, the OCV method is applicable for most Li-ion batteries. However, it works less well with batteries that have constant regions of OCV like LiFePO4 [18] or Li–S chemistries [19, 20].

6 Impedance Spectroscopy

A known technique for battery model parameterisation is electrochemical impedance spectroscopy (EIS) [21, 22]. The main advantage of this method is that the estimation can be done with dynamic currents, exploiting SOC changes when current is flowing. A full overview of this technique is presented in [23].

The results of a number of tests at different frequencies are plotted in the Nyquist plane which gives information on the ohmic resistance of the cell, their double-layer properties and mass transport effects when current is applied. It should be noted that the parameters are obtained under steady-state conditions that mean fixed SoC, temperature and SoH. So, the test should be repeated many times to cover various conditions.

For SoC estimation, however, this technique is rarely implemented due to advanced hardware needed, i.e. a frequency response analyser, and the effort and time needed to measure a whole range of different frequencies [1]. Furthermore, it is concluded in [23] that the marked temperature dependence of the impedance hinders a precise SoC estimation.

EIS has also been used for battery SoH investigations [24]. The impact of temperature, depth-of-discharge and cycling on discharge capacity were investigated as a function of cycle number for $Li_xMn_2O_4$-based lithium-ion cells. Both calendar life and cycling life are considered and tested, and the results demonstrated that an increase in temperature (from 25 to 45 °C) or depth-of-discharge (from 20 to 40%) leads to a higher capacity fade. The capacity fade was then obtained as a function of the square root of the cycle number that was a linear function in the cell's lifespan. Figure 4 shows cell voltage change measurements at various SoC levels, caused by current ($C/3$) interruptions: ΔV_1 is the voltage drop when the discharge load applies, ΔV_2 is the voltage relaxation when the current load is removed, and ΔV_3 is the total voltage relaxation after 20 min. These three measurable parameters were investigated in [24] to find a relationship between them and SoH. The results demonstrated that the voltage drop due to discharge (ΔV_1 that shows discharge resistance), at a fully charged state, has a linear relationship to the cycle number and can be used for SoH estimations.

The main advantage of techniques based on electrochemical analysis of the cell is the accuracy; however, the disadvantage is that they are difficult to apply for the user due to their complexity. So, they are not suitable for online applications. For further study, a full overview of the impedance spectroscopy method is presented in [23] and different battery modelling approaches are reviewed in [25].

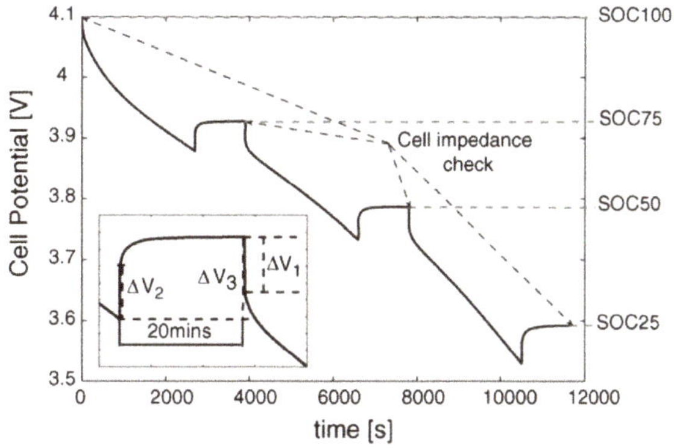

Fig. 4 Cell voltage change measurement under certain conditions to be used for SoH estimation [24]

7 Model-Based Approaches for Battery SoC and SoH Estimation

In a more advanced group of battery SoC estimation techniques, a cell model is constructed and used in support of the data resulting from the measurements to improve the estimation accuracy. In this approach, a mathematical cell model runs parallel to the real system, predicting the terminal voltage output of the battery for a current input. The error between the prediction and the real measurements is then used to correct the model. The behavioural changes of the cell over the whole discharge range, including OCV, internal resistances and transient behaviour, are used to estimate the SoC.

Different battery modelling approaches exist in the literature as reviewed in [25]. Available models are in a wide range: from complex electrochemical battery models to simpler equivalent circuit network (ECN) models [26]. The ECN models are mostly used for SoC estimation since they need less computational effort and can be relatively simple to parameterise in real time. The simplest ECN model, 'Resistance Model', predicts only the OCV and internal resistance (R_i) while more complex ones contain capacitors to model double-layer capacities or diodes to model hysteresis between charge and discharge. Each component of an ECN model represents parts of the battery response, as shown in Fig. 5, i.e. parts A, B and C.

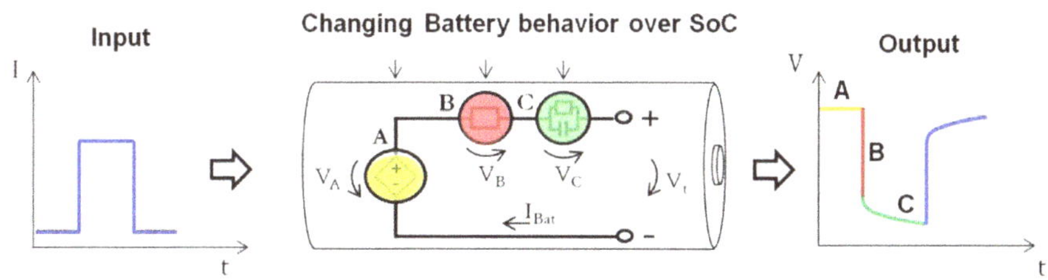

Fig. 5 Each component of an ECN model represents part of the battery response

Pulse discharge test is a common procedure for battery model identification; while the voltage before the pulse is seen as the OCV (component 'A'), the voltage drop is assigned to the internal resistance (component 'B') and the slow voltage change afterwards (component 'C') is determined by the double-layer capacitance of the electrode surfaces and is simplified by one parallel RC circuit [27]. Here it is worth to mention that the pulse test response also represents information about the battery behaviour similar to that of impedance spectroscopy. However, against EIS technique, here the model is parameterised in the time domain (rather than in the frequency domain) which is more intuitive for most users.

In Fig. 5, more RC networks (similar to component 'C') can be added to the model for more accuracy; however, this would add more complexity and computational effort as well. A proper trade-off between accuracy and complexity is necessary in cell modelling as discussed in [28, 29]. The battery model parameterisation is usually done using optimisation methods to minimise the difference between a model's output and experimental data.

7.1 Kalman Filter-Based Estimation Techniques

As mentioned before, battery parameters change over the SoC range which allows the adjustment of the estimated SoC based on differences between model's prediction and real measurement. As a rule of thumb, we can say the more pronounced the parameter variations over the SoC are, the more probable a precise estimation is. The principle of this model-based estimation technique is presented in Fig. 6. The example shows that the voltage prediction of the observer is constantly above the measurements. To minimise this error, the model's parameters and states (including SoC) are adjusted to fit the prediction to the measurements. The model in Fig. 6 can be represented by a standard state space representation described by

$$\begin{aligned} \dot{x}_t &= Ax_t + Bu_t \\ y_t &= Cx_t + D_t \end{aligned} \tag{11}$$

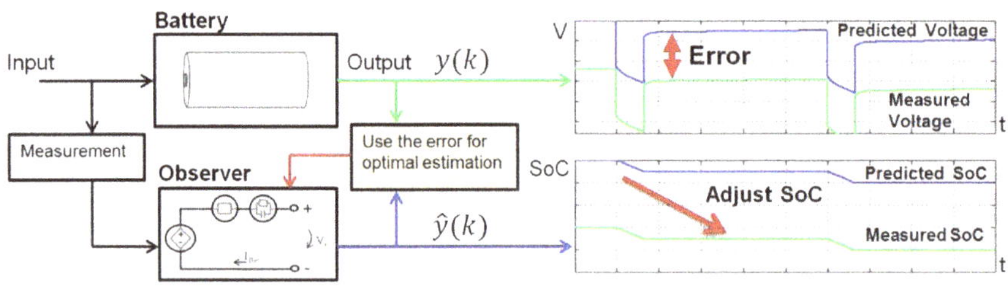

Fig. 6 Schematic of a model-based battery SoC estimation technique

Now the question is how to employ the estimation error mathematically for each time step, to get optimal state estimation results. As mentioned before, batteries are complex systems and a battery model always contains some level of abstraction and simplification. Methods that have been found to be robust against these imperfections are recursive filters that assign a stochastic variable to the observations y and states x, respectively. The principles of these filters are based on the assumption that both the measurements and the observer predictions are not perfect. While the measurement is influenced by random sensor noise, the imperfections of the observer are usually assigned to simplifications of the mathematical model.

For the propagations of these systems through time, the statistical framework of hidden Markov models (HMMs) of sequential data [30] is applicable, where all the past observations can be related to a state which is dependent on the inputs of the system. The state-dependent predicted measurements and the actual measurements can then be compared and the optimal state can be calculated combining both sources of information. For this, the recursive Bayes filter uses the evolution of the state transition probability, $P(x_t|x_{t-1}, u_t)$, and the measurement probability, $P(y_t|x_t)$, to propagate the probability densities from an initial value $P(x_0)$ over time. The algorithm possesses two main steps: (i) prediction and (ii) update [31].

Prediction:

$$\overline{\text{bel}}(x_t) = \int p(x_t|x_{t-1}, u_t)p(x_{t-1}|y_{1:t-1}, u_{1:t-1}) \tag{12}$$

where the actual probability density is calculated using the past measurements, the system's model and the control input u_t. Thereafter, the Bayes rule is used to estimate the posterior probability density bel (x_t) by employing the actual observation.

Update:

$$\text{bel}(x_t) = \eta^{-1}p(y_t|x_t)\overline{\text{bel}}(x_t) \tag{13}$$

where η is a normalisation factor and $p(z_t|z_{1:t}, u_{1:t})$ stands for certainty of the observations. An analytical solution of the Bayes filter for linear systems and Gaussian distributions is the Kalman filter (KF), minimising the error variance between true and estimated state. The introduced uncertainty allows the user to decide how trustworthy a parameter or measurement is. Less uncertainty leads to less correction and vice versa. The ECN model-based KF uses three general sources of information. The first one is the measurement, influenced by Gaussian white noise with zero mean. The second one are both system states of the Thevenin model (SoC, V_{RC}), which can be adjusted, and the third one are the model parameters (OCV, R_{i}, R_{p}, C_{p}) [19, 32]. Since the parameters of the model depend on the SoC, ideally a wrongly estimated SoC also leads to an imprecise terminal voltage prediction.

Since the KF algorithm is recursive, the continuous system has to be discretised as follows:

$$\begin{aligned} x_k &= Ax_{k-1} + Bu_{k-1} + w_{k-1} \\ y_k &= Cx_k + v_k \end{aligned} \tag{14}$$

where the additional terms w_k and v_k are random variables for the estimation process. The *prediction step* uses the input current and the previous states for estimation of the current states and probabilities.

$$\begin{aligned} \hat{x}_k^- &= A\hat{x}_{k-1}^+ + Bu_{k-1} \\ P_k^- &= AP_{k-1}^+A^T + Q. \end{aligned} \tag{15}$$

Using the new states and parameter predictions, prediction of the model's terminal voltage would be possible as well. The error between the terminal voltage prediction and measurement is then multiplied in the *correction step* by the Kalman gain to correct the states:

$$\begin{aligned} L_k &= P_k^- C^T \left(CP_k^- C^T + R\right) \\ \hat{x}_k^+ &= \hat{x}_k^- + L_k\left(y_k - C\hat{x}_k^-\right) \\ P_k^+ &= (I - L_kC)P_k^-. \end{aligned} \tag{16}$$

The mathematical background of minimisation and a good introduction to the KF can be found in [33, 34].

The Kalman gain dynamically weights the estimation in two opposite directions: the model's prediction and the real measurement. The performance of the filter depends heavily on the user's choices for the noise values of the states and measurements, combined in the matrices 'Q' and 'R'. While R is relatively easy to determine by testing the variances of the sensors or measurements, Q is guessed most of the time. Since models usually contain some level of abstraction, Q can also be used to indicate how imprecise the model predictions are. Although the user's choices seem quite difficult to make for the beginner, the parameterisation follows relative simple patterns. One example is the small variance assigned to the SoC. The implied small uncertainties emphasise the estimation on coulomb counting, which leads to less fluctuations of the SoC but also to a slower convergence with wrong initial states. Here the optimal decision depends on model's accuracy and application. As a rule of thumb, we can summarise: the less certain we are about a state, the more the Kalman filter uses the error between model and measurement for the corrections. One limitation of the Kalman filter is that it only works for linear systems with assumed Gaussian noise. Therefore, its nonlinear derivatives, the extended Kalman filter and unscented Kalman filter, are commonly used for estimations in the framework of batteries. Examples for both are given in [18, 35].

A dual KF architecture can be used to simultaneously compute both system states (e.g. SoC) and parameters (e.g. ohmic resistance). In such configuration, one filter is allocated to obtain the current system states using the previous parameters estimate, and one filter is designed to calculate the parameters using the previous states estimate. So, the state and parameter estimates are obtained together over time [36]. The battery parameter estimations can be used for battery SoH determination as well. For example, if we know the relationship between the ohmic resistance and SoH under various conditions (determined by SoC, temperature, etc.), KF can be used for identifying the values of ohmic resistance and SoC and then estimating SoH. For this purpose, a series of ageing tests should be performed under various conditions. There are many studies in the literature focusing on application of KF-based techniques for battery SoH estimation. For example, in [37], a new state variable was defined for internal resistance that is used as an indicator of battery SoH. In [38], EKF is used to obtain SoC and OCV in real time for SoH estimation. In [36], a model-based dual unscented KF was developed for cell SoC and SoH.

7.2 *Sliding-Mode Observer*

This technique comes from the sliding-mode control (SMC) theory [39–41]. SMC is a nonlinear control technique that changes the dynamics of a nonlinear system by using a discontinuous control signal that forces the system to 'slide' along some boundaries caused by control structures. One main advantage of SMC is its robustness. A system under SMC is considered as a variable structure system that moves through a continuous state space and also different discrete control modes. SMC uses very high gains to force the system's trajectory to slide along a restricted subspace. A geometrical locus consisting of the SMC boundaries is called 'sliding surface'. The system slides along the surface until it comes to rest at a desired equilibrium.

SMC theory can be used for design of state observers as well, called 'sliding-mode observer' (SMO). SMOs are able to bring the estimator's error dynamics to zero in a finite time period. Consider a linear time-invariant (LTI) system as follows:

$$\begin{cases} \dot{\mathbf{x}} = A\mathbf{x} + B\mathbf{u} \\ \mathbf{y} = [1\ 0\ 0\ \ldots].\mathbf{x} \end{cases}, \quad \mathbf{x} = (x_1, x_2, \ldots, x_n) \in \mathbf{R}^n \tag{17}$$

where $\mathbf{x}$ is the vector of system's states, $\mathbf{u}$ is the vector of inputs, and $\mathbf{y}$ is the output vector that is considered here as a scalar (equal to the first state of $\mathbf{x}$) for the sake of simplicity.

Assume a vector like $\hat{\mathbf{x}} = (\hat{x}_1, \hat{x}_2, \ldots, \hat{x}_n) \in \mathbf{R}^n$ containing state estimations of the system. The observer then takes the following form:

$$\dot{\hat{\mathbf{x}}} = \mathbf{A}\hat{\mathbf{x}} + \mathbf{B}\mathbf{u} + L.v(\hat{x}_1 - x_1) \tag{18}$$

where $e = \hat{x}_1 - x_1$ is the error between the estimated state and real state (which is the output y here as well), v is a nonlinear function ($R \to R$) of the error, and L is the observer gain. The nonlinear control law (i.e. function v) is designed in a way that produces a sliding surface like $0 = \hat{x}_1 - x_1$, which means the estimate $\hat{x}_1$ tracks the real state x_1 and the error goes to zero in a finite time. One possible form of function v that meets the required conditions of convergence is:

$$v = M.\text{sgn}(\hat{x}_1 - x_1) \tag{19}$$

where M is an enough large positive constant, and sgn is the sign function [39–41].

In [42], the SMO method was used for battery state estimation. In comparison to the Kalman filter-based estimation techniques, SMO also has attractive measurement noise resilience features. In addition, SMO does not need an operating environment with zero-mean noise condition, and its complexity is less relevant [42].

Like other model-based battery state estimation techniques, an ECN battery model has been used in [42]. However, the modelling errors, uncertainties and time-varying elements are considered as external disturbances. The modelling errors are expected to be compensated using SMO. It is discussed in [42] how the battery model parameters (including battery internal resistance) and states (SoC and SoH) are estimated using two SMOs running in parallel with exchanging information between them.

One limitation of such estimation techniques using an observer is the observability condition. As discussed in [42], the battery model should be observable and this depends on the OCV–SOC curve. For most of the battery types, including Li-ion batteries, an observable state-space representation of the model is achievable. However, this is not true for some other batteries like lithium–sulphur, as discussed in [43].

7.3 Online Battery Parameter Identification for SoC and SoH Estimation

A model-based technique of battery SoC estimation is presented in [43]. The generic framework of this idea is shown in Fig. 7 including real-time measurements, battery model parameterisation and a nonlinear mapping between the parameters and the battery SoC. The measurements consist of current, terminal voltage and temperature. Current and voltage measurements are used by the identification unit for online extraction of battery model's parameters. These parameters are then used by the estimation unit. The estimator is trained offline to find the relationship between the parameters and the SoC. The effect of temperature can also be taken into consideration in the estimation unit.

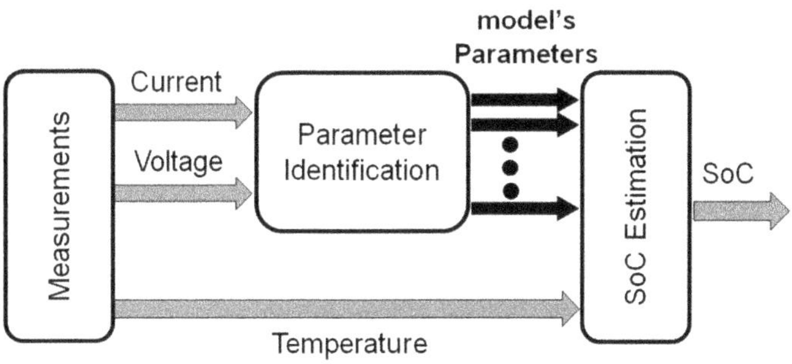

Fig. 7 Battery parameter identification for SoC estimation

The main advantages of this technique are: (i) in most cases, a simple model would be sufficient in this structure and there is no need to linearise the battery model, (ii) the technique is applicable to different types of battery, and (iii) the technique is fast enough for online applications. Number and type of the outputs of the identification unit are not predetermined; the number of parameters is chosen based on what is required for an effective state estimation that also depends on the battery chemistry [43]. In the estimation unit, a nonlinear mapping function (f) between the identification results and the SoC is used:

$$\mathrm{SoC} = f(P_1, P_2, P_3, \ldots) \quad (20)$$

where P_i is the ith identified battery parameter.

A similar idea can be developed for SoH estimation as well. This is a model-based SoH estimation technique, so a fast low-fidelity model, like ECN, is desired. The framework is shown in Fig. 8 including real-time measurements, battery model parameterisation (parameter identification) and a nonlinear mapping between the parameters and battery SoH (in Fig. 8, only one parameter is considered, i.e. ohmic discharge resistance). The SoH estimator is trained offline to find the relationship between SoC and SoH. The effect of temperature can also be taken

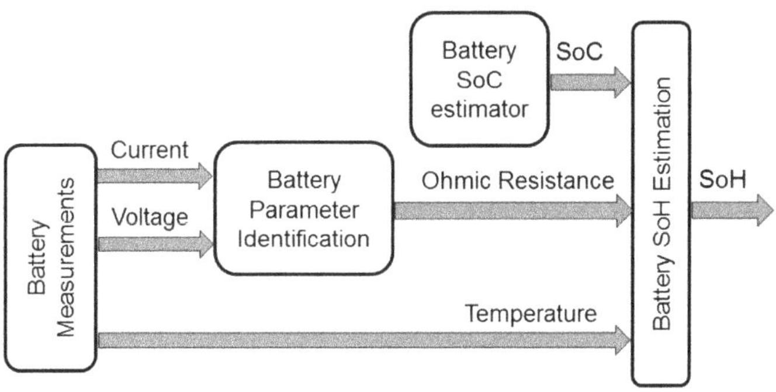

Fig. 8 Battery SoH estimation based on online parameter identification

into consideration in the estimation unit. In it, a nonlinear mapping function (g) between the identification results and SoH is constructed and used as follows:

$$\mathrm{SoH} = \mathrm{g}(P_1, P_2, P_3, \ldots) \tag{21}$$

where P_i is the ith identified battery parameter. In Fig. 8, only one parameter is considered, i.e. ohmic discharge resistance; however, a combination of parameters can be used in this framework. The disadvantage of this technique is the need of having enough test data for training the estimator under various conditions.

For example, in [13], Kalman filter was used to identify the battery's ohmic resistance online to be used for SoH estimation. In parallel, the battery ageing process was investigated to develop some lookup tables (graphs) including the relationship between ohmic resistance and SoH under various conditions. Finally, the whole system was constructed by combining the parameter identifier (KF) and the lookup tables using a concept similar to that shown in Fig. 8.

8 Battery SoH Estimation Using Mechanical Fatigue Theory

These methods come from the mechanical fatigue theory. Mechanical fatigue theory formulates the 'fatigue phenomenon' in a component subjected to a varying load. In some cases, a component might be subjected to a repeated on–off load. After a certain number of load cycles, the fatigue phenomenon might occur inside the component that usually leads to a collapse. Using the mechanical fatigue theory, the life of the component under such loading conditions can be estimated as a function of cycle number. For further studies in this area, the reader can refer to many textbooks [44–48]. Our discussion here is about the application of the fatigue theory in battery degradation analysis. In the following, some of the studies focused on this topic are reviewed.

One of the techniques of battery SoH estimation based on mechanical fatigue theory is called 'damage accumulation'. In this approach, battery ageing is estimated using a framework in which the age of a mechanical component is calculated using the Palmgren–Miner rule [44]. This rule formulates the life of a component under a sequence of variable loads. The component's life is calculated regarding the number of load cycles applied under a given condition. In such formulation, the component's EoL is defined by a life-reduction index (LRI), varying between 0 and 1, where the unit value means EoL. Experimental tests under various load conditions are necessary for each component in this approach. Assuming n_i as the number of cycles applied under certain load condition (L_i) and $N(L_i)$ as the number of cycles that can be applied to a new component under same load condition (L_i) before failure, the LRI under a sequence of variable loads (L_i, $i = 1\ldots s$) is defined as follows [49, 50]:

$$\mathrm{LRI} = \sum_{i=1}^{s} \frac{n_i}{N(L_i)} \tag{22}$$

$N(L_i)$ is obtained based on experimental test data and also the definition of EoL. As discussed in sect. 3, battery EoL is defined in different ways. Using the concept of capacity fade for battery EoL definition, a damage measure (ξ) is defined as follows:

$$\xi = \frac{C_{\mathrm{init}} - C_{\mathrm{batt}}}{C_{\mathrm{init}} - C_{\mathrm{final}}} \tag{23}$$

where the parameters are as explained in Sect. 3.3.

As discussed in [49], this approach gives a valid result if and only if the following factorisation would be possible.

$$\frac{\mathrm{d}\xi}{\mathrm{d}n} = \emptyset(\xi, L) = \emptyset_1(\xi)\sigma(L) \tag{24}$$

where $\emptyset_1(\xi)$ and $\sigma(L)$ are defined as the 'age factor' and the 'severity factor', respectively. The severity factor depends on parameters such as temperature, depth-of-discharge, current rate and is determined based on ageing experimental test data. For example, the severity factor is obtained for a lithium-ion battery pack to be used in a plug-in hybrid electric vehicle (PHEV) in [50].

Another technique of battery SoH estimation based on mechanical fatigue theory is called 'rainflow cycle counting' [51–54], a technique developed in the late 1960s for estimating fatigue damage due to an oscillatory load (mechanical stress) profile. The rainflow algorithm was originally developed to identify hysteresis loops in mechanical stress–strain space in order to estimate the fatigue damage. The fraction of fatigue life used, due to the application of a single hysteresis loop with a certain magnitude, is equal to the inverse of the number of cycles achievable before failure in the same condition. Rainflow cycle counting is a technique that formulates the equivalent effect of a varying load when it is replaced by a constant amplitude load. There are two well-known algorithms for rainflow cycle counting: three-point and four-point algorithms, both using the same concept. More explanations about these algorithms can be found in [54].

9 Conclusions

In this chapter, the concepts of battery SoC (representing the remaining energy inside the battery) and SoH (representing the remaining life of the battery) were discussed and various techniques of SoC and SoH estimation were reviewed. Coulomb counting or Amp-hour integration was discussed as the most common technique of battery SoC estimation but with some practical limitations. The Amp-hour integration leads to addition of small measurement errors that cause large errors over time and consequently imprecisions of that method in practice. Another issue can be the need for a precise initial condition, as this information might not be given at any point. Regarding these practical limitations, other advanced estimation techniques have been developed in the literature such as enhanced coulomb counting, SoC estimation based on OCV curve, online model identification and Kalman filter-based techniques. Selection of each of these techniques depends on the required accuracy and simplicity, so a trade-off between these parameters is required depending on the application.

Battery SoH and its estimation techniques were discussed in this chapter as well. As mentioned before, there is not a unique definition of battery SoH, and researchers define it in different ways but mainly according to capacity fade or power fade. In a common definition based on capacity fade, battery end-of-life is when the capacity falls to 80% of its initial value. Another definition of battery SoH is based on power fade that refers to how the battery power capability is decreased by the ageing process. The battery power directly depends on battery internal resistance. Battery ageing leads to an increase in internal ohmic resistance in almost all types of batteries. Consequently, this parameter can be used as an indicator of battery SoH. Depending on the definition that is considered for battery SoH, various techniques have been developed in the literature for its estimation, such as electrochemical impedance spectroscopy, online battery parameter identification, Kalman filter-based techniques, observer design using control theory (e.g. sliding-mode observer) and battery SoH estimation using mechanical fatigue theory, which all were reviewed in this study.

References

1. S. Piller, M. Perrin, A. Jossen, J. Power Sources **96**, 113 (2001)
2. L. Lu, X. Han, J. Li, J. Hua, M. Ouyang, J. Power Sources **226**, 272 (2013)
3. M.U. Cuma, T. Koroglu, Renew. Sust. Energ. Rev. **42**, 517 (2015)
4. D.U. Sauer, G. Bopp, A. Jossen, J. Garche, M. Rothert, M. Wollny, *State of Charge—What Do We Really Speak About*. International Telecommunications Energy conference (Copenhagen, Denmark, 1999)
5. W.Y. Chang, ISRN Appl. Math. Article ID 953792 (2013)
6. L. Serrao, Z. Chehab, Y. Guezennec, G. Rizzoni, *An Aging Model of Ni–MH Batteries for Hybrid Electric Vehicles*. IEEE Vehicle Power and Propulsion Conference (Chicago, USA, 2005)
7. J. Vetter, P. Novak, M.R. Wagner, C. Veit, K.C. Moller, J.O. Besenhard, M. Winter, M. Wohlfahrt-Mehrens, C. Vogler, A. Hammouche, J. Power Sources **147**, 269 (2005)
8. A. Hoke, A. Brissette, D. Maksimović, A. Pratt, K. Smith, *Electric Vehicle Charge Optimization Including Effects of Lithium-Ion Battery Degradation*. IEEE Vehicle Power and Propulsion Conference (Chicago, USA, 2011)
9. L.W. Juang, P.J. Kollmeyer, T.M. Jahns, R.D. Lorenz, *Implementation of Online Battery State-of-Power and State-of-Function Estimation in Electric Vehicle Applications*. Wisconsin Electric Machines and Power Electronics Consortium (WEMPEC) (Madison, USA, 2012)

10. P. Malysz, J. Ye, R. Gu, H. Yang, A. Emadi, I.E.E.E. Trans, Veh. Technol. **65**, 4512 (2016)
11. A. Zenati, P. Desprez, H. Razik, S. Rael, *A Methodology to Assess the State of Health of Lithium-Ion Batteries Based on the Battery's Parameters and a Fuzzy Logic System*. IEEE International Electric Vehicle Conference (IEVC) (Greenville, USA, 2012)
12. A. Pesaran, T. Markel, H.S. Tataria, D. Howell, *Battery Requirements for Plug-In Hybrid Electric Vehicles—Analysis and Rationale, EVS-23 International Electric Vehicle Symposium*, Anaheim, CA, NREL/CP-540-42240, 2007
13. D. Haifeng, W. Xuezhe, S. Zechang, *A New SoH Prediction Concept for the Power Lithium-Ion Battery Used on HEVs*. IEEE Vehicle Power and Propulsion Conference (Dearborn, MI, USA, 2009)
14. W. Peukert, Elektrotech. Z. **20**, 20 (1897)
15. L. Gao, S. Liu, R.A. Dougal, I.E.E.E. Trans, Compon. Packag. Technol. **25**, 495 (2002)
16. A. Hausmann, C. Depcik, J. Power Sources **235**, 148 (2013)
17. Y. Zheng, M. Ouyang, L. Lu et al., J. Power Sources **289**, 81 (2015)
18. W. He, N. Williard, C. Chen, M. Pecht, Microelectron. Reliab. **53**(6), 840 (2013)
19. K. Propp, D.J. Auger, A. Fotouhi, S. Longo, V. Knap, J. Power Sources **343**, 254 (2017)
20. A. Fotouhi, D.J. Auger, K. Propp, S. Longo, R. Purkayastha, L. O'Neill, S. Walus, I.E.E.E. Trans, Veh. Technol. (2017). https://doi.org/10.1109/TVT.2017.2678278
21. B.S. Haran, B.N. Popov, R.E. White, J. Power Sources **75**, 56 (1998)
22. E. Kuhn, C. Forgez, P. Lagonotte, P. Friedrich, G. J. Power Sources **158**, 1490 (2006)
23. F. Huet, J. Power Sources **70**, 59 (1998)
24. S. Brown, K. Ogawa, J. Power Sources **185**, 1444 (2008)
25. A. Fotouhi, D.J. Auger, K. Propp, S. Longo, M. Wild, Renew. Sust. Energy Rev. **56**, 1008 (2016)
26. V. Ramadesigan, P.W.C. Northrop, S. De, S. Santhanagopalan, R.D. Braatz, V.R. Subramanian, J. Electrochem. Soc. **159**, R31 (2012)
27. A. Jossen, J. Power Sources **154**, 530 (2006)
28. A. Fotouhi, D.J. Auger, K. Propp, S. Longo, IEEE Trans. Systems, Man Cyb.: Systems (2017) https://doi.org/10.1109/tsmc.2016.2599281
29. X. Hu, S. Li, H. Peng, J. Power Sources **198**, 359 (2012)
30. Y. Bengio, Neural Comput. Surv. **2**, 129 (1999)
31. S. Thrun, W. Burgard, D. Fox, *Probabilistic Robotics* (MIT Press, Cambridge, 2005)
32. L. Thevenin, CR Acad. Sci. **97**, 159 (1883)
33. G. Welch, G. Bishop, *An introduction to the Kalman filter* [Tech. Report]. University of North Carolina, Chapel Hill, NC, USA (1995)
34. A. Gelb, *Applied Optimal Estimation* (MIT Press, Cambridge, 1974)
35. G.L. Plett, J. Power Sources **134**, 262 (2004)
36. A. Mills, J. Zambreno, *Estimating State of Charge and State of Health of Rechargable Batteries on a Per-Cell Basis*. Workshop on Modeling and Simulation of Cyber-Physical Energy Systems (MSCPES) (Iowa, USA, 2015)
37. F. Zhang, G. Liu, L. Fang, *Battery State Estimation Using Unscented Kalman Filter*. IEEE International Conference on Robotics and Automation (Kobe, Japan, 2009)
38. M. Shahriari, M. Farrokhi, I.E.E.E. Trans, Ind. Electron. **60**, 191 (2013)
39. S.V. Drakunov, V.I. Utkin, Int. J. Control **55**, 1029 (1992)
40. C. Edwards, S. Spurgeon, *Sliding Mode Control: Theory and Applications* (Taylor and Francis, London, 1998)
41. V.I. Utkin, *Sliding Modes in Control and Optimization* (Springer, Berlin, 1992)
42. I.S. Kim, I.E.E.E. Trans, Power Electron. **25**, 1013 (2010)
43. A. Fotouhi, D.J. Auger, K. Propp, S. Longo, IET Power Electron. (2017). https://doi.org/10.1049/iet-pel.2016.0777
44. R.C. Juvinall, K.M. Marshek, *Fundamentals of Machine Component Design* (Wiley, New York, 2000)
45. J.A. Bannantine, J.J. Comer, J.L. Handrock, *Fundamentals of Metal Fatigue Analysis* (Prentice-Hall, Englewood Cliffs (NJ), 1990)

46. R.L. Mott, *Machine Elements in Mechanical Design (4th Ed.)* (Pearson Prentice Hall, Upper Saddle River, N J, 2004)
47. J.K. Nisbett, R.G. Budynas, *Shigley's Mechanical Engineering Design (8th Ed.)* (McGraw-Hill Higher Education, Boston, 2008)
48. Y.L. Lee, J. Pan, R. Hathaway, M. Barkey, *Fatigue Testing and Analysis: Theory and Practice* (Butterworth-Heinemann, Oxford, 2004)
49. M.T. Todinov, Comput. Mater. Sci. **21**, 101 (2001)
50. V. Marano, S. Onori, Y. Guezennec, G. Rizzoni, N. Madella, *Lithium-ion Batteries Life Estimation for Plug-in Hybrid Electric Vehicles*. IEEE Vehicle Power and Propulsion Conference (VPPC) (Dearborn, MI, USA, 2009)
51. S.D. Downing, D. Socie, Int. J. Fatigue **4**, 31 (1982)
52. I. Rychlik, Int. J. Fatigue **9**, 119 (1987)
53. C. Amzallag, J. Gerey, J. Robert, J. Bahuaud, Int. J. Fatigue **16**(4), 287 (1994)
54. C. McInnes, P. Meehan, Int. J. Fatigue **30**, 547 (2008)

Recycling of Batteries from Electric Vehicles

Tobias Elwert, Felix Römer, Kirstin Schneider, Qingsong Hua and Matthias Buchert

Abstract The introduction of electromobility will lead to a significant increase of waste traction batteries within the next decade. Recycling of these batteries is currently a huge challenge as the necessary legislative framework, logistic concepts, and recycling processes are in an early stage of development. In the first part of this chapter, the legal situation in the largest markets (European Union, People's Republic of China, and USA) is summarized and a forecast of traction battery return flows for cars and buses until 2025 is presented. The second part discusses the recycling chain including extraction of the batteries from end-of-life (EOL) vehicles, battery disassembly, and different approaches for cell recycling. The focus is on industrial efforts. In addition, economic and ecologic aspects are briefly addressed. The last part summarizes the main conclusions and highlights task fields to close the gaps in lithium-ion battery recycling.

1 Introduction

The environmental impact of a petroleum-based infrastructure, the finite nature of fossil fuel supply as well as advances in battery technology, fuel cells, electric motors, and power electronics have led to an increasing interest in hybrid electric

T. Elwert (✉) · F. Römer · K. Schneider
Department of Mineral and Waste Processing, Clausthal University of Technology, Walther-Nernst-Straße 9, 38678 Clausthal-Zellerfeld, Germany
e-mail: tobias.elwert@tu-clausthal.de

T. Elwert · Q. Hua
School of Mechanical and Electrical Engineering, Qingdao University, 308 Ningxia Road, Qingdao, Shangdong, People's Republic of China
e-mail: qihu@qdu.edu.cn

M. Buchert
Division Resources and Transport, Institute for Applied Ecology, Rheinstraße 95, 64295 Darmstadt, Germany
e-mail: m.buchert@oeko.de

G. Pistoia and B. Liaw (eds.), *Behaviour of Lithium-Ion Batteries in Electric Vehicles*, Green Energy and Technology,
https://doi.org/10.1007/978-3-319-69950-9_12

vehicle (HEV), plug-in hybrid electric vehicle (PHEV), fuel cell electric vehicle (FCEV), and battery electric vehicle (BEV). As these vehicle types, in the following summarized as xEVs, introduce new components, the spread of electromobility will lead to major changes in vehicle recycling practices within the upcoming years [1]. Especially the recycling of traction batteries, which are mainly lithium-ion batteries (LIB), is an enormous challenge, not only for vehicle recyclers and carmakers but also for subsequent battery recyclers, politics, and legislation. Challenges derive, among others, from a still-to-be-established recycling infrastructure, complex assemblies and material composites, currently low volumes with very different designs and chemical compositions as well as from high energy contents and weights [1, 2]. Nevertheless, in the case of a mass implementation of xEVs, sophisticated recycling technologies addressing these dynamic developments are of high importance, not only for environmental and safety reasons but also to secure the mid- and long-term supply of the required raw materials such as lithium and cobalt [1, 3].

In recent years, a considerable number of national and international efforts have been initiated by the private and public sectors in LIB recycling. The aim of this chapter is to present a general survey of the current developments and challenges in the recycling of traction batteries with a focus on (potential) industrial solutions. First, as background, the legislative framework and expected return flows are presented. Afterward, the recycling chain is covered from extraction of the traction batteries from end-of-life (EOL) vehicles to raw material production. Besides technical and economic aspects, also life cycle assessment (LCA) is addressed.

2 Background

The legislative framework and the estimated return flows of xEV batteries are determining factors for the emerging LIB recycling industry. Therefore, Sect. 2.1 summarizes the legislative framework in the largest markets: the European Union (EU), the People's Republic of China (PRC), and the United States of America (USA). In Sect. 2.2, a forecast of the traction battery return flows is presented.

2.1 Legislative Framework

2.1.1 European Union

In the EU, there are two important directives relevant to the recycling of traction batteries: Directive 2000/53/EC on EOL vehicles and Directive 2006/66/EC on batteries and accumulators and waste batteries and accumulators.

Directive 2000/53/EC addresses the prevention of waste from vehicles and the reuse and recycling of EOL vehicles and their components. These measures intend to reduce the disposal of waste and thereby ensure that EOL vehicles are discarded without endangering the environment. In 2015, the rate of reuse and recovery of EOL vehicles was set at a minimum of 95% by average weight per vehicle, the reuse and recycling to a minimum of 85%. According to Article 6, hazardous components, including waste traction batteries, must be removed and treated separately [4].

After removal, Directive 2006/66/EC regulates the further treatment of traction batteries. This directive addresses the manufacturing of batteries and accumulators as well as their disposal. Directive 2006/66/EC came into force in 2006 to harmonize national measures and to minimize the negative impact of EOL batteries on the environment. However, Member States retain some freedom how to transpose the directive's requirements in national laws [5].

According to Directive 2006/66/EC, traction batteries are industrial batteries. With regard to the recycling of industrial batteries, the most important regulations are the extended producer responsibility and the definition of minimum recycling efficiencies. Recycling processes of LIBs must achieve a minimum recycling efficiency of 50% by average weight. Battery producers, or third parties acting on their behalf, need to finance any net costs arising from the collection, treatment, and recycling of all waste industrial batteries collected [5].

Within the next years, a revision of the Directive 2006/66/EC is expected. Due to the growing electromobility sector, traction batteries will probably be addressed more specifically [6].

2.1.2 People's Republic of China

Since 2009, the Chinese government strongly promotes electromobility [7]. In that year, the government established the Thousands of Vehicles, Tens of Cities (TVTC) Program to demonstrate HEVs, BEVs, and FCEVs in public service vehicle fleets including buses and taxis [8]. The management of waste traction batteries, however, remains in an early stage despite the fact that the government has started to prepare the implementation of legislative regulations.

Building upon the TVTC Program, in 2012, the government formulated the "Energy-saving and New Energy Vehicles Industry Development Program (2012–2020)." Management policies on traction battery recycling are proclaimed, and the industry is encouraged to develop specialized battery recycling processes including a battery reuse cascade [9].

In 2014, the State Council issued the "Guiding Opinions of the General Office of the State Council on Popularization and Application of Electric Vehicles" to promote and establish a traction battery recycling policy [10]. This includes the investigation of a deposit system for traction batteries as well as the introduction of funds to finance the recycling. To implement this framework, the "Technology Policy on Power Battery Recycling for Electric Vehicles" was released in 2015 [11]. The policy postulates, among others, the following objectives:

- The establishment of a traction battery recycling network
- The industrial implementation of technologically advanced recycling processes should consider recycling efficiency, safety, personnel and environmental protection
- The establishment of a code for traction batteries to build up a traceability system
- The encouragement for battery design standardization to facilitate a reuse cascade of batteries.

2.1.3 USA

In the USA, the regulations on the recycling of LIBs are less stringent than in the EU. On federal level, the Universal Waste Regulations streamline the hazardous waste management for batteries (Title 40 of the Code of Federal Regulations Part 273), which were promulgated by the Environmental Protection Agency in 1995 [12]. The Universal Waste Regulations became effective in all 50 states through the Mercury Containing and Rechargeable Battery Management Act in 1996 [13]. The Universal Waste Regulations do not consider LIBs to be hazardous; therefore, these batteries are excluded [14].

Of 50 states, 30 have additional state battery recycling requirements in effect [15]. Only three states have rechargeable battery waste management regulations that incorporate LIBs: California's Rechargeable Battery Recycling Act of 2006 [16], New York State Rechargeable Battery Law of 2010 [17], and Minnesota Rechargeable Battery and Products Law of 1994 [18]. Legislation focuses on collection, storage, and transportation of waste batteries. Recycling is only encouraged, and no recycling or process efficiency targets are defined. LIBs are banned from landfills in California, New York, and Minnesota, but penalties for noncompliance are not existing or ineffective. Incineration is still possible [14]. Furthermore, the regulations do not explicitly address traction batteries.

2.2 Return Flows of Waste Traction Batteries

An important planning figure for the development of LIB recycling industry is the expected return flow of waste traction batteries. The mid- and long-term forecast for the return flows of waste traction batteries bears major uncertainties. Factors are, among others, the consumer behavior, the development of battery prices, politics regarding laws and subsidies, and battery lifetime [19]. Therefore, only a short-term analysis of the annual return flows of waste bus and car traction batteries is given until 2025. The forecast comprises BEVs and PHEVs and differentiates the three largest car markets: EU, PRC, and USA [20]. HEVs are not considered due to a lack of reliable statistics. Only the BEV and PHEV bus market of the PRC is taken into account due to its global market share of 99.6% [20].

Figure 1 shows the approach to this analysis. In the first part, the number of traction batteries entering the market by year is determined from annual sales of

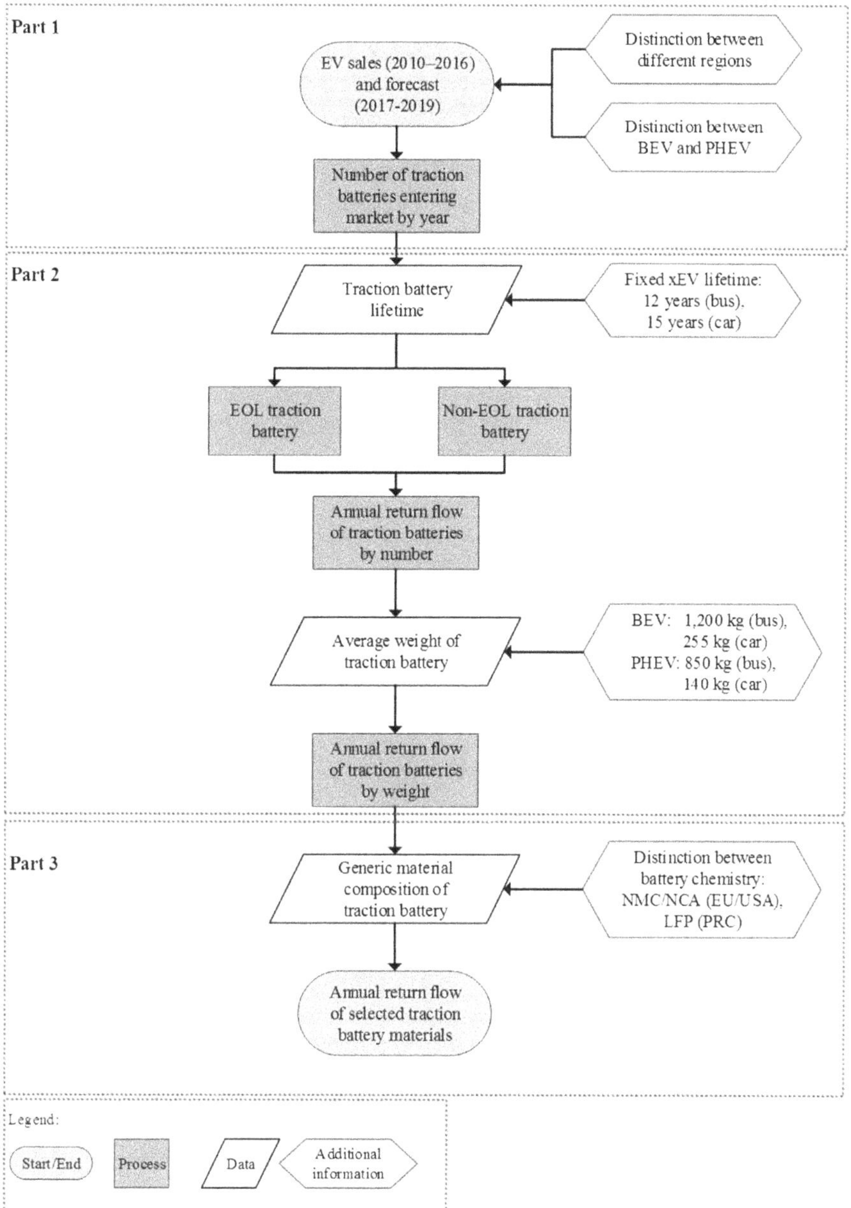

Fig. 1 Applied approach to forecast return flows of traction batteries in this *work*

BEVs and PHEVs. In the next step, the return flow of waste traction batteries is calculated based on battery and vehicle lifetimes. Used xEV exports are neglected as well as the effect of second life applications, e.g., stationary energy storage, which is not expected within the considered period [21]. Finally, the return flow of selected materials is calculated based on generic material compositions of traction batteries. In the following, the results of each part are presented.

2.2.1 Number of Traction Batteries Entering the Market

First, the annual inflow of traction batteries is needed for each considered market. Statistics of BEV sales are available from 2010 to 2016 and PHEV sales from 2011 to 2016 [20]. Based on these numbers, future sales are linearly extrapolated until 2019. The results are shown in Fig. 2.

Between 2010 and 2016, about 2.3 million BEVs and 1.6 million PHEVs (cars) entered the international market. It is estimated that from 2017 to 2019 further 6.9 million BEVs and 4.4 million PHEVs will be sold. The largest car market is the PRC, followed by the EU and the USA. From 2014 to 2015, car sales more than doubled in the PRC, because new subsidies became available for all types of

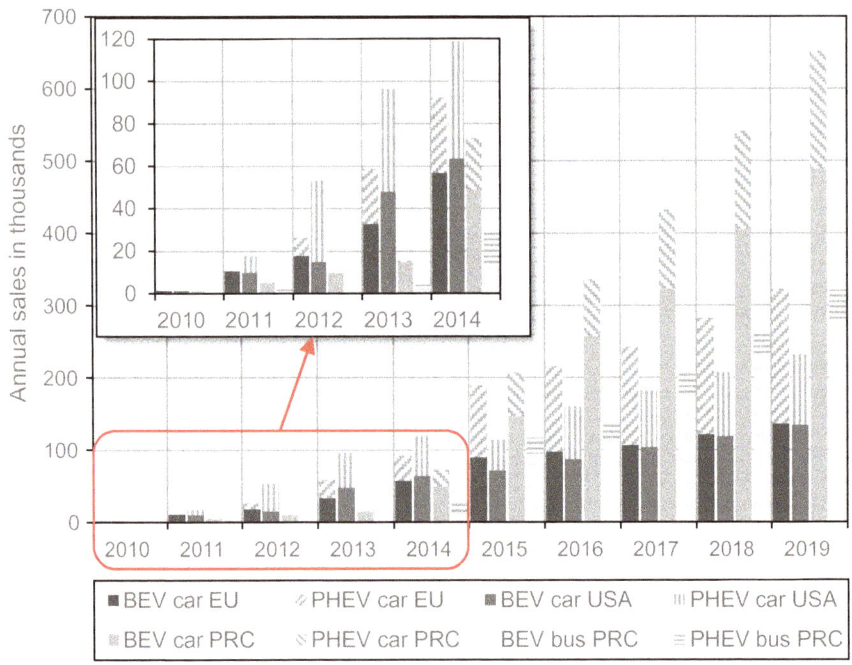

Fig. 2 Number of BEV and PHEV traction batteries entering the market from 2010 to 2019. Car sales based on IEA data from the Global EV Outlook 2017 © OECD/IEA 2017 [20]; bus sales based on [22, 23]

purchasers including private customers [24]. In this country, consumers prefer BEVs, which had a market share of about 76% in 2016. In contrast, BEVs and PHEVs have an almost equal share in the EU and the USA. Bus sales in the PRC are also steadily increasing. In 2018 and 2019, bus sales are expected to be as high as car sales in the EU.

2.2.2 Estimated Return Flows of Waste Traction Batteries

A traction battery is considered to have attained its EOL when it reaches about 80% of its original capacity [25]. The lifetime depends on various aspects such as charging and driving behavior as well as climatic conditions [26, 27]. Therefore, a lifetime distribution is assumed for car traction batteries according to the baseline scenario of [19] (Table 1). In case of bus traction batteries, no reliable data on lifetime distribution are available. Reported fixed lifetimes range between 5 and 12 years [28–30]. Therefore, a fixed lifetime of 8 years is assumed.

The lifetime of cars with internal combustion engine varies between 10 and 20 years, but so far no reliable data for xEVs exist [31]. Therefore, 15 years are assumed for cars and 12 years for buses based on [32, 33], respectively. As the calculations do not consider a second-hand battery market, it is assumed that batteries from EOL vehicles enter recycling regardless of the remaining battery capacity.

Figure 3 shows the results for the estimated return flows of waste traction batteries. Multiplying the return flows by number with average weights of different types of traction batteries (Table 2) results in the return flows by weight.

According to the calculations, a cumulative return flow of about two million waste traction batteries (80% relating to cars) will enter the recycling market until 2025, which corresponds to approximately one million metric tons (43% relating to cars). In this analysis, 64% of all waste traction batteries originate from BEVs.

With regard to the car market, the largest share of waste traction batteries will be allocated to the EU and the USA until 2020. This is a result of comparatively high annual sales of BEVs and PHEVs between 2010 and 2015 in these regions. Until the mid-2020s, the Chinese return flows will exceed the one of the Western regions by far.

In the sector of waste traction batteries from buses, a strong increase of the return flows is expected around the year 2023. Until 2025, the number of waste traction batteries from buses alone might be as large as that of the European BEV and PHEV car market. Due to the higher weights of traction batteries from buses, this corresponds to about 240,400 mt in comparison with 40,800 mt from the European

Table 1 Lifetime distribution of car traction batteries (according to [19])

Years	6	8	10	12
Traction batteries reaching EOL (%)	10	40	40	10

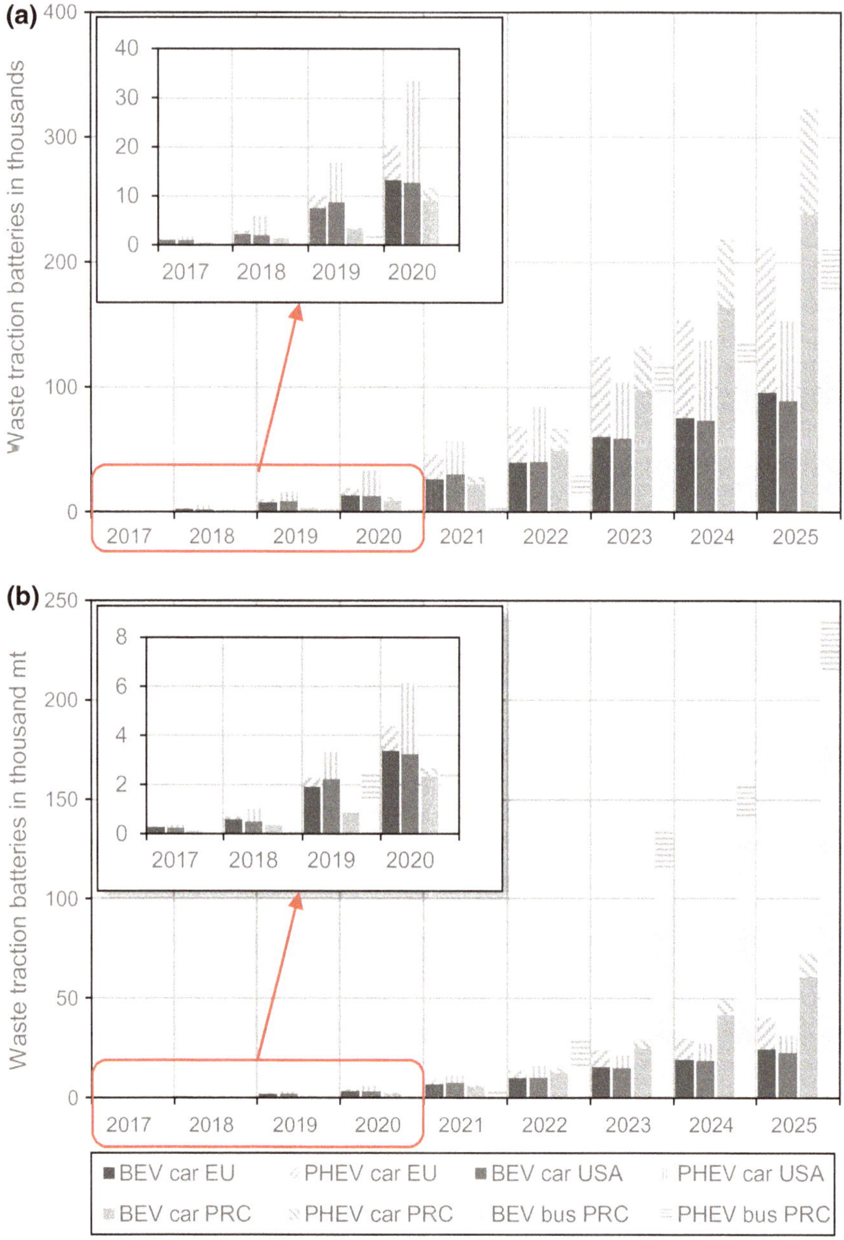

Fig. 3 Annual return flows of waste traction batteries from 2012 to 2025: **a** waste traction batteries by number; **b** waste traction batteries by weight

Table 2 Assumed average battery weights

	Bus [34]	Car [32]
BEV (kg)	1200	255
PHEV (kg)	850	140

car market. The return flows from buses also exceed the return flows from the Chinese car market by a factor of three.

In summary, a moderate increase of the return flows can be expected in the EU and the USA, whereas in the PRC, the implementation of a recycling infrastructure will be a much more demanding task due to the expected exponential growth.

2.2.3 Estimated Return Flows of Selected Battery Materials

In the last section of this forecast, an estimation of selected battery materials of the calculated waste flow is provided. In the EU and USA, the layered oxides lithium nickel manganese cobalt oxide (NMC) and lithium nickel cobalt aluminum oxide (NCA) are the preferred cathode materials [35]. Therefore, a mixture is assumed with a composition of 80% NMC and 20% NCA for these regions. In contrast, lithium iron phosphate (LFP) is currently the predominant cathode material in the PRC for cars and buses [36, 37]. Although a shift toward layered oxides is expected within the next years [38] and some car manufacturers such as BAIC and Zotye have started to use NMC [39], the return flow will be dominated by LFP in the considered period. In Fig. 4, generic material compositions of layered oxides and LFP car traction batteries are given.

About 35–40% of the traction battery accounts for the battery pack and module periphery, which consist of aluminum, cables, electronics, plastics, and steel. Until

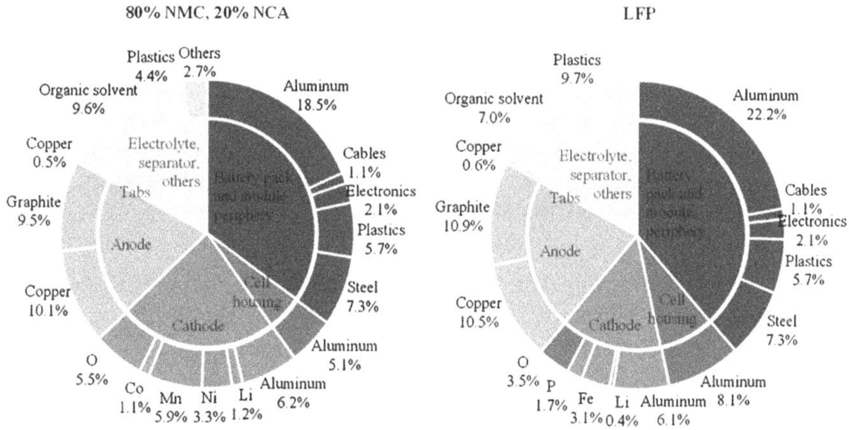

Fig. 4 Generic material composition of traction batteries: left: 80% NMC and 20% NCA, right: LFP (data based on [40])

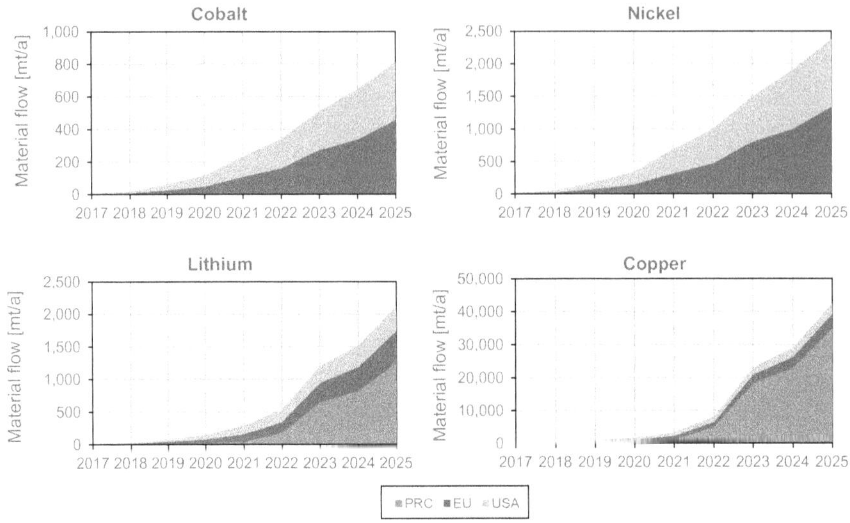

Fig. 5 Estimated recycling potential of cobalt, nickel, lithium, and copper from waste traction battery cells between 2017 and 2025; the scale of the ordinate varies

2025, the globally estimated return flows are about 221,000 mt aluminum, 11,000 mt cables, 21,000 mt electronics, 57,000 mt plastics, and 73,000 mt steel.

The cell material accounts for approximately 60–65% of the total traction battery weight. On the cell level, 55–65% relate to the anode and cathode. Depending on the battery chemistry, the constituent cathode materials vary substantially. From an economic point of view, cobalt, nickel, copper, and lithium are of highest interest. Hence, cell recycling processes focus on the recovery of these elements (see Sect. 3.4).

Figure 5 illustrates the estimated recycling potential of cobalt, nickel, lithium, and copper from waste traction battery cells between 2017 and 2025. Until 2025, the globally estimated return flows are about 2750 mt cobalt, 8100 mt nickel, 6000 mt lithium, and 110,200 mt copper (only from LIB cells).

3 Recycling of Traction Batteries

In order to direct the traction batteries into dedicated recycling processes, they must be extracted from EOL vehicles (see Sect. 3.2). Following the extraction, the batteries are usually disassembled down to module/cell level (see Sect. 3.3). With exception of the battery cells, there are established recycling routes for most components of the battery system. Therefore, the focus of Sect. 3.4 lays on different LIB cell recycling processes, which are currently under development. For all process steps, general safety measures need to be taken into consideration, which are presented in Sect. 3.1. The last two Sects. 3.5 and 3.6 present economic and ecological aspects of traction battery recycling.

3.1 General Safety Measures During Handling

When handling traction batteries, certain safety aspects must be taken into account due to the special risk potential of the batteries caused by their chemical and electrical energy content, high voltage as well as combustible and fluoride-containing electrolyte. Therefore, high safety measures as well as qualified personnel are required. The employment of unqualified personnel like in the manual pretreatment of electronic scrap is not permitted in Germany and several other European countries [3, 41].

The necessary safety requirements must at least cover the following [3, 41]:

- A fire protection concept that addresses the specific hazards of LIBs
- Occupational safety and health measures related to electrical hazards and high weight of batteries
- Separate storage of hazardous and non-hazardous battery scraps as well as storage locations for batteries with increased risk of fire (e.g., from cars involved in an accident)
- Training of employees
- Suction at workplaces.

3.2 Extraction of Traction Batteries

Traction batteries should be extracted from EOL vehicles prior to further processing because, on the one side, they pose a major environmental and safety risk and, on the other side, they contain valuable components and materials [32, 42]. European laws already regulate this and, presumably, similar rules will follow in other international markets. It can therefore be assumed that the batteries will be extracted from EOL vehicles if they enter a regulated recycling process [43, 44].

Due to high weights and voltages, the extraction of traction batteries requires qualified employees and specific tools. Results from different research projects show that the extraction of the battery pack is often time-consuming due to a conflict of interest between the vehicle design with respect to crash safety, center of gravity, space utilization, etc., and the realization of service-friendly installation positions. Additionally, a wide variety of designs exists. Standardization can hardly be expected, but as a recommendation, research projects suggest at least the installation of lifting points (e.g., eyelets, fishplates, or mounting threads) on all future battery packs to enable the disassembly by standard lifting tools [1, 3, 32, 42].

Prior to transportation and further disassembly, a check of the battery condition is highly recommended as several incidents have demonstrated the unpredictable fire risk of damaged traction batteries. Checks should include visual assessment (mechanical damages, signs of heat damage, electrolyte leakage) as well as evaluation of diagnosis data from the battery management system (voltage, state of charge, temperature sensors, etc.). The development of reliable diagnosis tools is therefore of high interest and part of research projects [1, 3, 32].

3.3 Disassembly of Traction Batteries

A comminution of entire traction batteries, for example, in shredders with subsequent sorting of the components, is currently not pursued because the process chain would have to be conducted in an inert atmosphere due to the reactivity of the electrolyte [3, 32, 45, 46]. Considering the benefits of pure material fractions after disassembly, the size of the battery packs, and the possible reuse of some battery components, instead, disassembly of the battery systems is usually conducted, followed by specific processing of the obtained material fractions [32, 42, 47, 48].

As explained in Sect. 2.2.3, the battery cells comprise approximately 60% of the battery weight. The rest of the weight is largely made up by materials for which established recycling routes are available (electronics, aluminum, steel, plastic, etc.). The battery cells must be directed to dedicated recycling processes which are described in the following Sect. 3.4 [3, 32, 45, 46].

Due to the current low return flows and the wide range of designs, only manual disassembly is industrially implemented at present. Robot-assisted disassembly is investigated in several research projects [42, 47, 49, 50]. Currently, most disassembly lines can be considered pilot plants. Umicore AG & Co. KG in Hanau, Germany, operates one of the most advanced disassembly plants. The plant has a daily capacity of 10 mt and can process batteries up to 1000 kg with maximum dimensions of 2.5 m × 1.5 m × 1.0 m [3]. In the following, the general disassembly procedure is briefly described. The steps might be adapted to individual battery pack designs or even left out. Figure 6 summarizes some of the disassembly steps and shows the contained material fractions.

After delivery, the batteries are inspected for damage. If the battery is undamaged, the battery is stored temporarily. If the battery shows signs of damage, special measures have to be taken. Prior to disassembly, the state of charge is measured and, if necessary, the battery is discharged and the energy is fed into the public power grid. After discharging, the casing is opened and the electrical connections between the cell modules and electrical components are interrupted in order to eliminate the risks of high voltage and prepare subsequent removal steps. In the following, all mechanical connections between the system components and the battery casing are undone and electronic components as well as battery modules are extracted. The battery modules are then disassembled, which yields further wiring

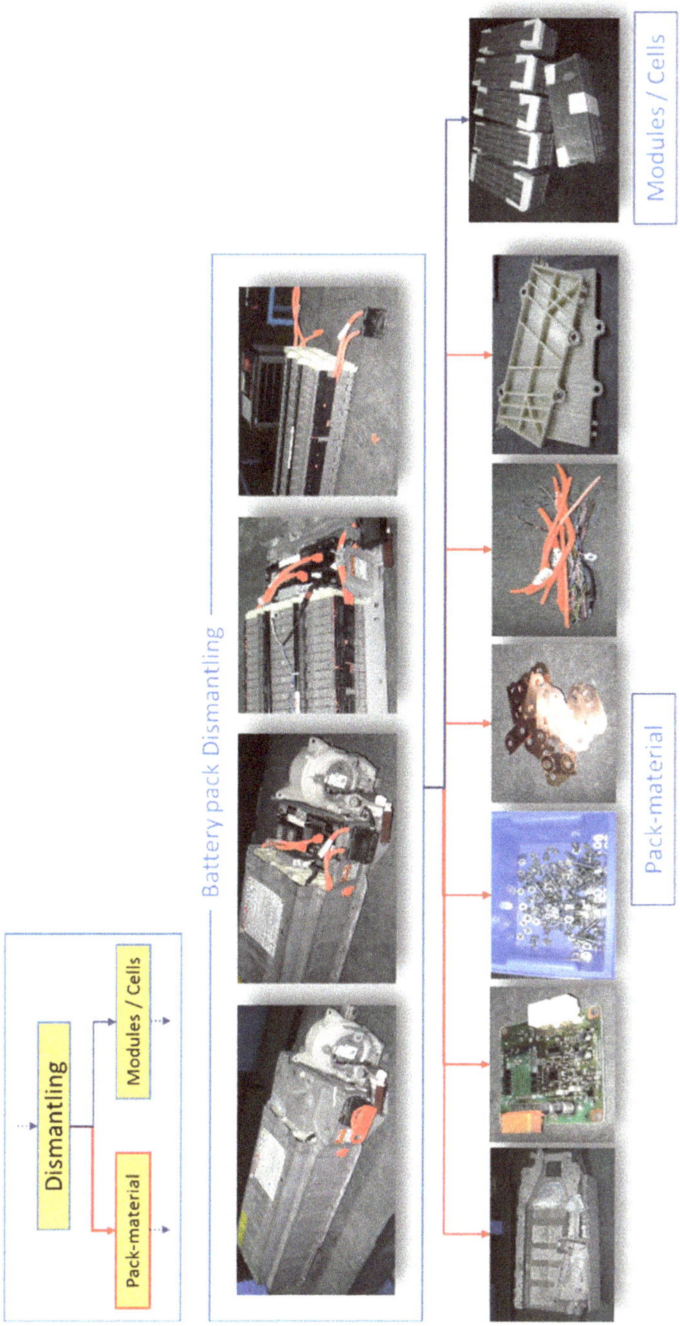

Fig. 6 Disassembly of traction batteries at Umicore Battery Recycling, Hanau, Germany. With permission of Umicore AG & Co. KG

and the battery cells. All fractions are sorted, stored, accumulated, and finally transported to their individual further processing [1, 3, 41].

Little information is available on the disassembly effort of specific battery systems. From the limited information publically available and from oral communication, it can be concluded that, due to high concerns about crash safety, many early generation battery systems are quite difficult to disassemble. However, due to the producer responsibility in Europe and the high costs of manual dismantling, design for recycling becomes more important with increasing sales numbers. Therefore, design for recycling is addressed in several research projects and improvements can be expected in future battery generations [1, 3, 32, 41, 51, 52].

3.4 Cell Recycling Processes

3.4.1 General Considerations Regarding Cell Recycling Processes

Within the last 15 years, several concepts have been developed for the recycling of LIB cells. From an economic point of view, the main drivers for recycling are cobalt (48,680 US$/mt, 99.8%) and nickel (10,124 US$/mt, 99.8%), followed by copper (5624 US$/mt, class A, LME, average prices from September 2016–August 2017) [53]. In the past, lithium was not a main driver due to its comparatively low price, difficult recoverability and its low content [53, 54]. However, since lithium prices have risen from approximately 4000–6000 US$/mt Li_2CO_3 (min. 99–99.5%) between 2007 and 2015 to about 13,000 US$/mt Li_2CO_3 after 2015 [53], lithium recycling gains more attention.

Nevertheless, most processes still concentrate on LIBs containing cobalt and nickel such as NCA and NMC [1, 45, 55, 56]. As LIB producers aim at a further cobalt reduction in layered oxides and other cobalt- and nickel-free cathode materials are introduced, declining levels of valuable materials must be expected, which makes an economical recycling of LIB cells more difficult [1, 45, 57, 58]. According to the extended producer responsibility principle implemented in Directive 2006/66/EC, battery producers or third parties acting on their behalf must finance the net cost of collecting, treating, and recycling in Europe. As producers will include these additional costs in their battery prices, price savings through cheaper raw materials like iron or manganese might be (over-)compensated [1, 56, 57, 59].

Also, regarding the environmental impact, the life cycle has to be analyzed critically. Due to high process complexity and long process chains, the recycling of specific materials does not necessarily offer an ecological advantage in comparison with the primary production of these raw materials (see Sect. 3.6). This is likely the case for raw materials with a comparatively low specific ecologic impact like iron. In this context, also the use of mass-centered recycling quotas as legislative steering instrument must be reviewed [45, 59, 60].

From a technological point of view, the processes vary significantly regarding scope of input batteries, applied process designs, and state of development. Due to the complex input material, all processes are characterized by long process chains containing combinations of mechanical and/or thermal and/or pyrometallurgical and/or hydrometallurgical unit operations [1, 55, 56, 61, 62].

A particular challenge is the high reactivity of the electrolyte, which reacts violently with oxygen, leading to fire hazards [1, 41, 55]. Due to the fluorine content of the electrolyte, the combustion fumes contain hydrogen fluoride and other highly toxic compounds and therefore pose both serious health and environmental risks [1, 55].

To solve these problems, different approaches have been developed including the direct melting of cells (Umicore NV/SA, Belgium), the vacuum thermal deactivation (Accurec Recycling GmbH, Germany), the mechanical treatment under a protective gas atmosphere (Recupyl SAS, France, Duesenfeld GmbH, Germany, and Batrec Industrie AG, Switzerland), and comminution in a liquid solution (Retriev Technologies Inc., USA) [46, 61].

Regarding process design, two contrary philosophies can be identified. On the one side of the spectrum, there are universal processes whose main goal is to process all different LIB types in one process. In addition, these processes are often capable of processing other materials containing similar target metals (e.g., nickel–metal hydride batteries or cobalt- and nickel-containing catalysts). Therefore, these processes tend to focus on the recovery of raw materials, high throughput, and are characterized by early-stage standardization. Usually, in these process routes, the battery cells enter a pyrometallurgical treatment with minimum or without mechanical pretreatment, followed by further treatment of the process outputs, i.e., alloy, slag, and flue dusts, mainly by hydrometallurgical processes to recover the individual metals [1, 51, 54, 56, 62, 63].

On the other side of the spectrum, the processes are individually adapted to each battery chemistry. These processes focus on recovering components for either reuse in LIB production, reuse in other applications, or recycling of materials. This possibly enables higher quality recycling regarding the waste hierarchy in comparison with more universal approaches like those aforementioned. However, it inevitably leads to high requirements regarding collection and sorting as well as knowledge about cell construction and chemistry. Typically, only business-to-business relationships can meet all these requirements. Consequently, these processes are unsuitable for mixed battery inputs, e.g., LIB charges from portable battery sorting [1, 51, 54, 62, 63].

In the following, three dedicated LIB recycling processes will be described. These are the Umicore Battery Recycling Process, the Duesenfeld process, and the Accurec process. They constitute different manifestations of the spectrum and were chosen due to a comparatively good knowledge of the authors regarding these companies. On the one side, the Umicore Battery Recycling Process represents a universal recycling approach, which can process all LIB types and nickel–metal hydride batteries within certain limitations. On the other side, the Duesenfeld process exemplarily shows a specific treatment for each LIB chemistry with a focus on compound recovery. The Accurec process lays in between and was chosen because of an alternative pretreatment and lithium enrichment technique. Additionally, the co-processing of LIB cells in primary and secondary nickel smelters is briefly described as it is currently one of the main recycling routes. For more existing recycling processes, see [41, 46, 54, 61, 63–65]. Figure 7 summarizes the main characteristics of the selected recycling processes, which are presented and discussed in the corresponding sections.

3.4.2 Processing of LIB Cells in Primary and Secondary Nickel Smelters

Besides the processing of LIBs in dedicated recycling processes, currently, they are also co-processed as secondary feedstock material in some extractive cobalt, nickel, and copper process routes. Thus, only the cobalt, nickel, and copper contents of spent LIBs are of interest. All other battery materials are slagged and/or used as energy source and as reducing agents. These processes include pyrometallurgical as well as hydrometallurgical process steps. As these processes were not designed for the recycling of LIBs, they can only co-process a certain share of batteries in the feed for metallurgical and technical reasons. Examples are Glencore plc and Nickelhütte Aue GmbH [63, 66–68]. In view of their limitation to mentioned target metals as well as potentially more stringent recycling regulations in the future, they, supposedly, either become less important within the next years or adapt their processes.

3.4.3 Umicore Battery Recycling Process

Umicore is one of the largest producers of cathode materials and a highly experienced metallurgical company. At the beginning of the 2000s, Umicore made the strategic decision to build up the recycling of nickel- and cobalt-containing batteries as a new business segment and has since invested at least a high double-digit million sum [51, 62, 69]. In comparison with most companies active in the emerging LIB recycling market, Umicore has a significant know-how advantage.

The Umicore Battery Recycling Process is one of the most advanced recycling processes for nickel–metal hydride and LIBs. It is based on a combination of pyrometallurgical and hydrometallurgical unit operations without mechanical pretreatment of battery cells. The process is mainly designed to recover nickel, cobalt, and copper as an alloy, which is further processed by hydrometallurgical methods. Lithium and rare earth elements can, in principle, be recovered from the slag fraction. A first pilot plant was run in Hofors, Sweden, for a couple of years. Based on experiences from this plant, a semi-industrial plant with an annual capacity of 7000 mt has been put into operation in 2011 in Hoboken, Belgium (investment: approximately 25 million €) [1, 45, 51, 61, 70, 71].

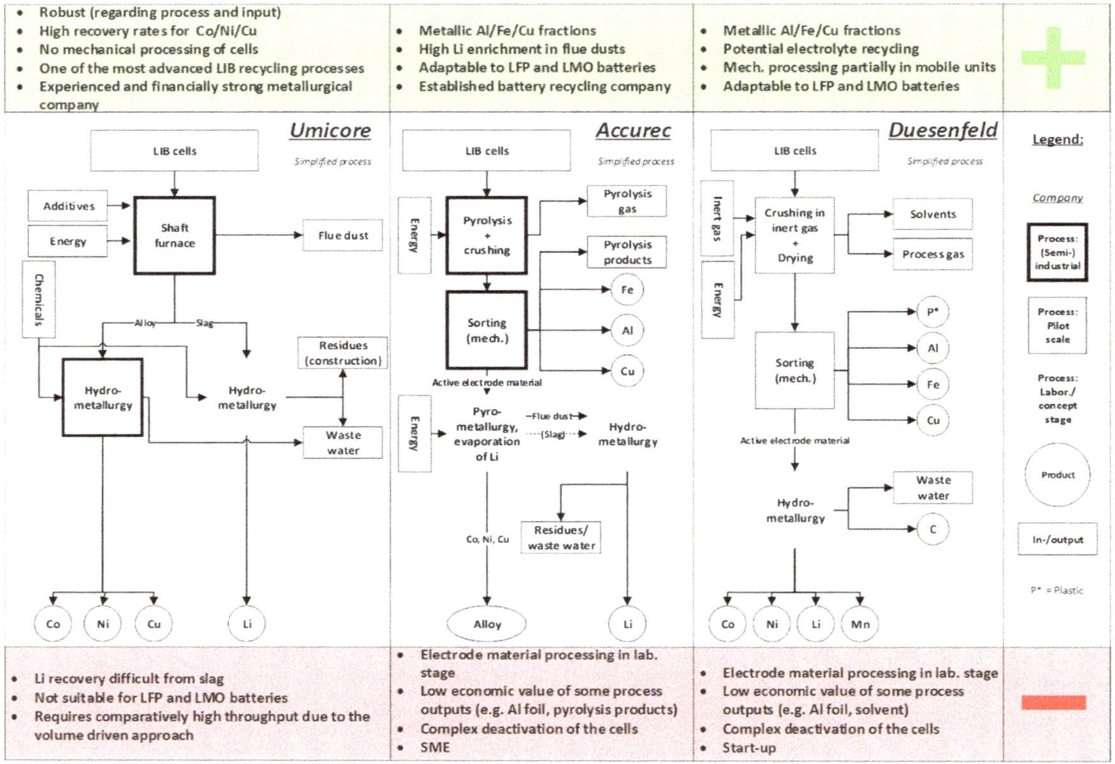

Fig. 7 Comparison of the introduced recycling processes for LIB cells

The process bears similarities to the co-processing in primary and secondary nickel smelters, but addresses the specific requirements of LIB cell recycling. These include, among others, the extensive gas treatment (fluoride capture and destruction of organics), corrosion resistance of materials (suitable refractory lining, fluoride-resistant alloys), concepts for lithium recovery, and tolerance of the process to undesired elements such as phosphate from LFP cells [1, 45, 51, 55, 62, 70].

Figure 8 shows a simplified flow sheet of the Umicore Battery Recycling Process. Without further pretreatment, the battery cells are filled into a shaft furnace with slag formers. The process is conducted with an Al_2O_3–CaO–Li_2O–MgO–MnO–SiO_2 slag system. Metallic aluminum, organics, and graphite of the LIBs are utilized as reductants. The redox potential is kept at a level where cobalt, nickel, and copper are completely transferred to the alloy fraction. Lithium, aluminum, and manganese are concentrated in the slag fraction as well as rare earth elements from nickel–metal hydride batteries. Volatile elements, especially halogens, are concentrated in the fine dusts [1, 3, 51, 70].

The alloy, containing mainly cobalt, nickel, and copper, is further refined by hydrometallurgical treatment including dissolution of the alloy, copper recovery, and nickel/cobalt separation by solvent extraction. After separation and refining, cobalt and nickel can be converted into precursor chemicals for new cathode materials [1, 51].

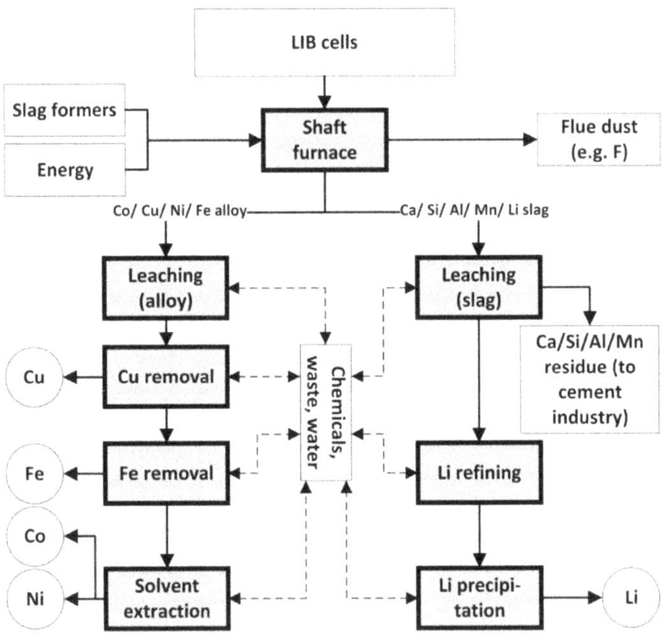

Fig. 8 Simplified flow sheet of the Umicore Battery Recycling Process (adapted from [51, 62])

The slag is completely inert and non-hazardous and is currently used as a construction material [51]. A lithium recovery from the slag was investigated in laboratory scale within the German national research project "Lithium Battery Recycling Initiative (LiBRI)". The investigations revealed that lithium can be leached from the slag using diluted sulfuric acid. After a subsequent precipitation of impurities from the leach liquor, lithium can be recovered as lithium carbonate. For details, see [70]. When processing pure nickel–metal hydride batteries, rare earth elements can also be leached from the slag and further refined to produce pure rare earth oxides [1, 51].

The gas emissions are treated by a gas cleaning installation, which ensures that no harmful dioxins or volatile organic compounds are released. It also captures fluorine and collects all possible dust carryover [1, 51].

The main advantages of the process are its robustness with regard to the battery input, the absence of a mechanical pretreatment of battery cells, and the high recovery rates for cobalt, nickel, and copper [1, 45, 56].

The main disadvantage of the process is that it is not an ideal solution for cobalt- and nickel-free batteries as only copper and, in the future, maybe lithium are recovered from these batteries. Furthermore, metallic aluminum, which is used as a reductant in the pyrometallurgical treatment, ends up as a low-value construction material [1, 45, 51].

3.4.4 Duesenfeld Process

The Duesenfeld process was developed within the German research projects LithoRec I (2009–2011) [32] and LithoRec II (2012–2015) [72]. A start-up company (Lion Engineering GmbH) was founded for the commercial implementation of some project results. In 2017, the Duesenfeld GmbH was founded as a spin-off, which is now continuing the implementation of the recycling activities. The competencies of the company lie in mechanical processing and mobile process implementation.

The original objective of the LithoRec projects was to recover components of the LIB cells for a reuse in battery production. Over the course of the projects, however, this concept was altered gradually because the quality requirements of the LIB industry are very high and difficult to fulfill. The commercial activities of Duesenfeld GmbH therefore now aim at raw material recovery. The recycling concept combines mechanical processing with subsequent hydrometallurgical processing of the active electrode materials. This concept is, in principle, also pursued by the companies: Recupyl SAS (France), Batrec Industrie AG (Switzerland), Retriev Technologies Inc. (formerly Toxco Inc., USA), as well as GEM High-tech Co., Ltd (PRC) and Guangdong Brunp recycling Technology Co., Ltd. (PRC) [45, 46, 58, 64, 65, 73]. In most cases, little is known about the process details. Therefore, the Duesenfeld process is presented as an example.

Figure 9 shows the patented process scheme for the thermo-mechanical recycling process [46, 74]. In the beginning, the battery cells undergo a shredding stage

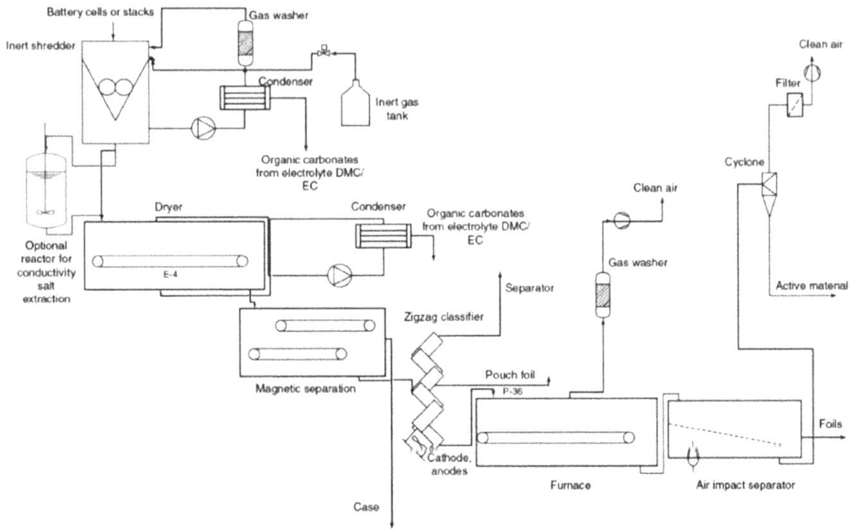

Fig. 9 Flow sheet of the thermo-mechanical recycling process developed within the LithoRec projects. Reproduced with permission from Hanisch [46]

with inert gas and a heating stage for solvent evaporation. These stages are ideally connected to condensers that allow the recycling of pure solvents and additional processing stages for conductive salt recycling (typically $LiPF_6$) [1, 44, 46, 75, 76]. A two-stage shredding improves the sorting significantly [76].

Thereafter, the shredded material enters physical separation which includes magnetic separation for the casings and density separation for copper and aluminum foils [77]. The cathodes and anodes are transferred to the high density fraction and are treated in a furnace (400–600 °C) as well as a special air jet sieve in order to disengage the active material and foil compound and to recover the fine-grained active material via a cyclone or sieve [1, 74, 76].

The process has been demonstrated in a simplified form in a small pilot plant in Braunschweig, Germany. In addition to an electrode active material concentrate, it produces a copper and an aluminum concentrate as well as a plastic and organic solvent fraction. It is planned to carry out the mechanical processing in mobile units (patented [78]) close to the customers. This would prevent the transport of dangerous goods.

The separated active material is to be processed further by hydrometallurgy. For this purpose, however, only laboratory processes have been developed within the framework of the LithoRec projects which are still far from an industrial implementation [76]. According to the final report, these hydrometallurgical processes are capable of recycling up to 85% of the lithium from LFP and up to 95% from NMC cathode materials. However, details are not published [32, 46, 74, 79]. Currently, Duesenfeld GmbH collaborates with the Department of Mineral and Waste Processing of Clausthal University of Technology in the further development of the hydrometallurgical processes and plans to build a pilot plant within the next years.

In contrast to the Umicore Battery Recycling Process, the Duesenfeld process is designed to win back more compounds. While other processes might suffer heavily from the trend that producers aim at reducing nickel and cobalt in LIBs, this process can also produce some other potentially valuable fractions from battery waste, i.e., solvents, conductive salts, and active materials. From the active cathode materials, an economic lithium recovery might be less challenging than from the slags of the Umicore Battery Recycling Process. Because of these advantages, the approach is of high interest for the development of recycling chains for cobalt- and nickel-free cathode materials such as LFP and lithium manganese oxide (LMO) for which the Umicore Battery Recycling Process is less suitable [44, 46, 60, 74, 75, 79, 80].

However, the aluminum concentrate is of relatively low value due to the grain size and shape (foils) and the recovered solvent cannot be recycled to a new electrolyte economically by now. Furthermore, the process is susceptible to changes in the input and requires complex measures (inert gas atmosphere, targeted evaporation) to minimize the risks of the electrolyte.

3.4.5 Accurec Process

Accurec Recycling GmbH is an established medium-sized battery recycling company. The competencies of the company lie mainly in the field of vacuum thermal treatment and mechanical treatment. In the field of metallurgy, the company cooperates with the Institute for Process Metallurgy and Metal Recycling of RWTH Aachen University. The company operates a disassembly line and the mechanical treatment of the pyrolyzed LIB cells in Krefeld, Germany, with an annual capacity of 5000 mt. The pyrolysis of the cells is conducted in collaboration with the Currenta GmbH & Co. OHG, Leverkusen, Germany [81].

The Accurec process has been developed in different research projects such as EcoBatRec, CloseWEEE, and "Rückgewinnung der Wertstoffe aus zukünftigen Li-Ion-basierten Automobil-Batterien" (Recovery of the valuable substances from future Li-Ion-based traction batteries) [82, 83]. As shown in Fig. 10, the developed process can roughly be divided into the three steps: pyrolysis of the organics, mechanical treatment of the cells, and metallurgical processing of the cathode material. For the last step, several process variants beyond the one depicted can be found in the literature [66, 68, 82, 84–86].

Regarding the target fractions prior to the metallurgical treatment, the Accurec process resembles the Duesenfeld process. The pyrolysis of the LIB cells produces a hydrocarbon mixture and, above all, destroys the electrolyte so that the subsequent mechanical treatment does not have to be conducted under protective atmosphere [66, 68]. Furthermore, fluorine is partly removed. Currently, the pyrolysis is carried out in a rotary kiln and the pyrolysis gas is used for thermal energy production [81].

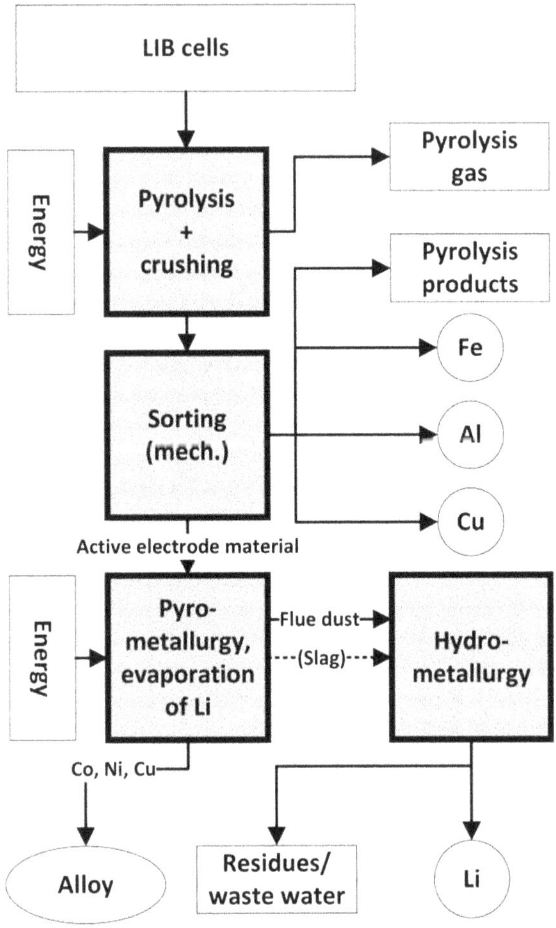

Fig. 10 Simplified process flow sheet for the recycling of LIBs of the Accurec Recycling GmbH, adopted and combined from [66, 68, 82]

The LIB cells are then comminuted, and different (metal) fractions are produced by sorting (density, magnetic, and eddy current separation) including steel, aluminum from casings, aluminum and copper foils, and active electrode materials [82].

For the treatment of the active cathode material, different approaches have been studied [66, 68, 82, 83]. These are based in most cases on a pyrometallurgical treatment with subsequent hydrometallurgical processing of the products (alloy, slag, flue dust). A special process development that distinguishes the Accurec process from the other processes is the selective transfer of the lithium into the flue dust by evaporation in a pyrometallurgical process. Only a few details are known about this process, but an important part of the process seems to be the proper slag composition as well as process technology [66, 82–84].

The great advantage of the lithium enrichment in the dust is that the subsequent hydrometallurgical recovery of the lithium might be simpler and more effective than the recovery from the slag phase in other processes. However, on the other hand, fluorine and other contaminants are also likely to be concentrated in the dust, which could interfere with further treatment. Georgi-Maschler et al. [66] report lithium carbonate products in demonstration runs that are suitable for high-quality glass production at high yields with sulfuric acid leaching and precipitation. Also, in contrast to the Umicore Battery Recycling Process, the produced Co–Ni alloy can be directly used as a master alloy and does not require additional hydrometallurgical treatment.

Both, the pyrolysis and the mechanical treatment, have been implemented and tested in the form of a demonstration plant at Accurec. According to the knowledge of the authors, the hydro- and pyrometallurgical experiments have hitherto been carried out only in laboratory or small pilot scale [82, 86].

3.5 Economic Assessment of the Contained Metal Values

Little information is publically available regarding the costs of LIB recycling. Furthermore, most processes are under development and no large-scale plants are in operation. Therefore, only an economic assessment of the contained metal values of the battery infrastructure and cells is given in this section.

Currently, only manual disassembly is employed in battery recycling (see Sect. 3.3). Therefore, the materials are recovered in separate fractions with high yield and in good quality. Based on scrap prices of September 2017 given in Table 3, the specific revenues of the battery pack scrap are about 225 €/mt batteries (without possible revenues from LIB cells). Approximately 50% of the specific revenues relate to the aluminum fraction. Therefore, a substitution of this material, e.g., by carbon fiber-reinforced polymers, could have a major impact on the economics.

An estimation of possible revenues from cell recycling is difficult because most processes are in an early stage of development and, so far, no data from industry have been published regarding yields and process costs. As presented in Sect. 3.4, current process developments aim at the recovery of aluminum, cobalt, copper,

Table 3 Scrap prices and specific revenues from battery pack disassembly

Material	Quality	Price (€/mt)	Source	Specific revenues (€/mt batteries)	
				NMC/NCA	LFP
Aluminum	Scrap with max. 5% adhesions	600	[87]	111	133
Cables	Copper cable scrap min. 38% copper	1400	[87]	15	15
Electronics	Printed circuit boards category II A	1530	[88]	32	32
Steel	Stainless steel scrap (V2A)	750	[87]	55	55
			Total	213	235

Scrap prices of September 2017

Table 4 Specific metal values of LIB cells based on stock market prices (average prices from September 2016 to August 2017) [53]

Material	Quality	Price (€/mt)	Specific metal values (€/mt cells)	
			NMC/NCA	LFP
Aluminum	High-grade primary	1651	316	381
Cobalt	High grade, min. 99.8%	43,979	820	–
Copper	Grade A	5081	913	916
Lithium carbonate	Min. 99–99.5%	10,399	1471	443
Nickel	Min. 99.8%	9146	512	–
		Total	4031	1739

lithium, and nickel. The comparison of the specific metal values based on stock market prices (Table 4) shows a significant difference between layered oxides and LFP. Considering the complexity of the cell recycling processes, an economic recycling of LFP is a very challenging task.

3.6 *Life Cycle Assessment*

The described process developments were accompanied by LCA to determine the ecological impact of the recycling processes in comparison with the primary production. In this section, selected results of the latest LCAs are presented, which were conducted within the EcoBatRec (Accurec process) [89] and LithoRec II (Duesenfeld process) [90] research projects. For the Umicore Battery Recycling Process, no LCA results are presented as the last LCA was conducted in 2011 [80], which does not represent the current state of development.

Both LCAs were conducted according to ISO 14040/44 including a critical review by an independent external expert. The following impact categories were assessed in both LCAs: global warming potential (GWP), acidification potential (AP), elementary abiotic resource depletion potential (ADP_{elem}), eutrophication potential (EP), photooxidation potential (POCP), and cumulated energy demand (CED).

Figure 11 shows the system boundaries of the LCA for the Accurec process. The LCA is based on test runs of a demonstration plant and results of laboratory experiments. The functional unit was defined as recycling of one metric ton batteries (definition of representative battery composition of the cathode type NMC). The Accurec process aims to yield the following products:

- Suitable starting material for cathode production (Co/Ni/Mn/Li)
- Further valuable products (Cu, Al, stainless steel, plastics, etc.).

In Fig. 12, the LCA results of the Accurec process for the environmental impact category GWP are shown. The net result (black column) demonstrates a decisive

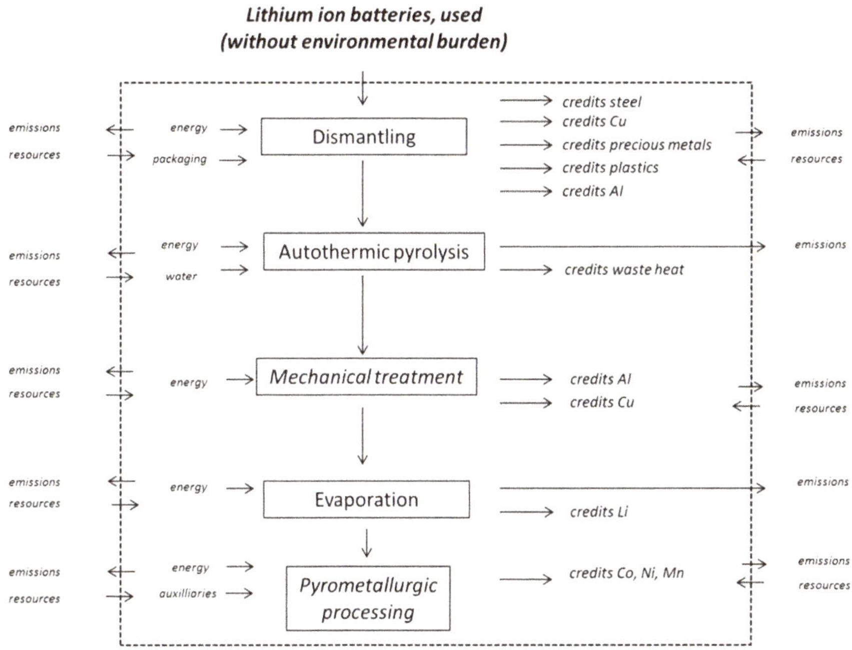

Fig. 11 Accurec process: system boundaries of the LCA [89]

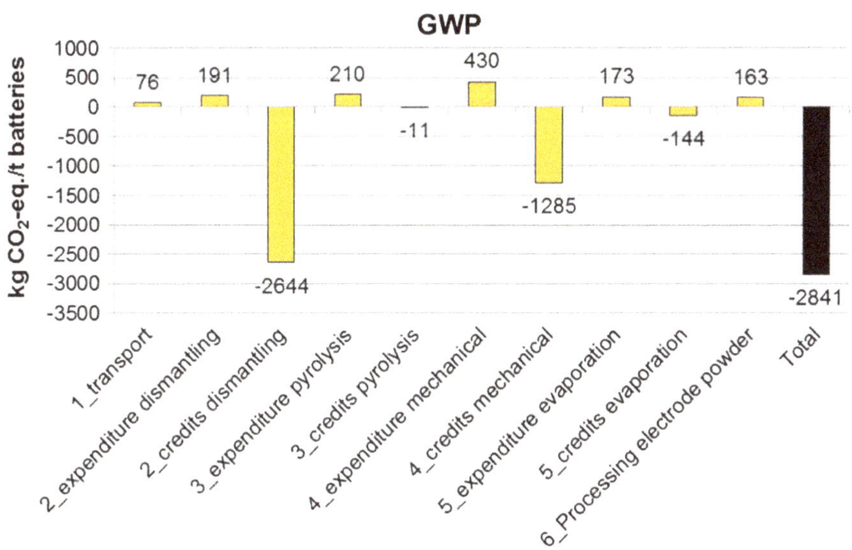

Fig. 12 Accurec process: LCA results for the impact category global warming potential (GWP) [89]

advantage of this recycling process. The main credits derive from dismantling of the battery packs (aluminum, precious metals, copper, steel, and plastics), followed by the mechanical processing of the cells (aluminum, copper). The metallurgical processing shows a slight disadvantage compared to the primary production. The environmental advantages of the Accurec process are also confirmed for the other impact categories.

Figure 13 shows the system boundaries of the LCA for the Duesenfeld process. The LCA is based on pilot plant and laboratory-scale results. The functional unit was again defined as recycling of one metric ton batteries (definition of representative battery composition of the cathode type NMC). The LithoRec II Process aims to yield the following products:

- Suitable starting material for cathode production (Co/Ni/$MnSO_4$ in mixed dilution, LiOH)
- Further valuable products (Cu, Al, stainless steel, plastics, etc.).

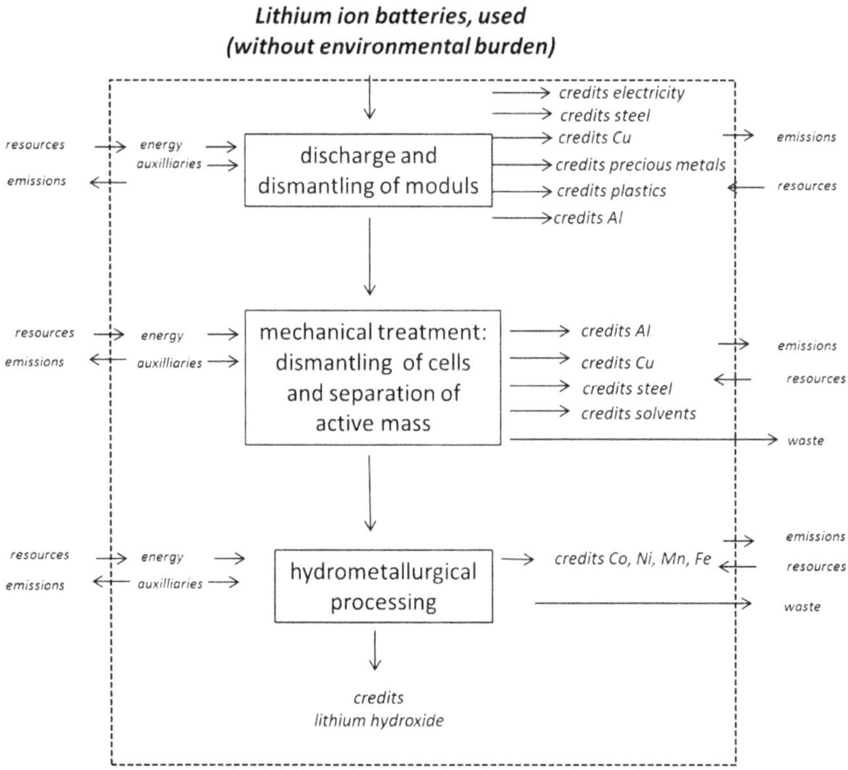

Fig. 13 Duesenfeld process: system boundaries of LCA [90]

In Fig. 14, the LCA results of the Duesenfeld process for the environmental impact category GWP are shown. The results resemble those of the Accurec process. The net result shows a significant excess of the credits compared to the burdens for this recycling route. The credits derive again mainly from dismantling and mechanical processing, whereas the hydrometallurgical processing shows a disadvantage compared to the primary production. The results of the other impact categories point in the same direction.

When evaluating the results of both LCAs, one has to consider that especially the metallurgical data are based on laboratory experiments. Therefore, data uncertainties are high and considerable improvements can be expected during the further development of the processes. Furthermore, lithium carbonate is currently mainly produced from high-grade brines. In the production process, lithium is enriched by solar evaporation [91, 92]. Hence, the ecological burden of this process is comparatively low. However, due to the sharp rise in lithium demand, deposits showing less favorable production conditions are currently put into operation (low-grade brines, hard rock deposits containing spodumene) [92, 93]. As a result, that ecological burden of primary lithium will increase and recycling likely becomes more favorable from an ecological point of view.

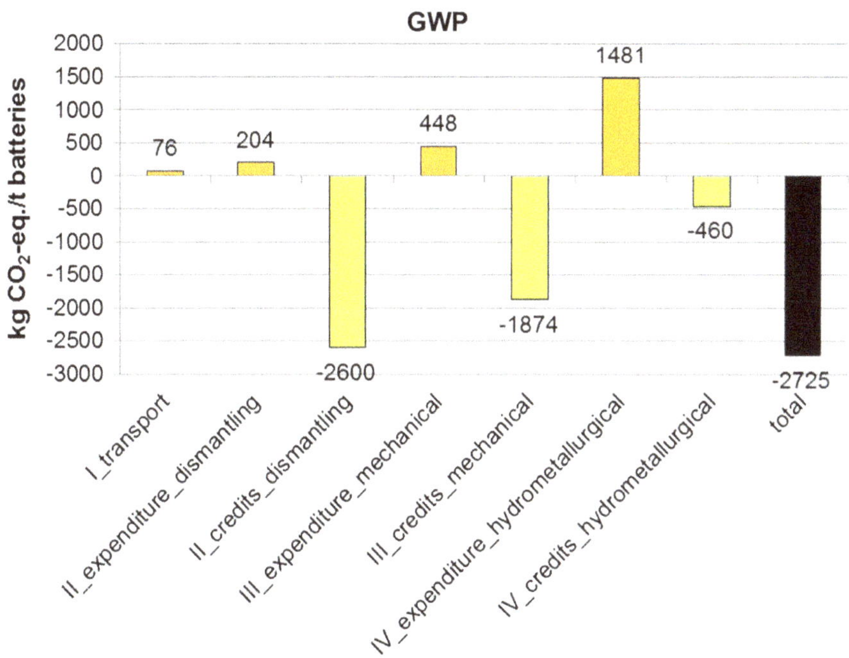

Fig. 14 Duesenfeld process: LCA results for the impact category global warming potential (GWP) [90]

4 Conclusions and Outlook

In the past decade, and parallel to the rise of xEVs, a considerable number of national and international efforts have been initiated by the private and public sectors in LIB recycling. Although general trends can be identified in different fields relevant to the recycling of traction batteries, high dynamics and new developments in this young market lead to unavoidable uncertainties.

In the field of legislation, currently, no specific regulations for traction batteries are implemented in the main markets: EU, PRC, and USA. Furthermore, the regulations on waste battery treatment in general differ significantly. Whereas some legislations still allow landfilling, others impose strict recycling regulations and requirements. At present, the EU regulations seem to be most suitable to ensure an organized recycling of batteries, although specific regulations regarding xEVs and traction batteries are missing. Key elements include a landfill ban, the extended producer's responsibility principle, and recycling efficiencies. In view of the increasing return flows, the EU and the PRC are currently working on specific regulations. Other regions are expected to follow.

The xEV and electric bus markets are characterized by high growth rates. Annual global sales of PHEVs and BEVs already exceeded 0.8 million in 2016 and are expected to reach about one million in 2017. This will inevitably lead to high return flows of traction batteries in the next decade. Additionally, growing return flows from other applications such as electric two-wheeled vehicles have to be expected.

The biggest market by far is the PRC. Besides cars and in contrast to the EU and USA, also electric buses play a major role in the PRC. Another important difference is the dominance of LFP in contrast to layered oxides in the Western markets. Therefore, the requirements especially for battery cell recycling processes differ considerably.

Regarding the technical realization of the recycling, all recycling routes start with the extraction of the battery from the EOL vehicle, followed by disassembly down to module/cell level. This allows to direct the different material fractions of the battery infrastructure, about 40% of the battery weight, to dedicated recycling processes. At present, these steps are conducted manually. Against the background of sharply increasing return numbers, safety issues, and high labor costs, the development of automated extraction and disassembly technologies is of high interest. Prerequisites are design for recycling and standardization in LIB construction.

Today, layered oxides containing LIB cells are typically co-processed in primary and secondary cobalt–nickel plants where cobalt, nickel, and copper are recovered. However, specific recycling processes for LIB will raise in importance due to increasing return flows, more stringent regulations, decreasing cobalt and nickel contents, growing market share of LFP, and increasing interest in lithium recovery. This is especially true for LFP because of the low specific metal value.

In principle, two recycling approaches have been developed and partially implemented. The first approach resembles the above-mentioned co-processing with pyrometallurgy and subsequent hydrometallurgy. Although the process has been adapted to and optimized for the processing of LIB cells, its focus still lays on the recovery of cobalt, nickel, and copper. Lithium recovery from the slag phase is possible but has not been industrially implemented so far. The second approach comprises a mechanical processing of the cells, followed by a metallurgical processing of the separated active electrode materials. The mechanical processing requires measures to control the hazard potential, but yields different material fractions from the cells such as aluminum casings, electrode foils, etc. Another advantage of this concept is the theoretical possibility to recover compounds for reuse and the adaptability of the metallurgical operations, which is especially relevant for LFP. However, due to economic reasons, most developments aim at a raw material recovery with focus on cobalt, nickel, and lithium from layered oxides.

Evidently, there are solutions for the recovery of cobalt, nickel, and copper, which are partially implemented industrially. Recovery concepts for lithium exist, but their industrial implementation has not been profitable so far. However, considering increasing return flows and lithium prices, lithium recovery will probably become lucrative and therefore implemented. In the case of LFP, the specific metal value is particularly low. Hence, political steering seems to be required to ensure an adequate recycling.

First LCA results for different battery recycling routes demonstrate promising overall results, especially for the recycling of NMC batteries. Also, the results for the recycling of LFP batteries reveal positive overall results due to the high credits for the battery casing, cables, circuit boards, etc. However, especially the further development of the metallurgical processes should be accompanied closely by LCA as these processes currently show in some impact categories unfavorable results in comparison with the primary production.

References

1. T. Elwert, D. Goldmann, F. Römer, M. Buchert, C. Merz, D. Schueler, J. Sutter, Recycling **1**, 25 (2015)
2. G. Pistoia (ed.), *Lithium-Ion Batteries: Advances and Applications*, 1st edn. (Elsevier, Amsterdam, 2014)
3. F. Treffer, Entwicklung eines realisierbaren Recyclingkonzeptes für die Hochleistungsbatterien zukünftiger Elektrofahrzeuge. Gemeinsamer Abschlussbericht des Konsortiums, 2011, http://edok01.tib.uni-hannover.de/edoks/e01fb12/727409611.pdf. Accessed 30 Aug 2017
4. European Parliament and the Council of the European Union, Directive 2000/53/EC of the European Parliament and the Council of the European Union of 18 September 2000 on end-of life vehicles, 2000, http://eur-lex.europa.eu/resource.html?uri=cellar:02fa83cf-bf28-4afc-8f9f-eb201bd61813.0005.02/DOC_1&format=PDF. Accessed 26 July 2017
5. European Parliament and the Council of the European Union, Directive 2006/66/EC of the European Parliament and of the Council of 6 September 2006 on batteries and accumulators

and waste batteries and accumulators and repealing Directive 91/157/EEC, 2006, http://eur-lex.europa.eu/LexUriServ/LexUriServ.do?uri=OJ:L:2006:266:0001:0014:en:PDF. Accessed 26 July 2017
6. Öko-Institut e.V., Strategien für die nachhaltige Rohstoffversorgung der Elektromobilität. Synthesepapier zum Rohstoffbedarf für Batterien und Brennstoffzellen. Studie im Auftrag von Agora Verkehrswende, 2017
7. H. Gong, M.Q. Wang, H. Wang, Mitig. Adapt. Strat. Glob. Change **18**, 207 (2013)
8. Ministry of Science and Technology of the People's Republic of China, Notice on Implementing Energy-saving and New Energy Vehicles Pilot Project, 2009, http://www.most.gov.cn/fggw/zfwj/zfwj2009/200902/t20090224_67588.htm. Accessed 31 July 2017
9. The Central People's Government of the People's Republic of China, Energy-saving and New Energy Vehicles Industry Development Program (2012–2020), 2012, http://www.gov.cn/zwgk/2012-07/09/content_2179032.htm. Accessed 31 July 2017
10. State Council General Office, Guiding Opinions of the General Office of the State Council on Popularization and Application of Electric Vehicles, 2014, http://www.gov.cn/zhengce/content/2014-07/21/content_8936.htm. Accessed 31 July 2017
11. National Development and Reform Commission, Technology Policy on Power Battery Recycling for Electric Vehicles, 2015, http://www.ndrc.gov.cn/gzdt/201601/W020160128605285191658.pdf. Accessed 31 July 2017
12. U. S. Government Publishing Office, Electronic Code of Federal Regulations Part 273, 2017, https://www.ecfr.gov/cgi-bin/text-idx?SID=0501d91ec562faafa833c60c2404d806&mc=true&node=pt40.27.273&rgn=div5. Accessed 30 Aug 2017
13. United States Environmental Protection Agency, Implementation of the Mercury-Containing and Rechargeable Battery Management Act, 1997, http://www.call2recycle.org/wp-content/uploads/ImplementationoftheMercury-ContainingandRechargeableBatteryManagementAct.pdf. Accessed 30 Aug 2017
14. K. Richa, C.W. Babbitt, G. Gaustad, J. Ind. Ecol. **21**, 715 (2017)
15. call2recycle, Recycling Laws By State, http://www.call2recycle.org/recycling-laws-by-state/. Accessed 30 Aug 2017
16. United States California, AB 1125: Rechargeable Battery Recycling Act, 2006, http://www.leginfo.ca.gov/pub/05-06/bill/asm/ab_1101-1150/ab_1125_bill_20051006_chaptered.pdf. Accessed 21 Sept 2017
17. United States New York, Rechargeable Battery Recycling Law, 2010, http://www.dec.ny.gov/docs/materials_minerals_pdf/batterylaw.pdf. Accessed 21 Sept 2017
18. The Office of the Revisor of Statutes, 115A.9157: Rechargeable batteries and products, https://www.revisor.mn.gov/statutes/?id=115A.9157. Accessed 21 Sept 2017
19. K. Richa, C.W. Babbitt, G. Gaustad, X. Wang, Resour. Conserv. Recycl. **83**, 63 (2014)
20. International Energy Agency, Global EV Outlook 2017, 2017, https://www.iea.org/publications/freepublications/publication/GlobalEVOutlook2017.pdf. Accessed 18 Aug 2017
21. A. Thielmann, A. Sauer, M. Wietschel, Gesamt-Roadmap: Stationäre Energiespeicher 2030, 2015, http://www.isi.fraunhofer.de/isi-wAssets/docs/t/de/publikationen/GRM-SES.pdf. Accessed 21 Sept 2017
22. Clean Technica, China: 100% Electric Bus Sales grew to ~115,700 in 2016, 2017, https://cleantechnica.com/2017/02/03/china-100-electric-bus-sales-grew-115700-2016/. Accessed 29 Aug 2017
23. Y. Zhou, M. Wang, H. Hao, L. Johnson, H. Wang, Mitig. Adapt. Strat. Glob. Change **20**, 777 (2015)
24. S. Ou, Z. Lin, Z. Wu, J. Zheng, R. Lyu, S.V. Przesmitzki, X. He, A Study of China's Explosive Growth in the Plug-in Electric Vehicle Market, 2017, https://www.researchgate.net/publication/313083799_A_Study_of_China%27s_Explosive_Growth_in_the_Plug-in_Electric_Vehicle_Market. Accessed 30 Aug 2017
25. R. Spotnitz, J. Power Sources **113**, 72 (2003)
26. T. Yuksel, S. Litster, V. Viswanathan, J.J. Michalek, J. Power Sources **338**, 49 (2017)
27. S. Pelletier, O. Jabali, G. Laporte, M. Veneroni, Transport. Res. B-Met. **103**, 158 (2017)

28. K. Nørregaard, B. Johnsen, C. Hedegaard Gravesen, Battery Degradation in Electric Buses, 2016, https://www.trafikstyrelsen.dk/~/media/Dokumenter/06%20Kollektiv%20trafik/Forsogsordningen/2013/Elbusser/Battery%20degradation%20in%20electric%20buses%20-%20final.pdf. Accessed 29 Aug 2017
29. BYD, World's First Mass Produced 40ft Long Range Battery-electric Bus, http://www.byd.com/na/ebus/ebus.html. Accessed 29 Aug 2017
30. YUTONG, New Energy Buses & Solutions, http://en.yutong.com/z/newenergybus/. Accessed 29 Aug 2017
31. A. Thielmann, A. Sauer, M. Wietschel, Produkt-Roadmap: Energiespeicher für die Elektromobilität 2030, 2015, http://www.isi.fraunhofer.de/isi-wAssets/docs/t/de/publikationen/PRM-ESEM.pdf. Accessed 21 Sept 2017
32. A. Kwade, G. Bärwaldt, *LithoRec: Recycling von Lithium-Ionen-Batterien*, 1st edn. (Cuvillier, Göttingen, 2012)
33. S. Rothgang, M. Rogge, J. Becker, D. Sauer, Energies **8**, 6715 (2015)
34. S. Sturn, Die E-Buslinie 63 in Mannheim, 2016, https://www.swr.de/swr1/rp/programm/e-buslinie-63-in-mannheim/-/id=446640/did=17431390/nid=446640/1uxa3cu/index.html. Accessed 29 Aug 2017
35. E. Maiser, S. Michaelis, D. Müller, A. Kampker, C. Deutskens, H. Heimes, N. Sarovic, N. Klusmann, A. Thielmann, A. Sauer, Roadmap Batterie-Produktionsmittel 2030, 2014, https://prod.vdma.org/documents/105963/5521486/Roadmap%20Batterie-Produktionsmittel%202030/f422a080-dfaf-4620-acc3-ed747d33fae1. Accessed 21 Aug 2017
36. G. Masiero, M.H. Ogasavara, A.C. Jussani, M.L. Risso, RAI Revista de Administração e Inovação **13**, 3 (2016)
37. F. Thorpe, Global Battery Electric Bus Market, 2016, http://www.cte.tv/wp-content/uploads/2016/12/5_Thorpe.pdf. Accessed 30 Aug 2017
38. Research in China, Global and China Lithium Iron Phosphate ($LiFePO_4$) Material and Battery Industry Report, 2015-2018, http://www.researchinchina.com/Htmls/Report/2015/10133.html. Accessed 21 Aug 2017
39. China Auto Web, Chinese-Made Electric Cars, http://chinaautoweb.com/electric-cars/. Accessed 21 Aug 2017
40. M. Buchert, J. Sutter, Ökobilanzen zum Recyclingverfahren LithoRec II für Lithium-Ionen-Batterien, 2015, http://www.erneuerbar-mobil.de/sites/default/files/publications/endbericht-kobilanzen-zum-recyclingverfahren-lithorec-ii-fuer-lithium-ionen-batterien_1.pdf. Accessed 18 Aug 2017
41. K. Wegener, S. Andrew, A. Raatz, K. Dröder, C. Herrmann, Procedia CIRP **23**, 155 (2014)
42. K. Wegener, W.H. Chen, F. Dietrich, K. Dröder, S. Kara, Procedia CIRP **29**, 716 (2015)
43. R. Holzhauer, *Recycling und Rohstoffe* (TK Verlag Karl Thomé-Kozmiensky, Neuruppin, 2014), p. 149
44. C. Hoyer, *Strategische Planung des Recyclings von Lithium-Ionen-Batterien aus Elektrofahrzeugen in Deutschland* (Springer Fachmedien Wiesbaden, Wiesbaden, 2015)
45. L. Gaines, J. Dunn, in *Lithium-Ion Batteries: Advances and Applications*, 1st edn., ed. by G. Pistoia (Elsevier, Amsterdam, 2014), p. 483
46. C. Hanisch, J. Diekmann, A. Stieger, W. Haselrieder, A. Kwade, in *Handbook of Clean Energy Systems*, ed. by J. Yan (John Wiley & Sons Ltd, Chichester, 2015), p. 2865
47. D.A. Dornfeld, B.S. Linke (eds.), *Leveraging Technology for a Sustainable World* (Springer, Berlin Heidelberg, Berlin, Heidelberg, 2012)
48. C. Hoyer, K. Kieckhäfer, T.S. Spengler, in *Glocalized Solutions for Sustainability in Manufacturing*, ed. by J. Hesselbach, C. Herrmann (Springer, Berlin Heidelberg, 2011), p. 79
49. P. Ay, J. Markowski, Pempel, Harry, M. Müller, *Recycling und Rohstoffe* (2012), p. 443
50. N. Natkunarajah, in *3. Symposium Rohstoffeffizienz und Rohstoffinnovationen: 05/06. Februar 2014, Neues Museum Nürnberg*, ed. by U. Teipel (Fraunhofer-Verl, Stuttgart, 2014), p. 475
51. J. Tytgat, F. Treffer, *Proceedings of the Belgian Platform on Electrical Vehicles*, 31 Mar 2011
52. B. Friedrich, Rückgewinnung der Wertstoffe aus zukünftigen Li-Ion-basierten Automobil-Batterien: Lithium-Ionen-Batterie (LIB2015): Abschlussbericht zum

Verbundvorhaben im Rahmen des FuE-Programms Innovationsallianz "Lithium Ionen Batterie", 2013, http://edok01.tib.uni-hannover.de/edoks/e01fb13/769010806.pdf. Accessed 14 June 2016
53. Deutsche Rohstoffagentur DERA, Preismonitor August 2017, 2017, https://www.deutsche-rohstoffagentur.de/DE/Themen/Min_rohstoffe/Produkte/Preisliste/cpl_17_08.pdf?__blob=publicationFile&v=2. Accessed 26 Sept 2017
54. J. Ordoñez, E.J. Gago, A. Girard, Renew. Sustain. Energy Rev. **60**, 195 (2016)
55. A. Sonoc, J. Jeswiet, V.K. Soo, Procedia CIRP **29**, 752 (2015)
56. K. Kotaich, S.E. Sloop, *Encyclopedia of Electrochemical Power Sources* (Elsevier, 2009), p. 188
57. X. Wang, G. Gaustad, C.W. Babbitt, K. Richa, Resour. Conserv. Recy. **83**, 53 (2014)
58. L. Gaines, Sustain. Mater. Technol. **1–2**, 2 (2014)
59. C. Hoyer, K. Kieckhäfer, T.S. Spengler, in *Re-engineering Manufacturing for Sustainability*, ed. by A.Y.C. Nee, B. Song, S.-K. Ong (Springer Singapore, Singapore, 2013), p. 543
60. M. Buchert, W. Jenseits, C. Merz, D. Schüler, Ökobilanzen zum Recyclingverfahren LithoRec II für Lithium-Ionen-Batterien, 2011, http://www.erneuerbar-mobil.de/de/projekte/foerderprojekte-aus-dem-konjunkturpaket-ii-2009-2011/batterierecycling/abschlussberichte-recycling/lca-analyse-lithorec.pdf. Accessed 15 May 2016
61. A. Vezzini, in *Lithium-Ion Batteries: Advances and Applications*, 1st edn., ed. by G. Pistoia (Elsevier, Amsterdam, 2014), p. 529
62. J. Tytgat, Li-Ion and NiMH Battery Recycling at Umicore: Strategic Choices (Brussels), 19th June, 2009
63. B. Swain, Sep. Purif. Technol. **172**, 388 (2017)
64. X. Zeng, J. Li, L. Liu, Renew. Sustain. Energy Rev. **52**, 1759 (2015)
65. F. Gu, J. Guo, X. Yao, P.A. Summers, S.D. Widijatmoko, P. Hall, J. Clean. Prod. **161**, 765 (2017)
66. T. Georgi-Maschler, B. Friedrich, R. Weyhe, H. Heegn, M. Rutz, J. Power Sources **207**, 173 (2012)
67. M. Neumann, *Proceedings* (GDMB Verlag GmbH, Clausthal-Zellerfeld, 2017)
68. R. Weyhe, in *Recycling und Rohstoffe*, ed. by K.J. Thomé-Kozmiensky, D. Goldmann (TK Verlag Karl Thomé-Kozmiensky, Neuruppin, 2013)
69. Umicore Invests in Recycling of Rechargeable Batteries. CP-2009-31-R, 2009
70. T. Elwert, D. Goldmann, T. Schirmer, K. Strauß, *Proceedings of the European Mineral Resources Conference*, 19–21 Sept 2012, p. 575
71. J. Vliegen, Umicore, Battery Recycling, 18 Nov 2010
72. TU Braunschweig, Recycling von Lithium-Ionen-Batterien: Abschlussbericht der TU Braunschweig für das Projekt Lithorec, Braunschweig (2016)
73. L. Li, Current Situation of Recycling and Reusing for Spent Batteries in China (2011)
74. C. Hanisch, W. Haselrieder, A. Kwade, Verfahren zum Wiedergewinnen von Aktivmaterial aus einer galvanischen Zelle und Aktivmaterial-Separationsanlage, insbesondere Aktivmetall-Separationsanlage (2013)
75. M. Grützke, X. Mönnighoff, F. Horsthemke, V. Kraft, M. Winter, S. Nowak, RSC Adv. **5**, 43209 (2015)
76. J. Diekmann, C. Hanisch, L. Froböse, G. Schälicke, T. Loellhoeffel, A.-S. Fölster, A. Kwade, J. Electrochem. Soc. **164**, A6184 (2016)
77. C. Hanisch, T. Loellhoeffel, J. Diekmann, K.J. Markley, W. Haselrieder, A. Kwade, J. Clean. Prod. **108**, 301 (2015)
78. C. Hanisch, W. Haselrieder, M. Schoenitz, B. Westphal, Verfahren zum Behandeln gebrauchter Batterien, insbesondere wieder aufladbarer Batterien und Batterie-Verarbeitungsanlage (2015)
79. C. Hanisch (Lion Engineering GmbH), Recycling of Lithium-Ion Batteries, 14 June 2014
80. M. Buchert, W. Jenseit, C. Merz, D. Schüler, Verbundprojekt: Entwicklung eines realisierbaren Recyclingkonzepts für die Hochleistungsbatterien zukünftiger Elektrofahrzeuge—LiBRi:

Teilprojekt: LCA der Recyclingverfahren. Endbericht, 2011, https://www.oeko.de/oekodoc/1499/2011-068-de.pdf. Accessed 21 Sept 2017
81. Accurec Recycling gmbH, Accurec Battery Recycling, http://accurec.de/. Accessed 26 Sept 2017
82. R. Weyhe, B. Friedrich, Demonstrationsanlage für ein kostenneutrales, ressourceneffizientes Processin ausgedienter Li-Ionen Batterien aus der Elektromobilität: Abschlussbericht zum Verbundvorhaben EcoBatRec (2016)
83. B. Friedrich, Rückgewinnung der Wertstoffe aus zukünftigen Li-Ion-basierten Automobil-Batterien, Bernd Friedrich (2009)
84. R. Weyhe, 1. Accurec Recycling GmbH, Amsterdam (2014)
85. R. Weyhe, in *Recycling und Rohstoffe*, ed. by K.J. Thomé-Kozmiensky, D. Goldmann (TK Verlag Karl Thomé-Kozmiensky, Neuruppin, 2010)
86. EU Recycling, Akkumulatoren-Recycling durch Accurec, 2016, http://eu-recycling.com/Archive/13624. Accessed 17 May 2017
87. Schrottankauf Gouchev, Schrottpreise, http://www.schrottankauf-bitterfelderstr23.de/schrottpreise.php. Accessed 11 Sept 2017
88. ESG Edelmetall-Service GmbH & Co. KG, Ankaufspreise und Sortierkriterien für Elektroschrott/E-Schrott, 2017, https://www.scheideanstalt.de/elektroschrott-preise-und-sortierkriterien/. Accessed 11 Sept 2017
89. M. Buchert, J. Sutter, Aktualisierte Ökobilanzen zum Recyclingverfahren EcoBatRec für Lithium-Ionen-Batterien: Stand 09/2016, 2016, http://www.erneuerbar-mobil.de/sites/default/files/2017-01/EcoBatRec-LCA-Update%202016.pdf. Accessed 21 Sept 2017
90. M. Buchert, J. Sutter, Aktualisierte Ökobilanzen zum Recyclingverfahren LithoRec II für Lithium-Ionen-Batterien: Stand 09/2016, 2016, http://erneuerbar-mobil.de/sites/default/files/2017-01/LithoRec%20II-LCA-Update%202016.pdf. Accessed 21 Sept 2017
91. U. Wietelmann, M. Steinbild, in *Ullmann's Encyclopedia of Industrial Chemistry*. 7 edn. (Wiley-VCH, Weinheim u.a.) 2010, 1
92. USGS National Minerals Information Center, Mineral Commodity Summaries 2017: Lithium, https://minerals.usgs.gov/minerals/pubs/commodity/lithium/mcs-2017-lithi.pdf. Accessed 21 Aug 2017
93. British Geological Survey, Lithium, 2016, https://www.bgs.ac.uk/downloads/start.cfm?id=3100 Accessed 30 Aug 2017

Business Models for Repurposing a Second-Life for Retired Electric Vehicle Batteries

Na Jiao and Steve Evans

Abstract The rapid development of electric vehicles (EVs) has caused a problem for the industry: what happens to the batteries at the end of their useful life in EVs? Repurposing those batteries for a less-demanding second-life application, e.g. stationary energy storage, could provide a potential solution to extract more value than just recycling or disposal. This paper explores the current battery second use (B2U) business models and the key challenges of implementing B2U. Based on empirical interview data from stakeholders involved in B2U, this paper presents a typology of current B2U business models—standard, collaborative and integrative business models—and offers implications for designing business models that incorporate sustainability at the core. The findings also show that innovative business model is a key to addressing the B2U challenges and overcoming the 'inferiority' of second-life batteries as used products.

Keywords Electric vehicles · Battery second use · Business models
Sustainability · Value

1 Introduction

Electric vehicles (EVs) hold great promises for a more sustainable transportation in the future. Governments of several European countries such as France and Britain have set ambitious goals for no more petrol or diesel cars to be sold by 2040.[1] Overtime, when the batteries are no longer able to provide sufficient power and

[1]See https://www.theguardian.com/politics/2017/jul/25/britain-to-ban-sale-of-all-diesel-and-petrol-cars-and-vans-from-2040 (read 2017-08-20).

N. Jiao (✉) · S. Evans
Institute for Manufacturing, University of Cambridge, Cambridge, UK
e-mail: nj268@cam.ac.uk

S. Evans
e-mail: se321@cam.ac.uk

G. Pistoia and B. Liaw (eds.), *Behaviour of Lithium-Ion Batteries in Electric Vehicles*, Green Energy and Technology,
https://doi.org/10.1007/978-3-319-69950-9_13

range due to their ageing characteristics, there will be millions of tons of batteries coming out of the cars. If not properly treated, those retired batteries could place tremendous burden on the environment.

In general, an EV battery could retain 70–80% of its original capacity intact upon reaching the end of its vehicular life [1], and replacement is recommended in order to satisfy the driving range demand of EV owners [2]. Upon retirement, there would still be sufficient capacity left in the batteries to support less-demanding applications such as load shifting, renewable energy storage and backup power [1, 3–6]. Compared with recycling which entails costs, energy and wastes, recapturing the residual value from retired EV batteries for non-vehicular applications could generate alternative revenue streams to help overcome EV cost-hurdles and create synergic value for energy storage [7]. Major automotive companies such as Nissan, BMW and Daimler have started initiatives to investigate or even commercialise battery second use (B2U).[2]

This paper explores the current B2U business models (BMs) and the key challenges of implementing B2U to understand how to improve business models to extract more value from second-life batteries. In-depth semi-structured interviews with major players in the B2U industry provide substantial data to draw lessons for business models of second-life batteries. First, a typology of current B2U business models generated from empirical case studies is presented. Second, critical challenges of implementing B2U are discussed. Third, innovative business models are identified as the key to addressing the B2U challenges and overcoming the 'inferiority' of second-life batteries as used products. Finally, we present three critical business model design elements that will help companies better extract value from B2U.

2 Case Study Data

In the nascent stage of B2U, only a handful of cases can provide substance to studying business models at the commercial level. Most of the B2U projects are still in the planning, piloting or demonstration phases and are more focused on technical or economic aspects. In this study, the unit of analysis is the business model of B2U. A firm could have several B2U business models in parallel which are treated as separated cases.

Seven case studies were selected from the existing B2U markets that have passed the phase of demonstration or pilot projects to reach the early commercialisation stage. In-depth semi-structured interviews were carried out with experts from both automotive and energy industries. Each interview lasted 0.5–3 h and

---

[2]See http://news.vattenfall.com/en/article/used-e-car-batteries-get-second-life-hamburg;
http://media.daimler.com/marsMediaSite/en/instance/ko/Worlds-largest-2nd-use-battery-storage-is-starting-up.xhtml?oid=13634457;
https://www.theguardian.com/business/2017/may/04/nissan-launches-british-made-home-battery-to-rival-teslas-powerwall (read 2017-08-21)

followed a pre-defined interview protocol to gain an in-depth understanding of the B2U business models. The interviews were recorded, transcribed and coded for analysis according to the qualitative case study methodology [8]. A list of the case study interviews is presented in Table 1.

Table 1 Case study interviews

Case no.	Company	Stakeholder role	Region	Interviewees' position	Reference code
I	A	OEM	North America	General Manager	O-1-1
	B	Energy storage/ B2U system provider	California	CEO and Co-founder	O-1-2
	C	Lifecycle management	US	President and Founder	L-1
	D	B2U joint venture	Japan	President General Manager, Planning General Manager, R&D	E-1
II	A	OEM	North America	General Manager	O-1-2
	E	Energy storage/ B2U system provider	California	COO	E-2
III	A	OEM	Europe	General Manager, Zero Emission Strategy; Manager, V2G and Stationary Storage; Expert Leader, Technology Planning, Advanced Engineering	O-1-3
	F	Energy management	Europe	Vice President of EMEA Marketing	E-3
IV	G	OEM	North America	Manager, Connected eMobility	O-2-1
			Germany	Program Leader for Battery Second-Life; Head of Development Stationary Storage Systems	O-2-2
V	I	Energy storage	Germany	Managing Director	E-3
VI	J	OEM	France	Program Manager, Energy Services	O-3
VII	L	OEM	Japan	Project General Manger, New Business Planning; Project Manager, Environmental Affairs; Group Manager, Planning	O-4

3 A Typology of Current B2U Business Models

The complete set of interviews shows that the automotive OEMs are creating and capturing value from B2U in different ways. In most cases, OEMs are adopting battery selling as their main business model. However, due to the lack of knowledge and resources in applying the batteries to the energy market, the OEMs interact, in different ways and to different extents, with stakeholders from the energy sector to help develop and deliver the final solutions to the end-customers. In this paper, the key stakeholder from the energy sector who offers the final solution to the end-customers is termed ‘B2U solution provider’. The major stakeholder roles generalised from the seven case studies are the automotive OEM, the B2U solution provider and the end-customer.

The data show that the main differences between the business models in the case studies originated in the various relationships and interactions between the OEM and the B2U solution provider. The degree that the OEM integrates B2U into their businesses varies from nearly zero to full integration. Compared across cases, it was found that the value generated for the OEMs from B2U increases as the degree of integration raises. For example, in case V where the OEM retains the battery ownership, they benefit from the energy services provided by the batteries during their entire second-life while in case I the OEM only get additional revenues through selling the batteries at a very cheap price ($85/kWh as referred to by one of the interviewees). This indicates the degree of integration as a key factor in the value creation and capture for the OEMs. Based on the business model analysis in the seven individual case studies, a typology of existing B2U business models is presented to illustrate how key stakeholders across the automotive and energy sectors interact to create and capture value from B2U in different ways.

The business models examined from the seven empirical case studies can be categorised into three types: standard business model, collaborative business model and integrative business model. These categories correspond to the various relationships between the cross-sector stakeholders, namely (a) pure supplier–customer relationship, (b) collaboration and (c) the OEM internalising the role of the B2U solution provider. Accordingly, the degree of integration increases from standard to collaborative and to integrative business models. Within the collaborative business model type, there are subtypes, depending on the degree of integration and the relative dominance of stakeholders in the final solution development. Examples from the case studies are given for each subtype.

A schematic illustration of the typology is shown in Fig. 1. The degree of integration in the business model types increases from left to right. The boxes represent key stakeholders involved in B2U: the OEM, the B2U solution provider and the end-customer. The height of the OEM box represents the relative degree of integration. The red and green arrows represent the flow of battery ownership and the knowledge and information flow between the OEM and B2U solution provider, while the purple arrow represents the delivery of final solutions to the end-customers.

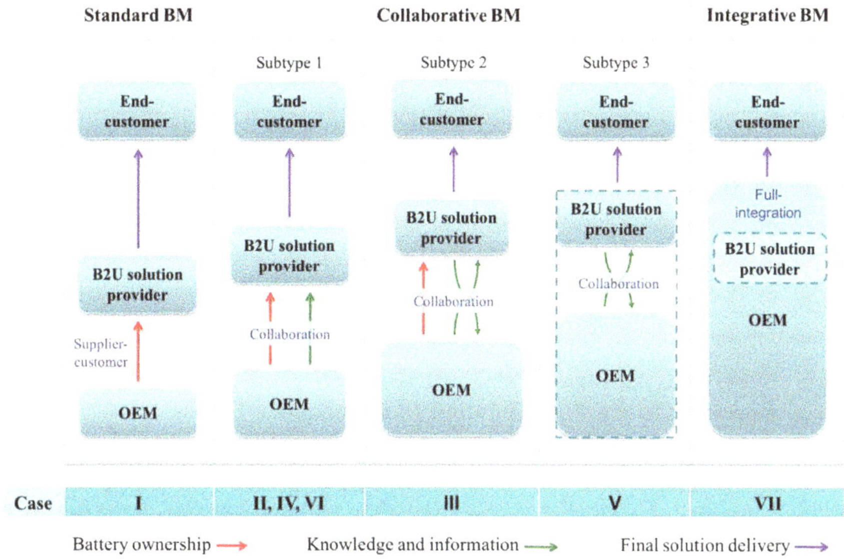

Fig. 1 Schematic typology of B2U business models: how cross-sector stakeholders interact to develop business models for second-life batteries

3.1 *Standard Business Model*

The standard business model is where the OEM simply sells second-life batteries to the B2U solution provider (e.g. case I). The B2U solution provider develops the final solutions for the batteries and delivers that solution to the end-customers. The interaction between the OEM and the B2U solution provider is just supplier–customer relationship, like in most business models under the 'sell-and-disengage' logic. In that case, the OEMs involve in B2U to generate additional revenues from selling the battery property in the free market. The OEM's degree of integration is nearly zero. They are not engaged in the final B2U solution development, and there is almost no knowledge and information flow between the OEM and the B2U solution provider. This type of business model requires little OEM engagement, but the value captured by the OEM is also small. In addition, the standard business model is very vulnerable to competitors.

3.2 *Collaborative Business Model*

Most of the manufacturing firms developing B2U fall into the second type—collaborative business model (case II, IV and VI). Under this type, the OEMs collaborate with B2U solution providers and involve in the final solution development. Instead of just selling the battery asset, OEMs under this type collaborate with B2U solution providers in different ways and to different extents to add to the value of second-life batteries and capture more benefits than just selling the batteries. Three subtypes of collaborative business models are generalised, depending on the relative dominance of the final solution development between the OEM and B2U solution provider. The three subtypes are (a) assistant collaborative where OEMs assist B2U solution providers in the final solution development; (b) OEMs co-develop the final B2U solution with B2U solution providers; and (c) B2U solution providers develop the final B2U solution for the OEMs. These subtypes are discussed in turn below.

3.2.1 Subtype 1. Assistant Collaborative—OEMs Assist B2U Solution Providers in the Final Solution Development

In this subtype, the final solution is still developed and delivered by the B2U solution provider. Unlike the standard business model, however, the OEM also collaborates with the B2U solution provider to share knowledge and resources that contribute to the final solution development, in addition to selling the batteries. In case II and VI, for example, the OEMs collaborate with the B2U solution providers to share their expertise and information on the batteries (e.g. battery historical data and remained performance evaluation) to make second-life batteries better fit into the storage systems developed by the B2U solution providers. In case IV, the OEM provides consultancy services and tailor-made batteries to the B2U solution providers to help them better develop the final solutions. In this subtype, the OEM's degree of integration is higher than the standard business model but is still very low. The B2U solution provider is dominating the final solution development. The OEMs provide knowledge and information on the battery side to assist the B2U solution provider but are not actually engaged in developing and providing the final solutions. The value captured by the OEM in this business model type is higher than that in the standard business model but is still small because the OEM does not benefit from the final solutions.

3.2.2 Subtype 2. Co-development Collaborative—OEMs Co-develop the Final Solution with B2U Solution Providers

In this subtype, the final solution is co-developed by the OEM and the B2U solution provider. Each stakeholder has its own set of knowledge and resources. Through the collaboration, the two stakeholders integrate complementary capabilities to design and optimise the battery systems, as well as develop and deliver the final products and solutions. The OEM's degree of integration is higher than in subtype 1 because the OEM is engaged in integrating the capabilities of the two parties to co-develop the final solution, in addition to selling batteries. There are bidirectional and interacting flows of knowledge and information between the two stakeholders because the final solutions are their mutual objectives and outcomes. The value captured by the OEM in this business model type is higher than the previous two business model types because the OEM shares the benefits from the final solutions together with the B2U solution provider. A typical example of this business model subtype can be found in case III.

3.2.3 Subtype 3. Integration-Collaborative—B2U Solution Providers Develop the Final Solution for the OEMs

In this subtype, the OEM retains the ownership of the battery and gather its partners (including the B2U solution provider) together to develop and market the final solutions for them. Retaining the ownership of the batteries allows the OEM to continuously engage in B2U to extract value throughout the second-life of the batteries. The degree of integration is higher than in the previous two subtypes because the OEM created a joint venture where they exploited the B2U solution provider to help them develop and deliver the final solutions to the end-customers. The OEM dominates the final solution development in this subtype. There are also bidirectional and interacting knowledge and information flows between the two stakeholders. The value captured by the OEM in this business model subtype is higher than the previous business models because the OEM retains the battery ownership, which enables them to continuously optimise and extract value from second-life batteries. A typical example of the business model subtype can be found in case V.

3.3 Integrative Business Model

The integrative business model is where the OEM internalises B2U into their own business, developing and delivering the final solutions for the end-customers. In this case, the OEM leverages its own networks to maximise the value that they can obtain from delivering that final product or service to the end-customers. The integrative business model requires very high OEM engagement and diversified resources and capabilities of the company and is likely to be restricted to certain applications due to OEMs' limited access to certain markets (e.g. grid-scale applications). The OEM's value capture portion is the highest among all the business model types because the OEM internalises the role of B2U solution provider which enables them to obtain all the potential value delivered by the final solution. However, the value captured from second-life batteries might not be the highest because of the OEM's limited access to the energy market.

In summary, a typology of B2U business models is proposed. The typology compares existing B2U business models in practice to illustrate how automotive OEMs interact in different ways with stakeholders from the energy sector to create and capture value from second-life batteries. The standard and integrative business models are the two extremes of the existing B2U business models. Evidence from the case studies shows that the standard business model requires little OEM engagement but is very vulnerable to competitors. The integrative business model, on the other side, allows the OEM to capture higher value portion from B2U than other business models but is restricted in terms of markets and applications.

4 Challenges of Implementing B2U

Despite the envisioned benefits of repurposing retired EV batteries for stationary storage, there are manifold challenges regarding B2U that could significantly reduce the value of second-life batteries. Four critical challenges are found from the seven case studies: competitiveness, uncertainty, design and regulation. The four critical challenges are refined and generalised from the individual case studies to reflect the more general nature of the cross-case findings. For example, the 'uncertain flow of batteries', the 'uncertain second-life battery performance' and 'customers' concerns over second-life batteries' are grouped into the challenge of 'uncertainty'.

4.1 Competitiveness

The competition that comes from new batteries was commented on by interviewees from all seven case studies. In four out of the seven cases (case I, II, III and IV), the competition from increasingly cheap new batteries was described by the OEMs as one of the most critical B2U challenges. The existing data show that currently, the relatively cheap price of second-life batteries compared with new batteries is regarded as the main motivation for many companies to develop B2U. However, it is also expected that by the time the 8- or 10-year-old batteries are taken out of the cars, there will be new generations of batteries in the market with not only cheaper price but also much better quality and performance, which would make the life of second-life batteries more difficult. In that case, the cost competitiveness and thus the attractiveness of second-life batteries would be diminished. The OEMs are now trying to reduce the cost of battery repurposing, for instance, using the whole battery pack as it is to avoid costs regarding opening the pack, so as to keep the cost competitiveness of second-life batteries (case IV, V and VI).

4.2 Uncertainty

4.2.1 Uncertain Flow of Second-Life Batteries

The other critical challenge most clearly mentioned by the OEMs is managing the flow of second-life batteries (case I, III, VI, VII). Unlike new batteries, the volume of second-life batteries that will come back to the OEMs is somewhat out of control because it depends on the customer's behaviour—when they retire the batteries, and whether they will trade their old EVs into the second-hand car market. This adds to the uncertainty in terms of the volume of batteries available for the OEMs. And if the OEMs sell second-life batteries to their customers, the uncertain return flow of the batteries also causes anxiety for the purchasers of the battery due to a lack of steady supply. This would make the business more difficult, especially for large-scale applications which require a steady supply of batteries with a large volume.

In case I, for example, the energy storage start-up (Company B) expressed their concerns about the amount of second-life batteries available from the OEM. They are not too concerned now because the scale of their business is small. However, they said they need to be sure about the volume in order to scale up their businesses; otherwise, they will stop using second-life batteries in the future. This is not mentioned in case II, and it might be because the core business of Company E is using new batteries and second-life battery is just an option for them. And in case V, the business model is service-based rather than selling the batteries so the OEM does not need to make their customers feel 'secure' about the battery supply. The OEM in case IV did not indicate the challenge, but they said they have sold out of second-life batteries which imply a lack of steady battery flow.

In the author's view, it seems that the scale of second-life batteries would solve this problem in the future but it might also generate more competition for the volume out there. When second-life battery supply scales up, it becomes a competitive issue where the best models that offer the highest price will be able to obtain the most batteries.

4.2.2 Uncertain Second-Life Battery Performance

The uncertainty in the remained battery lifetime and performance degradation in various energy storage applications is perceived as another B2U challenge for both the second-life battery providers and buyers (case I, III, V and VI). Unlike new batteries that are designed for energy storage, the lifetime and degradation of second-life batteries are quite uncertain, depending on both how they were used in their first life in the EVs and how they are going to be used during their second-life in stationary storage applications.

This uncertainty is said to be caused by (a) the lack of systemic and sophisticated data collection onboard; (b) the lack of effective data analysis; and (c) the lack of sharing with downstream stakeholders on the battery health over its first life in vehicles. One of the OEMs said: '*We have the global data centre so the data is enough. What matters is how we can analyse the data. For the time being, we can't precisely predict the remained lifetime but it is under the way...*' (O-1-1). However, the interviewee of one of the downstream energy companies said: '*They (the OEM) don't track all the information we would like to have...at least not in such an efficient way as we would like it to be...If they track all the information, we would get much more proper picture about the quality of the battery at the end of life...*' (E-4). In addition, one of the interviewees commented on current battery testing: '*This testing is not going to tell you the performance forward because that requires a lot more information from the OEM*' (L-1). It can be seen from the data that there is information asymmetry between the battery provider and downstream stakeholders, which indicates the need for closer communication and information sharing between B2U stakeholders.

Besides, there is a lack of understanding of second-life battery ageing behaviours in specific energy storage applications and a data sharing platform among stakeholders. Since each battery ages differently under the varying historical operating conditions in the EVs and complex usage profiles in energy storage, it is difficult to predict the exact ageing behaviour of the batteries during their second-life. Without a proper tracking of the historical usage data of the batteries and evaluation of the second-life battery performance, it is not very likely that the in the long-term OEMs could persuade customers to buy their second-life batteries.

4.2.3 Customers' Concerns Over Second-Life Batteries

The data show that from the customer's side, the major challenge of acquiring second-life batteries is their concerns over second-life batteries (case III and IV). As one of the interviewees said: '*Another challenge is that people often have low price expectation for second-life batteries, so we have to tell them the value of the system...*' (E-3). In general, customers have a bias against used products—they feel insecure about second-life batteries and they have low price expectations. Besides, customers have poor understanding of the value of second-life batteries. They do not have the knowledge about the functionalities of the battery storage system, and they lack the skills or experience for operating the system. Those concerns could strongly influence customers' buying decisions and impair the real value of second-life batteries.

4.3 Design

From the B2U repurposer's perspective, one of the challenges of B2U is in regard to the initial battery design. As commented by one of the interviewees in case V: '*Currently the car manufacturers design the batteries only for being used in the car*' (E-4). The data show that the battery repurposing cost is significantly affected by how the batteries were initially designed. For example, if the BMS does not properly track and collect the battery usage data over its first life in EVs, the battery state of health (SOH) could not be evaluated, and thus batteries would need to be sent to a third party for testing. This incurs extra cost in battery transportation and testing. Besides, if the components inside the battery pack such as the BMS are not compatible with stationary storage applications, a new BMS needs to be built and implemented to the battery system which brings additional costs for battery repurposing.

It is obvious that the car manufacturers will always prioritise the battery design for the EVs. However, a systemic design thinking that incorporates second-life repurposing into the initial battery design would greatly smooth the whole repurposing process and reduce/avoids relevant costs. And as one of the interviewees commented: '*It is a matter of consideration, not cost*'. Some OEMs are aware of the importance of design for repurposing and have taken measures either in the form of battery architecture redesign or the improvement of battery control and data tracking system (e.g. case IV).

4.4 Regulation

In terms of regulation, the data show that there are three major challenges existing, depending on the countries and regions. The first is the waste and transportation regulation. Currently, second-life batteries are not clearly defined in the regulation in most countries. As the interviewee of case VI said: '*Because the battery is considered to be dangerous goods, the transportation is very expensive…the second-life battery is not really defined…it shouldn't be regarded as waste, otherwise there will be other regulations and complicated stuff*'.

The second regulatory challenge is regarding battery storage for the energy market. As commented by one of the interviewees in case VI, for example, '*If the regulation is not open, the business model could not fly*'. The large amount of confirming data from other interviews also shows that the electricity market regulations in most regions are not open and transparent yet, which might kill many potential business cases. Stakeholders in the case studies expect that the regulators could understand more about the role and value of batteries in the energy market and accept battery storage as equal to other means of storing or creating energy to support the power grid.

The third challenge is that in some regions such as California, there are incentive programmes that only subsidise new batteries but not second-life batteries, which is unfavourable for second-life batteries to compete in the energy market.

In summary, the four critical challenges identified across the seven case studies, namely competitiveness, uncertainty, design and regulation, show that B2U is still at its emerging stage with multiple challenges facing B2U players. A summary of the B2U challenges is presented in Fig. 2.

5 Role of Business Models in B2U

Many of the critical challenges confronted by B2U stakeholders are still perceived as operating under the 'business-as-usual' scenario with the traditional product selling model. Offering the repurposed battery as a discounted product (compared

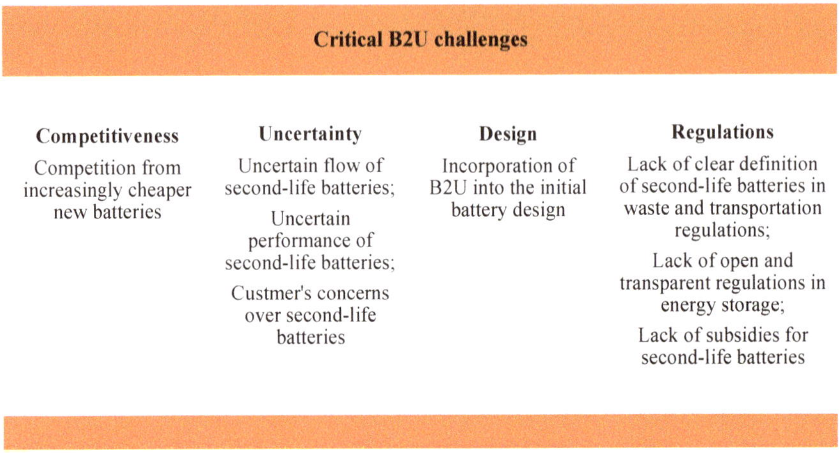

Fig. 2 Summary of critical B2U challenges

with new batteries) is pushing the 'inferior' second-life batteries into increasingly fierce competition with new batteries, which reduces the real value of the battery and is not sustainable. Under the 'sell-and-disengage' logic, the only way for second-life batteries to compete is to lower the price continuously with the decreasing cost of new batteries. The reward from selling the battery asset is thus very low. And one day when the cost advantage of second-life batteries becomes negligible, the 'inferior' aged batteries will be driven out of the market. The data from the seven case studies suggest that the 'sell-and-disengage' business model is not helping stakeholders extract the potential value of second-life batteries in energy storage.

To achieve the potential benefits that can be delivered from B2U, a new perspective of perceiving the value of second-life batteries more than a physical product is needed. As one of the interviewees in case I commented: '*One important thing to keep in mind is that it (B2U) doesn't diminish the utility of the battery, the battery is just as good as any energy storage device. And regardless of what the price point is comparatively, it still has a good function and capability. It is, how can you create a structure that makes it worthwhile to pursue that matters*' (O-1-1). Though somewhat degraded in terms of capacity, the value of the storage capacity of second-life batteries should not be discounted. When applied in certain applications, a second-life battery could deliver just the same functions and services as a new one. The key is to best utilise the remained capacity of second-life batteries in the right energy storage applications to generate value.

However, a second-life battery itself does not have a value proposition, and it is the business model that creates value for second-life batteries and helps stakeholders capture the benefits. Business models and the logic of value exchange were a constant part of the case study interviews even when speaking to the technical people. One of the interviewees, for example, emphasised the importance of a good business model on B2U: '*A good business model is the key so the customer can pleasingly accept the storage system*' (BJV-1). In some cases, customers do not really care whether the battery is new or old, they only want the power or capacity services delivered by the batteries. As captured in the following quotes: '*A battery doesn't do anything—it is what you built around the battery to solve what pain points for your customers. Our customers don't care whether you use second-life batteries, as long as it does what they tell us to do*' (E-1).

A shift to services in the B2U business models was found in some case studies. The data suggest that the 'inferiority' of second-life batteries could be overcome by delivering the solutions customers want rather than selling the physical asset. In that case, what matters most is not how fancy the battery itself is, but the value of the solutions delivered by the battery through the business models. The value that stakeholders capture from delivering that solutions could also be optimised because they can continuously engage in and benefit from the various services provided through the battery rather than the one-off product selling.

Due to the very nature of second-life batteries, which are used products, the business model plays a pivotal role in achieving the potential value of the batteries. The data indicate that a good business model could help address the challenges of B2U to overcome the 'inferiority' of second-life batteries as used products.

6 How to Better Design Business Models for B2U

Three aspects, namely lifecycle thinking, system-level thinking and the shift to services are proposed as helpful perspectives for stakeholders to better design their BMs so as to achieve the potential value of second-life batteries. The three aspects are discussed in turn below.

6.1 Lifecycle Thinking for Analysing the Potential Value of Second-Life Batteries

In the light of the earlier discussion of the critical challenges of B2U (Sect. 4) that might impair the real value of second-life batteries, we can now ask the question what is the potential value of second-life batteries? At the nascent stage of B2U, the value of second-life batteries is still poorly understood and B2U stakeholders are not very good at extracting value from second-life batteries. All seven case studies show that there are actually more B2U benefits available that manufacturing firms are not accessing. Across the data, it is found that there are value opportunities existed in various stages of the battery life cycle. In this section, a battery lifecycle thinking perspective is proposed to help analyse the potential value of second-life batteries and identify opportunities for improved value creation along the battery life cycle.

Second-life batteries, by definition, are 'inferior' to new batteries in terms of performance, lifetime and functionalities for some specific applications. However, the potential value of second-life batteries could be as high as or even higher than that of a new battery if equipped with a good business model. At this emerging stage of B2U, it is important for stakeholders to understand the potential value of second-life batteries so as to identify value opportunities to better design business models for increased value creation. Based on the value creation and capture analysis in the seven case studies, it is found that a whole lifecycle thinking, which integrates the battery's first life in EVs, second-life in storage applications and end-of-life recycling, is helpful to understand the value created for various stakeholders and potential value opportunities along the whole battery life cycle.

Unlike new batteries designed for stationary storage, the data show that second-life batteries involve many different stakeholders at various stages of its life cycle. B2U itself is considered as an end-of-life strategy for vehicle batteries and a circular approach to creating value from 'waste'. However, repurposing a second-life for the batteries also means that those once scrapped batteries will start a new life cycle in a different application. For second-life batteries, the initial battery conditions depend on how they were designed and used during their first life in the vehicles. In other words, the battery's first life partially determines the performance and remained value of second-life batteries. On the other hand, the value analysis should also include the final EOL when the batteries could not be utilised anymore, for example, the value of recycling.

Therefore, analysing the value of second-life batteries should be embedded in considering its whole life cycle in a broader sense that includes multiple lives: (a) the first life in the EVs, (b) the second-life in, for example, stationary storage applications and (c) the EOL when the batteries are recycled/disposed. The key stages of the battery life cycle are illustrated in Fig. 3 to help analyse how second-life battery value could be improved by considering the entire battery life cycle.

(a) **Battery first life**

As shown in Fig. 3, the battery first life includes battery design and manufacturing, vehicle use and return for collection. The battery is designed and manufactured for the EVs by the automotive OEMs and/or battery producers. Then the batteries are

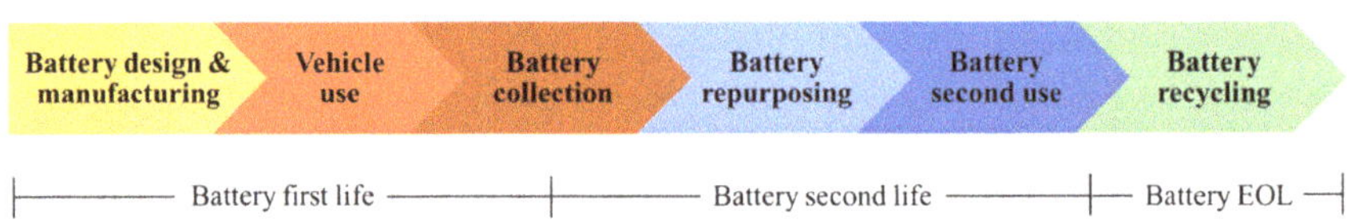

Fig. 3 Key stages of the lifecycle thinking for increased value of second-life batteries

used as the vehicle traction by the EV customers. After 8–10 years when the batteries could not satisfy EV drivers' demands such as driving range, acceleration and charging rate, the EV owners will return the old batteries to the OEMs. The initial status and thus the remained value of the second-life batteries depends on how they were initially designed and used during their first life. The data indicate that three aspects of the battery first life, namely initial battery design, EV battery ownership and education for consumers, could be considered to facilitate B2U and increase the battery value.

Initial battery design

As discussed in the previous section, incorporating B2U into the initial EV battery design through, for example, better data tracking and collection, as well as improved reusability and durability of the battery pack components, could greatly reduce the battery repurposing cost and smooth the process of B2U.

EV battery ownership

The second is the EV business models regarding battery ownership. One challenge discussed above is the management of the second-life battery flow. In most cases, once the OEMs sell the EVs they do not have the ownership of the batteries anymore. They might have the liability for battery recycling, depending on regions, but they actually have little control over the battery flow in terms of when the retired batteries will come back, for example. Interestingly, in one of the case studies (case VI) where 80 to 90% of the EV batteries are on the leasing mode, the OEM remains the ownership of the battery and, thus, has much more control over the volume and quality of the batteries coming back. In addition, because the OEM owns the battery property, they provide various maintenance services to the EV customers to keep the battery running under the best possible conditions. When the battery capacity drops to a certain level, the OEM swaps the batteries and keeps the old ones for repurposing or recycling depending on the battery conditions. In that case, the battery quality is also more unified which enables more efficient and profitable B2U.

Education for the consumers

The third aspect of improving the residual value of second-life batteries is to educate the EV customers on better utilising the batteries. OEMs could give advice to the EV drivers in terms of how to take care of the batteries during EV driving and help them understand the value of their batteries after the vehicle life. Rewarding mechanisms could also be built where customers get a better price if they return batteries with higher quality. Besides, the OEMs could offer maintenance services on a regular basis to check the batteries and repair any degraded components if necessary to avoid further deterioration of the batteries. In that case, the customer relationship is strengthened which also brings value for the OEMs in terms of more valuable EV offerings for the customers.

(b) **Battery second-life**

As shown in Fig. 3, the battery second-life includes battery collection, repurposing (e.g. testing, grading, system integration) and second use in various energy storage applications. After the batteries are retired from the EVs, the automotive OEMs collect the batteries through their car dealers, test them and decide whether to repurpose or recycle them. For batteries that could be further utilised, they will be graded according to their remained capacity and then sorted and repackaged. Depending on the applications, the batteries are integrated to build the energy storage systems by the automotive OEMs and/or the energy companies. The storage system composed of second-life batteries is then sold to the customers in the energy market or operated to provide various energy services. Since the batteries are repurposed for a different application (energy storage) than in the automotive industry, multiple stakeholders across sectors might be involved at different stages and it is essential to coordinate among stakeholders to optimise the cost structure and improve the total value creation. The data suggest that four aspects of the battery second-life, namely battery redemption, battery repurposing strategies, and battery testing and grading should be taken into consideration to increase the value of second-life batteries.

Battery redemption

In terms of battery collection, normally the EV customers will return the old batteries to the car dealers who then send the all the batteries, good or bad, back to the OEMs. Interestingly, in one of the case studies (case VII) the interviewees proposed a fast testing plan where the batteries are tested at the dealers to quickly check their conditions. Only the batteries qualified for further utilisation will be transported to the OEMs while the bad quality ones will be sent directly for recycling which saves cost in battery transportation. In another case study, the OEM outsourced a third party to do the logistics who collect the batteries for them from their car dealers. The OEMs need to weigh the cost of battery collection against their specific situations to decide a most cost-efficient plan.

Battery repurposing strategies

Based on the seven case studies, the data show that there are generally two different battery repurposing strategies: (a) to disassemble the battery pack into modules and repackage the modules and (b) to reuse the whole battery pack as it is. According to the case studies, both strategies are used in practice by different OEMs. Most of the OEMs adopt the latter strategy to reuse the whole battery pack in that the costs regarding opening the battery pack, testing individual modules and repackaging can be avoided. In addition, key components such as the BMS and cooling functions could also be reused to avoid additional cost. However, some of the OEMs insist that reassembling the modules of similar conditions could extend the lifetime of the battery and thus increase its residual value. Currently, there is no consensus on which strategy is more economically viable but both will require the incorporation of second use into the initial battery design. For example, if you want to disassemble the battery pack into modules for reuse, the battery should be designed for easy disassembling. On the other hand, if you want to reuse the pack as it is in stationary storage applications, you need to ensure the reusability and durability of the battery components needed for second-life applications so that they could also be reused as a whole in a more sustainable way.

Battery testing and grading

The batteries are then tested, graded and sorted for different second-life applications. In some cases, battery testing and grading are done by external parties which incur extra costs. How much the cost can be internalised depends on how much efforts the OEMs make in tracking the battery data during its first life in EVs and being able to analyse that data. Depending on the capacity remained, the batteries are then graded and sorted for different stationary storage applications. However, knowledge in energy storage is also required to determine what is the best usage profile for each battery to better utilise the battery value. In most cases, the OEMs bring in partners from the energy sector to develop or assist them in designing the final solutions that commercialise second-life batteries into the right energy market.

(c) **Battery EOL**

Depending on the battery conditions after its first life and its usage profiles during second-life, second-life batteries could be used in stationary storage applications for another 5–15 years. After that when the batteries could not be further utilised, they will be recycled. Currently, the recycling system for lithium-ion batteries is not established yet, so the cost of recycling EV batteries could be quite high in the near term. According to the case studies, battery recycling incurs cost nowadays but it is possible that in the future, recycling will bring profits instead of incurring expenses. Through deploying second-life batteries in stationary storage applications for another 5 or 10 years, for example, the OEMs could defer the recycling phase and turn the cost into revenue opportunities. As one of the OEMs said: '*This (recycling) is important to follow up because recycling cost will always change. Today there might be a cost to it, tomorrow it might be a benefit*' (O-3). In B2U, the stakeholders should also make clear the battery recycling responsibilities for the very end of the battery life.

In summary, thinking about B2U from the lifecycle perspective is helpful to analyse the potential value of second-life batteries and identify opportunities where possible for increased value creation. At the emerging stage of B2U, it is essential that stakeholders understand the potential value of second-life batteries at this system level so as to better design their business models to achieve that value. The lifecycle thinking helps integrate resources and knowledge from cross-sector stakeholders to optimise the cost structure over the entire battery value chain.

6.2 System-Level Design for Achieving the Potential Value of Second-Life Batteries

The value of second-life batteries in energy storage is delivered to the end-customers either in the form of the battery products or the services provided by the storage systems. How to best utilise the value of the batteries requires the integration of knowledge and expertise from both the automotive and the energy sectors, as well as a good business model that helps deliver that value to various stakeholders involved. The business model analysis in the seven case studies shows that the system-level perspective which transcends the firm boundaries is helpful to analyse the total value creation for second-life batteries. This section aims to present how system-level thinking could help achieve the potential value of second-life batteries.

It can be seen from the data that if the OEM only looks at the benefits of B2U from the firm perspective, the perceived value of second-life batteries is segmented because the full value of the battery is not achieved until the final battery solution is delivered to the end-customers. From the OEM's perspective, for example, in cases where they sell the batteries, the value of second-life batteries for them is just the sales of the battery asset. However, there are much more benefits delivered through the batteries, for example, the value of various energy services, that the OEMs are not accessing. The system-level thinking is helpful for the B2U stakeholders to analyse the full potential value of the batteries, identify value opportunities and design business models to better achieve that value.

On the other hand, thinking about B2U only from the firm perspective is not helping extract the potential value from second-life batteries. The data show that under the 'selling and buying' logic, some of the systems are badly designed and they are fragile. In that case, people just want to sell the batteries to obtain additional revenues without trying to optimise the value system. They seem to work for a period of time when they can predict second-life battery price is lower than new batteries, but they are not sustainable business model in the medium term because other things are changing. For example, one of the interviewees of Company B complained: '*If we can't get a warranty then we will stop using Company A's second-life batteries... In order to scale up, we need to be 100% certain that Company A is going to keep providing us batteries, but I can't be sure about that, not today...If the OEMs are too difficult to work with, then we will simply move to other suppliers... and because battery price is falling so quickly, we think in the long run our main suppliers would be new battery manufacturers*' (E-2).

When Company B is small and only doing business on a small scale, it is not so concerned about the transparency of the schedule, but if they want to scale it, they have to be sure about the battery supply and they need proper contracts and warranties. However, the OEM (Company A) in this case is not helping make the life easier for Company B. They are not concerned about how Company B creates value for second-life batteries, and they are not helping optimise the total value creation for the batteries. The consequence of that is their reward from B2U is very small, and their business model is very vulnerable to competitors.

At the system level, the total value creation for second-life batteries determines the 'overall size of the value pie', which is also the upper limit of the value that stakeholders can capture from. The data suggest that if one is only trying to create and capture value from the firm perspective, they are partially optimising the value without increasing the 'overall size of the value pie'. In order to increase the total value creation for the batteries and thus their value capture potential, stakeholders should also consider the value creation of other players and design business models to facilitate value creation for the whole value system. The system-level thinking that considers value creation and capture of various stakeholders as well as the synergies between them help stakeholders better understand how to increase the system-level value creation for second-life batteries to enlarge the 'value pie'.

In summary, system-level thinking is helpful for stakeholders to analyse the potential value of second-life batteries and identify value opportunities. It could also help stakeholders increase the total value creation for the batteries and thus optimise the 'overall size of the value pie'. B2U stakeholders should take the system perspective into their business model design to enlarge the 'value pie' so as to achieve the potential value of the batteries and enable more value capture for themselves.

6.3 *Shift to Services*

As discussed in Sect. 6.1, business model is a key in overcoming the B2U challenges and achieving the potential value of second-life batteries. The data suggest that the traditional 'selling and buying' logic is no longer suitable for B2U, and a shift to services has been observed in all seven case studies. The section further synthesises the previous findings and discusses how the concept of service could help achieve the battery value and how stakeholders could better integrate this concept into their business models.

In terms of providing the final battery solutions to the end-customers, the business model analysis shows that in most cases, the B2U solution providers are offering substituting or adapting services that either extend the value propositions of the battery or replace the purchase of a product altogether. The interview data suggest that integrating services into the business models changes the perceived value of second-life batteries. Interestingly, in three out of seven cases, the B2U solution providers are offering energy storage as a service without selling any physical products. Customers in these cases are not so concerned about how good the battery is, and they are not comparing the prices because they do not own the battery asset. What matters to them is the energy storage solutions and the value of the services provided through the batteries. For example, one of the interviewees in case I commented: '*Our customers don't care whether you use Company A's old batteries, as long as it does what they tell us to do*' (E-1). One of the interviewees in case II also said: '*Customers won't care that it's used batteries because they can get more savings. In our case, it becomes our risk where we own the asset*' (E-2). The data suggest that offering substituting services allows companies to take full advantage of the remained value of the batteries to design differentiated value propositions that satisfy customers' demands in energy storage. Furthermore, it reduces the risks on the customers, which makes it easier and faster to enter the market. It proves to be useful, especially in the ferment stages of B2U, when customers are not familiar with the technology and feel unsecured about used products.

In terms of the OEM's business model, the data show that apart from case V and VII where the OEMs retain the battery ownership, in all the other five cases the OEMs are selling the batteries to the B2U solution providers. Most of them provide smoothing services such as warranties and technical support that complement their battery offers. They are not separately providing and benefiting from the services, but they obtain higher revenues from selling the more 'premium' battery product compared with the pure selling model. In case III, the consultancy service is also an important part of the OEM's value proposition. The OEM profits from providing consultancy services apart from selling the batteries. In those cases, the OEMs benefit more or less from providing the services. However, they are still selling the batteries and once they sell the batteries they stop profiting from the potential value of the various energy services provided by the batteries. In case V, the OEM retains the ownership of the battery and brings the batteries into the joint venture. The B2U solution provider is providing services to the OEM to help them develop and deliver the final solutions to the end-customers. The OEM shares the revenues from the energy services provided by the batteries, and they are able to continuously capture value from the batteries during the entire second-life of the battery.

The data suggest that integrating the concept of service could help OEMs generate more value from B2U than the traditional selling model. If the main value proposition for the OEM is the sales of the battery, there are various transaction costs involved and the OEM also fails to profit from the potential value of the energy services provided through the batteries. As commented by one of the interviewees: '*If OEMs sell the battery they are in huge competition because there will always be someone who sells cheaper. As the most valuable asset, it doesn't make sense for OEMs to sell the battery*' (E-4). With new battery price dropping rapidly, the selling model for second-life batteries would put OEMs in increasingly fierce competition in the battery market. The interviewee continued: '*As an OEM, you know the value, how long the battery can last and so on...they should provide the battery and they also know the battery best. If they sell the battery, the customers ask for warranty for several years and so on and these are all the cost factors that you pay for.*' (E-4)

In summary, in the nascent phase of B2U, there is no established market for second-life batteries. B2U stakeholders are still exploring how to approach potential customers—whether to just sell the batteries, or add some services to the battery offers, or just offer services. The findings from this research suggest that in this early stage of the industry characterised by high uncertainties of both the supplier and customer, B2U stakeholders could either provide complementary services or retain the ownership of the battery to reduce risks on the customers. In particular, providing energy storage as a service instead of selling the physical product enables stakeholders to differentiate their value propositions and overcome the 'inferiority' of second-life batteries as 'used product'. Result-oriented services also lead to opportunities for B2U stakeholders to develop innovative business models for second-life batteries.

7 Conclusions

In summary, this paper has provided insights concerning the value system of B2U. Through seven in-depth case studies with multiple B2U stakeholders, a typology of current B2U business models has been proposed that show how B2U stakeholders are interacting with each other to generate value from second-life batteries. Four critical B2U challenges have been identified (competitiveness, uncertainty, design and regulations) that help understand the factors that might impair the potential value of second-life batteries. The findings also show that innovative business model is a key to addressing the B2U challenges and overcoming the 'inferiority' of second-life batteries as used products. Three critical business model design elements, namely lifecycle thinking, system-level design and shift to services, have been proposed as helpful aspects for B2U stakeholders to consider for better designing their business models and extract more value from B2U.

References

1. J. Neubauer, A. Pesaran, J. Power Sources **196**, 10351 (2011)
2. E. Cready, J. Lippert, J. Pihl, I. Weinstock, P. Symons, Technical and Economic Feasibility of Applying Used EV Batteries in Stationary Applications, Sandia National Laboratories, Albuquerque, New Mexico, 2003 (http://www.rmi.org/Content/Files/Technicalandfeasible.pdf)
3. P. Wolfs, An Economic Assessment of "Second Use" Lithium-Ion Batteries for Grid Support. Universities Power Engineering Conference (AUPEC), 20th Australas. IEEE, 2010
4. V.V. Viswanathan, M. Kintner-Meyer, IEEE Trans. Veh. Technol. **60**, 2963 (2011)
5. S. Beer, T. Gómez, S. Member, D. Dallinger, I. Momber, C. Marnay, M. Stadler, J. Lai, I.E.E.E. Trans, Smart Grid **3**, 517 (2012)
6. M. Knowles, A. Morris, Br. J. Appl. Sci. Technol. **4**, 152 (2013)

7. B. Gohla-Neudecker, M. Bowler, S. Mohr, Battery 2nd life: Leveraging The Sustainability Potential of Evs and Renewable Energy Grid Integration. 5th International Conference Clean Electrical Power, Taormina, Italy, 2015
8. M.B. Miles, A.M. Huberman, J. Saldaña, Qualitative Data Analysis: *A Methods Sourcebook* (3rd Ed.), Sage Publishing, 2014

GPSR Compliance
The European Union's (EU) General Product Safety Regulation (GPSR) is a set of rules that requires consumer products to be safe and our obligations to ensure this.

If you have any concerns about our products, you can contact us on

ProductSafety@springernature.com

In case Publisher is established outside the EU, the EU authorized representative is:

Springer Nature Customer Service Center GmbH
Europaplatz 3
69115 Heidelberg, Germany

www.ingramcontent.com/pod-product-compliance
Ingram Content Group UK Ltd.
Pitfield, Milton Keynes, MK11 3LW, UK
UKHW022214051026
15624UKWH00001B/162
* 9 7 8 3 3 1 9 6 9 9 4 9 3 *